T0331789

Structural and Phase Stability of Alloys

Structural and Phase Stability of Alloys

Edited by

J. L. Morán-López

Universidad Autónoma de San Luis Potosí
San Luis Potosí, S.L.P.
Mexico

F. Mejía-Lira

Late of Universidad Autónoma de San Luis Potosí
San Luis Potosí, S.L.P.
Mexico

and

J. M. Sanchez

The University of Texas
Austin, Texas

Springer Science+Business Media, LLC

Library of Congress Cataloging-in-Publication Data

Structural and phase stability of alloys / edited by J.L. Morán-López, F. Mejía-Lira, and J. M. Sanchez.
p. cm.
Includes bibliographical references and index.
ISBN 978-0-306-44211-7 ISBN 978-1-4615-3382-5 (eBook)
DOI 10.1007/978-1-4615-3382-5
1. Alloys--Congresses. 2. Phase rule and equilibrium--Congresses. 3. Order-disorder in alloys--Congresses. 4. Alloys--Thermal properties--Congresses. 5. Alloys--Metallography--Congresses. I. Morán-López, J. L. (José L.), 1950- II. Mejía-Lira, F. (Francisco) III. Sanchez, J. M.
TN689.2S74 1992
669'.9--dc20 92-8513
CIP

ISBN 978-0-306-44211-7

Originally published by Plenum Press, New York in 1992

Preface

This volume contains the papers presented at the Adriatico Research Conference on Structural and Phase Stability of Alloys held in Trieste, Italy, in May 1991, under the auspices of the International Centre for Theoretical Physics. The conference brought together participants with a variety of interests in theoretical and experimental aspects of alloys from Argentina, Belgium, Bulgaria, Czechslovakia, France, Germany, Italy, Japan, Mexico, People's Republic of Congo, Portugal, Switzerland, United Kingdom, United States, U.S.S.R., and Venezuela.

The conference was purposely designed to succinctly cover experimental and theoretical aspects of magnetic and non-magnetic alloys, surfaces, thin films and nanostructures. The Conference opened with an overview of a select class of advanced structural materials, with a potential in engineering applications, for which the conventional "physics" approach, both theoretical and experimental, should have a significant impact. A number of papers were dedicated to the use of phenomenological approaches for the description of thermodynamic bulk and surface properties. It was clear from these presentations that the phenomenological models and simulations in alloy theory have reached a high degree of sophistication. Although with somewhat limited predictive powers, the phenomenological models provide a valuable tool for the understanding of a variety of subtle phenomena such as short-range order, phase stability, kinetics and the thermodynamics of surfaces and antiphase boundaries, to name a few.

Several papers in the conference were dedicated to "first-principles" theories of alloy phase stability which, by their very nature, should eventually overcome the predictive limitations of the phenomeological models. Here, once again, the presentations underscored the remarkable developments over the last several years in our understanding of alloy properties on the basis of their electronic structure. Although much work remains to be done in order to achieve an quantitatively accurate, finite temperature theory of alloys, we have, as of now, a solid basis to build on, and a workable blueprint. Local density functional theory allow us to describe, from a strictly microscopic view point, the properties of compounds with relatively complex structures. The reach of these first principles calculations can be considerably extended using, for example, the "bond-order" theory which allows for the description of the energy of both topologically and configurationally disordered systems. Properly coupled with a sound statistical mechanics and kinetic theory, such as Monte Carlo, Cluster Variation or the Path Probability methods discussed at some length in this volume, we come one step closer to a comprehensive theory of real materials.

We thank the participants for all the efforts they have put on the presentations and the manuscripts. We greatefully acknowledge the technical assistance of J.M.

Montejano-Carrizales. Sponsorship was also received from the Mexican National Council of Science and Technology.

F. Mejía-Lira
J.L. Morán-López
San Luis Potosí, S.L.P. México

J.M. Sanchez
Austin, Texas

August, 1991

Contents

HIGH TEMPERATURE MATERIALS

MAGNETIC PROPERTIES

STATISTICAL MECHANICS AND THERMODYNAMICS

ELECTRONIC THEORIES

LOW DIMENSIONAL SYSTEMS

SUMMARY

The Frontiers of High Temperature Structural Materials

J. K. Tien, G. E. Vignoul, E. P. Barth, and M. W. Kopp

Strategic Materials Research & Development Laboratory
The University of Texas
Austin, Texas 78712
U.S.A.

Abstract

The science and technology fronts of advanced high temperature structural materials are defined and discussed. Current research is focussing on monolithic intermetallic and ceramic compounds, as well as metal, intermetallic, and ceramic matrix composites. An emerging field, in addition to these, is intelligent materials. These composite structures are either self diagnostic, self attenuated, or both. Challenges being addressed pivot about mechanical integrity, interdiffusion and phase stability, interface engineering, and material status sensing and signal processing.

I. Introduction

Since the advent of the jet age, the evolution of high temperature structural materials has been at the forefront of defining economic and military strength as a result of the critical nature of these materials in aircraft and aerospace systems. In order to meet the demands imposed by both military and commercial end users, the development and commercialization of advanced high temperature structural materials must keep pace with, and in fact anticipate, the requirements of the advancing propulsion technologies.

To date, the nickel and iron-base superalloys have been the materials of choice in critical high-temperature structural applications. The evolution of these materials, as measured in terms of increased effective use temperature has largely kept pace with the demands of the jet turbine industry (Figure 1). However, as a consequence of the conservative nature of the jet turbine industry, no new class of high temperature

Structural and Phase Stability of Alloys
Edited by J.L. Morán-López *et al.*, Plenum Press, New York, 1992

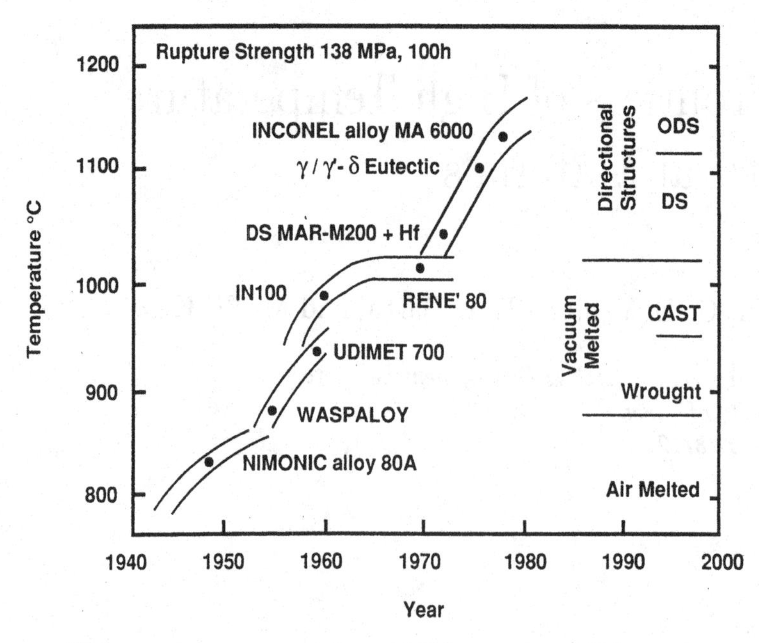

Figure 1. High temperature materials capabilities as measured by use temperature.

structural materials have found acceptance within the jet turbine engine since its inception.

While significant effort and resources have been devoted to the development of a new generation of high temperature structural materials, these endeavors have largely been focused on improving the performance of existing technologies. The classic example of this trend has been the extensive work devoted to the research and development of monolithic Ni_3Al as a replacement for the structural superalloys, which already contain upwards of 60% of this phase as strengthening precipitates. However, it has been known that this particular intermetallic compound, either by itself or alloyed with ternary or quaternary elements, offers little advantage over that of the superalloys. Indeed, the beneficial effects of boron and hafnium alloying additions have long been recognized in the superalloy literature.[1–3] Further, even with the alloying additions that improve ductility, the level of ductility attained is generally considered insufficient. While this effort on Ni_3Al has resulted in a body of scientific knowledge, its future as a monolithic intermetallic is not in the aerospace sector. The titanium, iron and nickel aluminides, however, do offer certain near term gains for special turbine applications.[4]

Marginal improvements in current high temperature material performance may be sufficient to carry the turbine industry through the next decade, but the driving force imposed by the desire for and needs of hypersonic flight, for example, implicitly demand a quantum leap in philosophy and technology. In light of the projected temperatures that will be encountered during hypersonic flight, it is clear that the materials represented in Fig. 2 present the future of high temperature materials. In essence, the future of high temperature structural materials is encompassed by intermetallics whose melting points exceed those of any being actively studied today and composite materials that are metallic-based, intermetallic-based, ceramic-based

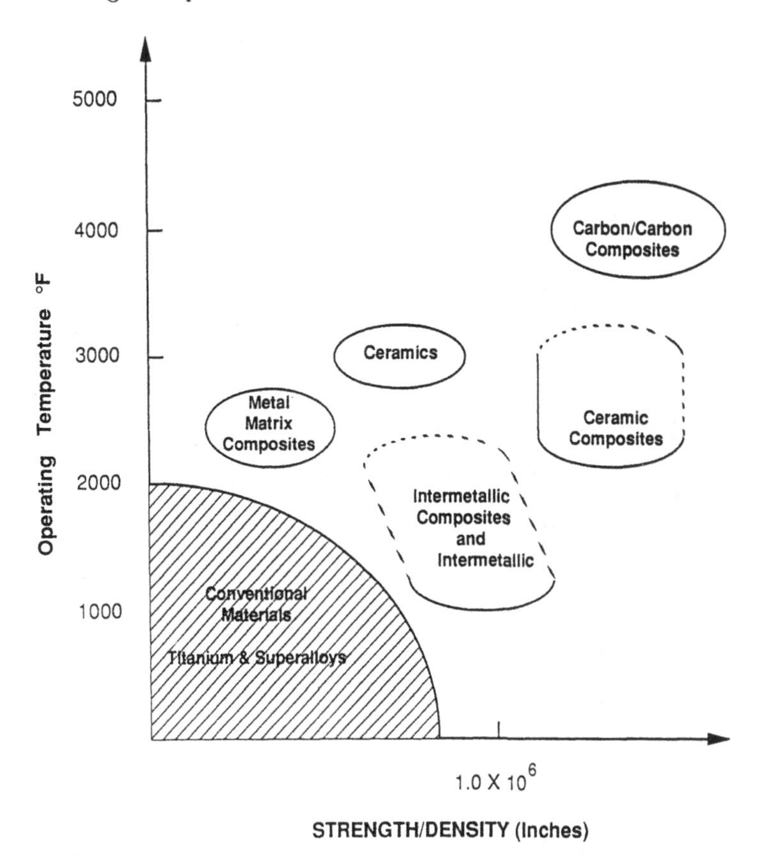

Figure 2. Plot of strength to density ratio versus temperature for a variety of high temperature material classes.

and carbon-based. Clearly, the brute-force temperature capabilities of carbon/carbon composite materials make them the optimal solution in terms of future high temperature applicability once the oxidation problem is addressed more satisfactorily. Ceramics, and especially ceramic/ceramic composites ductillized by appropriately weak interfaces, is another class of materials which have elicited significant interest and effort. Consequently, this paper will deal exclusively with the present and future of high temperature intermetallics and ceramics, speak on the status of metallic, intermetallic, and ceramic matrix composites, and introduce the status and outlook of the emerging field of intelligent materials.

II. Intermetallics

Ordered intermetallic compounds have garnered increased attention as potential high temperature structural materials because they have a number of properties that make them intrinsically more appealing than other metallic or ceramic systems for high temperature use. Specifically, intermetallic compounds tend to (1) be inherently very strong and maintain this strength to high temperatures, (2) have high stiffness that

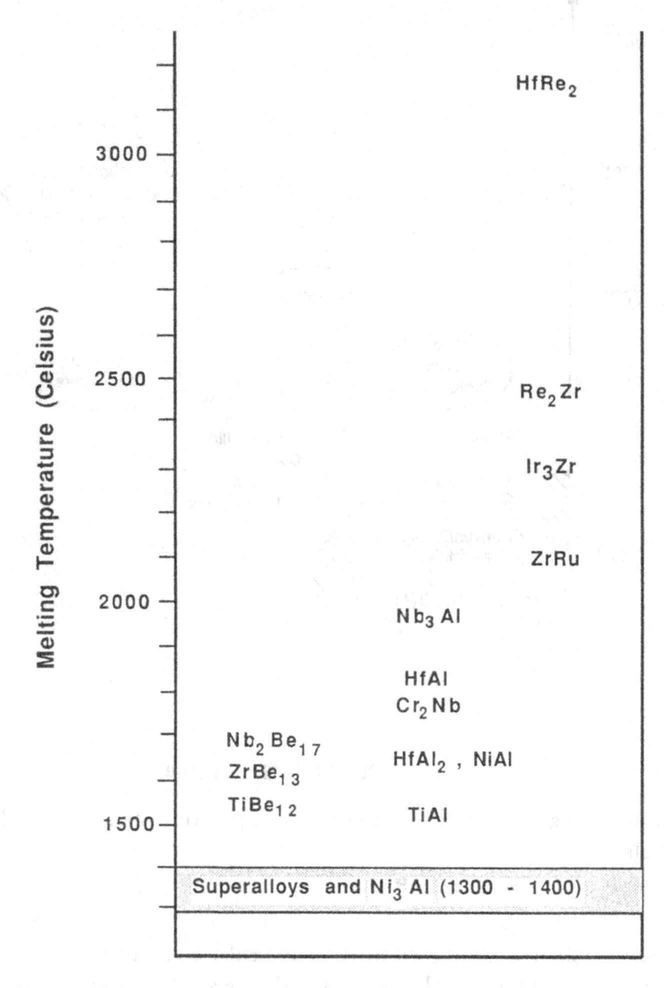

Figure 3. Melting point of a variety of intermetallic compounds relative to superalloys.

decreases slower than disordered metals with increasing temperature, and (3) posse low self-diffusion coefficients, leading to excellent microstructural stability and improved creep strength. In addition to these advantages, intermetallics can be easily manufactured by a variety of current production techniques and have better thermal conductivities than ceramics.

As has been recognized for some time, a material's melting point is a useful first approximation of its high temperature performance, since a variety of high temperature mechanical properties (*i.e.* strength, creep resistance) are limited by thermally assisted or diffusional processes and thus tend to scale with the melting point of the material. Therefore, intermetallics can be crudely ranked in terms of their melting points to indicate their future applicability as high temperature structural materials. As may be seen in Fig. 3, metallic materials (intermetallics or otherwise) which are currently in use or being studied melt at temperatures much lower than 1650° C. If these materials are discounted from consideration, the remaining intermetallics in

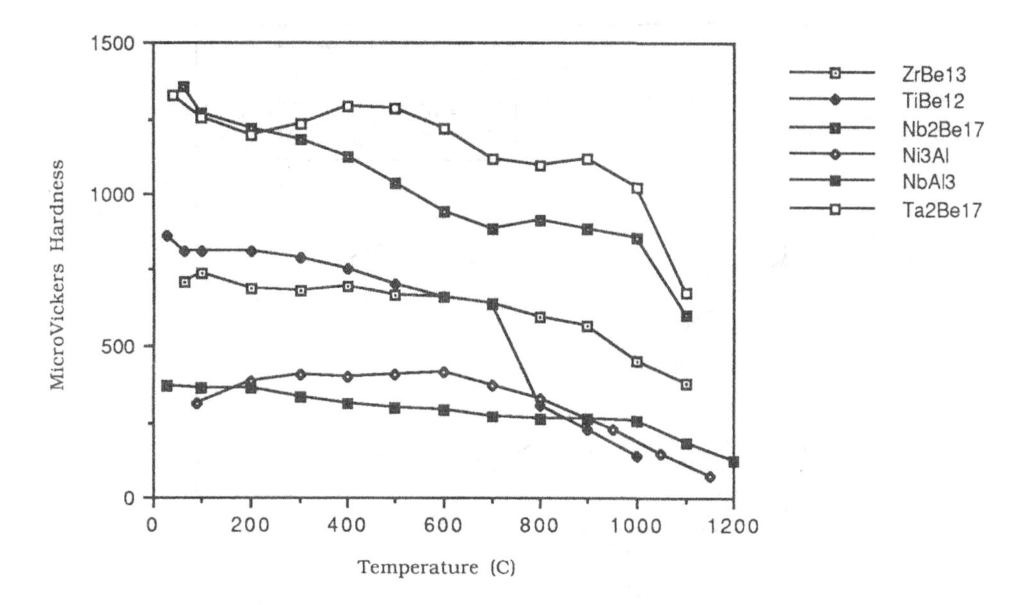

Figure 4. Micro-Vickers hardness versus temperature for selected intermetallics.

Fig. 3 may be roughly divided into two groups; those that fall in the temperature range just above 1650° C and those whose melting points extend to much higher temperatures.

Of those compounds whose melting points are near 1650° C, intense interest and effort has recently been directed at refractory metal beryllide intermetallics. Several of the transition metal beryllides such as Nb_2Be_{17}, Ta_2Be_{17}, and $ZrBe_{13}$, for example, have properties that make them ideal for certain applications. Specifically, these compounds possess extremely low densities and high coefficients of thermal expansion. Although these materials have high σ/E ratios, they are rather brittle and thus eventual incorporation into composite structures has formed the basis for the majority of current research efforts. Of course, before a realistic understanding of composite form and behavior can be obtained, basic studies of the component monoliths is required and in fact is underway. As illustrated in Fig. 4, the high temperature strength (here represented by microhardness) shows the clear advantage that the beryllides can have over more conventional intermetallics.[5] However, the single largest disadvantage of the beryllides is an inherently low ductility and lack of toughness that can be attributed to the very low symmetry of their crystal structures.

Many of the second group of intermetallic compounds (those with MP> 1650° C) belong to a group of intermetallics which are predicted on the basis of the Engel-Brewer phase stability theory and encompass compounds whose melting temperatures approach the incipient solidus of structural ceramics ($\geq$ 3000° C). In fact, it can be argued that these intermetallics have effective melting points which are greater than those of either SiC or Si_3N_4, since these ceramics generally utilize glassy phases as binders that have critical softening temperatures that are on the same order as, or in fact lower than, the melting temperatures of the higher melting Engel-Brewer intermetallic compounds. Further, these compounds are predicted to be very stable.[6–8] Indeed, extremely large negative free energies of formation have been es-

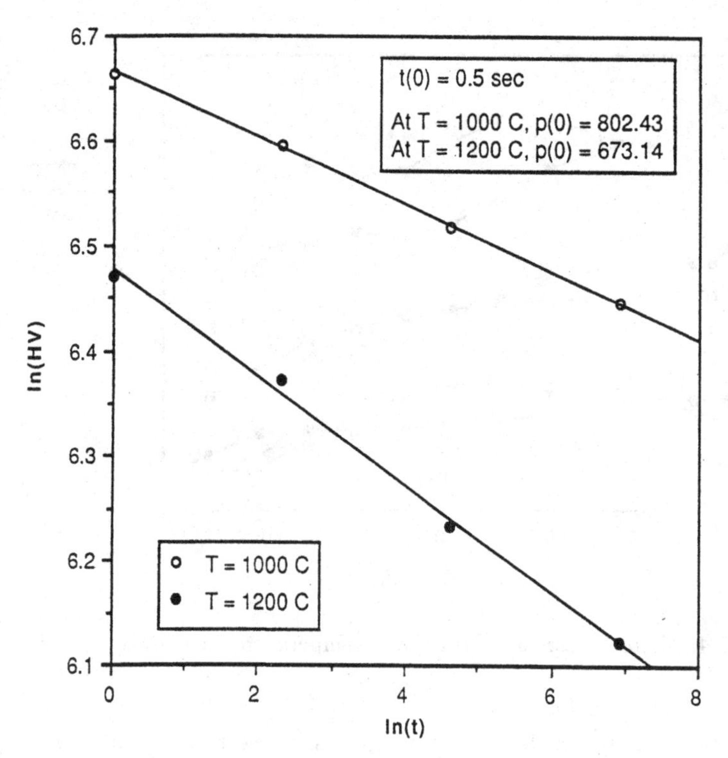

Figure 5. Plot of hardness versus time on load for Cr_2Nb at 1000 and 1200° C.

tablished for several such systems. For example, ZrRu has a free energy of formation of −21.5 kcal/g-atom, which is about three times as negative as the −7.4 kcal/g-atom reported[9] for Ni_3Al.

It should be noted that large negative free energies of formation translate not only into very high melting points, but potentially also into inherent oxidation resistance because the constituent compound atoms may prefer each other more than they prefer oxygen, even at elevated temperatures. Of course, kinetics will be the determining factor with respect to this issue.

Within the group of intermetallics with melting points greater than 1650° C, but not so high as the typical Engel-Brewer intermetallic, recent studies have indicated that there are several systems which are attractive candidate high temperature structural materials. For example, the Cr_2Nb binary system has been examined in some depth.[10] As may be seen in Fig. 5, the results of this study have shown that this intermetallic system possesses strength that is up to three times greater than that found for Ni_3Al at comparable temperatures. In fact, the strength of the Cr_2Nb system at 1300° C is comparable to or greater than the peak strength exhibited by a Ni_3Al(B, Hf) compound.

Further, the microindention creep behavior of the Cr_2Nb system was studied by varying time on load at $T = 1000$ and 1200° C. The results of this aspect of the study are shown in Fig. 5. Analysis of the data showed that $m = 24$ and $Q_{app} = 478$ kJ/mole. These unusually high values are indicative of the existence of an effective resisting stress against creep. This is somewhat surprising, given that resisting stresses against

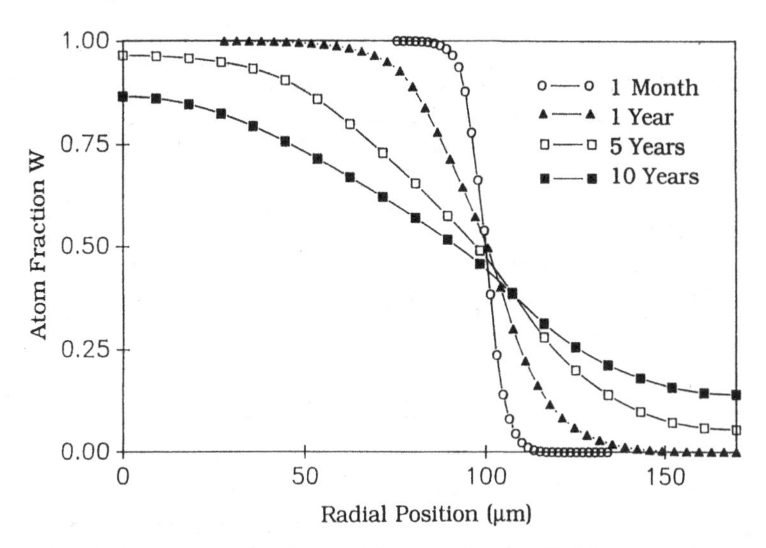

Figure 6. Predictions of composition-position profiles for W/Nb composites at 1500 K.

creep are normally associated with such complex multi-phase systems as high-volume fraction γ' superalloys and oxide dispersion strengthened (ODS) alloys. When the data were fit against a microindention creep deformation law which was modified to incorporate an effective resisting stress term, it was determined that $m = 4.5$, $Q_{\text{creep}} = 357$ kJ/mole and the resisting stress term $\sigma_r = 300$ MPa. While the actual mechanistics for this apparent creep resistance have yet to be fully explored, it is believed that the resistance to creep is, at least in part, due to the lack of active glide planes in the C15 crystal structure and the resultant complex dislocation-dislocation interactions that occur during deformation.

Studies of the ZrRu Engel-Brewer intermetallic have also been initiated recently. The crystal structure of this compound is B2 which, a priori, should allow for a reasonably simplified view of dislocation slip systems, dislocation configurations and dislocation core transformations. However, it should be noted that the tight binding of this intermetallic may play a large role in deformation behavior and thereby preclude such simplification. Work in progress on this system is focusing on the basic characterization and elucidation of high temperature mechanical properties and behavior (*i.e.* yield, creep, deformation mechanisms).

III. Metal and Intermetallic Matrix Composites

III.1 Metal Matrix Composites

The predominant factor affecting the implementation of metal matrix composites (MMCs) for elevated temperature applications are the degree of chemical interaction between the fiber and matrix components and thermal-mechanical stability issues due to mismatch of the coefficients of thermal expansion (CTE) for fiber and matrix. Interdiffusional phenomena can take the form of fiber dissolution, fiber properties being "poisoned" by matrix element influx, fiber/matrix reactions and fiber

Table I

Reaction zone growth kinetics for selected TFRS composites annealed at 1093° C.

Matrix	Fe	Fe+Co	Ni	$K_{rz}^{1/2} K_f^{1/2}$ ($\times 10^{-12}$ cm^2/sec)
FeCrAlY	71	71	0	3.5
SS316	70	70	12	2.9
Incoloy 907	57	70	25	1.7
Incoloy 903*	42	57	38	0.8
Waspaloy	0	13	56	0.3
Alloy 89	0	0	66	0.05

* annealed at 1100° C.

coarsening. Also, CTE mismatch can result in thermal fatigue and fiber/matrix load transfer problems due to debonding at the interface. Efforts in recent years to address these problems directly have focussed on developing methodologies for assessing these phenomena, as well as developing diffusion barrier and compliant interface layers to combat them.

Previous work on very simple W/Nb single phase composites (*i.e.* complete solid solution) have resulted in a methodology that allows long term prediction of interdiffusional behavior.[11] The first priority was determining composition dependent interdiffusion coefficients. This was easily accomplished by Boltzmann-Matano analysis of planar interface diffusion couples. Having determined the interdiffusion coefficients for the temperatures of interest, composition profiles were calculated using numerical solutions to Fick's second law. This finite difference computer code, which was adapted from the program of Tenney and Unnam,[12] calculates diffusion profiles for diffusion couples with planar, cylindrical, or spherical geometry with finite boundary conditions. Using this method, forecasts the level of interdiffusion for W/Nb composites for very long term exposures are possible. For this effort, the ability of the numerical solutions utilized to handle finite boundary conditions (*i.e.*, overlapping diffusion fields) was crucial. Fig. 6 illustrates radial diffusion profiles for a 40 volume percent 200 micron diameter fiber reinforced composite seeing 1500 K service exposures for from 1 month to 10 years. Clearly, significant degradation of fiber properties can be expected at the very long times.

It must be noted that a great deal of care is necessary in determining diffusion coefficients. Often accuracy in determining these coefficients is limited to within a factor of 2 or 3. This uncertainty can lead to large systematic errors when predicting composition profiles for very long times.

Tungsten fiber reinforced superalloys (TFRS) have long been candidates for high temperature composites allowing for significant increases in operating temperatures through increased creep resistance and strength. TFRS composites present several concerns including diffusion induced recrystallization of the tungsten fibers and reaction zone formation at the fiber/matrix interface. Recrystallization has a pronounced effect on the strength, creep resistance, and toughness of the tungsten fibers. Fiber/matrix diffusional reactions in TFRS have been shown to produce brittle intermetallic phases that, as they continue to grow, may adversely affect mechanical strength, due to loss of fiber cross-section, and toughness, from the defect sensitive intermetallic itself. Thus, this system presents a more difficult set of problems than the simple W/Nb system.

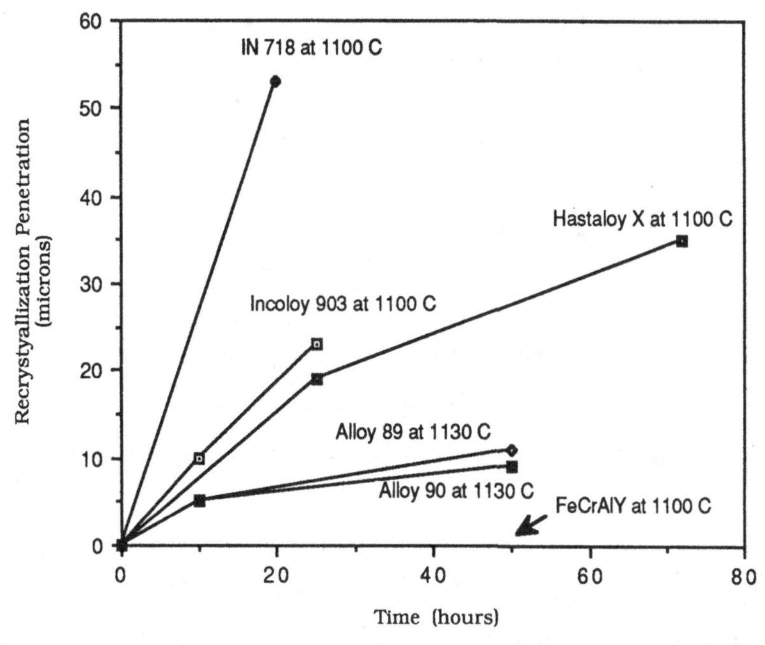

Figure 7. Recrystallization of W-fibers in various superalloy matrices as a function of time.

Reaction zone growth has been determined to be rate controlled by interdiffusion across the reaction zone phase.[13] The interdiffusion coefficient for these systems using a pseudobinary approximation can be expressed as being proportional to the product of the roots of parabolic rate constants for the growth of the entire reaction zone and the growth of the portion of the reaction zone that displaces the fiber. Several matrix alloys have been ranked according to the product of the roots of these rate constants. The ranking of these matrix alloys by reaction zone growth kinetics is shown in Table I. Alloys 89 (19.05 wt%Cr, 8.70 wt% W, 3.01 wt% Ti, 2.57 wt% Al, bal. Ni) and 90 (17.75 wt%Cr, 16.25 wt% W, 2.81 wt% Ti, 1.22 wt% Al, bal. Ni) are experimental alloys developed in an effort to minimize reaction zone growth kinetics. As can be seen, reductions of iron and cobalt decrease interdiffusion across the reaction zone and thereby decrease the kinetics of reaction zone formation.

Recrystallization of the tungsten fibers reinforcement is also a primary concern for TFRS composites. Recrystallization of ThO_2 doped tungsten wires normally occurs at about 2000° C. A number of studies have shown that a number of elements, most notably nickel, cause this recrystallization temperature to drop dramatically when they are in contact with the fibers. In fact conventional wisdom to date has been that any increases in nickel content of the matrix alloy in TFRS composites would be accompanied by increased recrystallization kinetics of the fibers in that composite system. Since very pure tungsten wires recrystallize at about the same temperature as the poisoned ThO_2 doped wires, it appears that the infusion of the poisoning elements affects the recrystallization inhibiting nature of the dopant. A complication has been some uncertainty regarding the nature and distribution of dopant particles. Perhaps the most notable theory to explain the diffusion-promoted recrystallization of doped tungsten wires is that the poisoning species, which diffuse primarily along grain boundaries through short circuit paths, lowers the interfacial energy of the grain

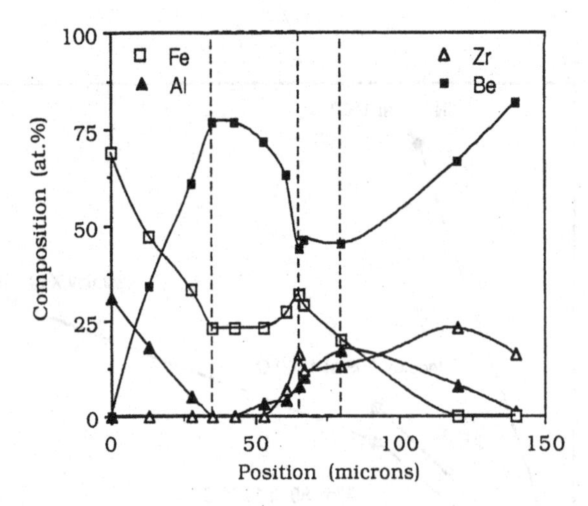

Figure 8. Composition–position profile of $ZrBe_{13}$/Fe–40 %Al.

and subgrain boundaries, thereby overcoming the effect of the pinning dispersoids. However, the effect of the poisoning species on the dispersoid/bulk tungsten energy may also be playing a significant role.

As stated above, the conventional wisdom with regard to matrix nickel content has been that increases in nickel content should promote accelerated tungsten fiber recrystallization. The purpose of this investigation pivoting about alloys 89 and 90 is to attempt to elucidate the previously reported matrix chemistry effect.

Unexpected results were found when considering the level of fiber recrystallization for these various matrix materials.[14] The FeCrAlY matrix composites did not exhibit significant recrystallization at the temperatures and exposure times studied. Figure 7 illustrates recrystallization penetration data for several other matrix materials annealed at 1100° C and 1130° C, including some of the 1100° C data of L. O. K. Larsson.[15] The earlier composites of Larsson show more accelerated recrystallization than the higher and lower nickel containing alloys of the present study. Further, the very high nickel alloys 89 and 90 were annealed 30° C higher than the others. Although considerably more nickel is available in the matrix of these alloys to source diffusion induced recrystallization, these alloys were specifically designed to minimize diffusion across the anticipated reaction zone. Thus, there is apparently some level of competition between total nickel availability and interdiffusional kinetics that provide nickel to the fiber surface.

III.2 Intermetallic Matrix Composites

As discussed above, beryllide intermetallic have been of great interest for composite application. This has lead to attempts to employ them as reinforcing element in intermetallic matrix composites. One of the prime matrix candidates has been FeAl due to its mechanical and environmental properties. Four beryllide reinforced Fe-40% Al matrix systems have been studied.[16] The beryllides studied included $TiBe_{12}$, $ZrBe_{13}$ Nb_2Be_{17}, and Ta_2Be_{17}. A similar dual phase reaction zone evolution was observed for each system. This growth behavior took the form of a decreasing rate of growth from

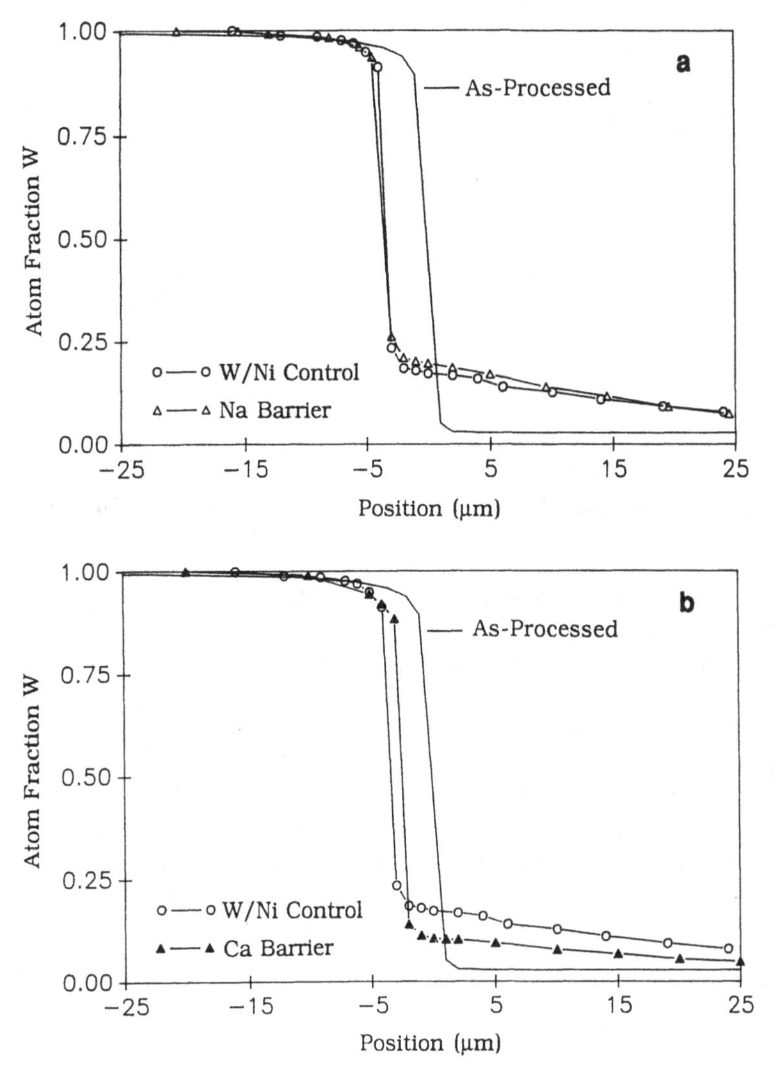

Figure 9. Effect of (a) Na and (b) Ca ion implanted barrier layers in W/Ni diffusion couples.

the initially observed parabolic rate law. The shape of the interphase growth behavior of these beryllide/Fe-40% Al composites implies that some mechanism is operating that increasingly retards diffusion across the reaction zone as annealing time increases.

Examination of the composition-position plots for these systems has revealed some very interesting phenomena that are believed to account for this decrease in reaction zone growth kinetics. As a typical example, the plot for the $ZrBe_{13}$/Fe-40% Al system is shown in Fig. 8. The most striking features of these plots are the concentrations of beryllium and aluminium from the terminal phases through the reaction zone phases. For both these elements, concentration drops in the immediately adjacent reaction zone phase from the element rich terminal phase, climbs in the second and more distant phase, then drops to zero in the opposite terminal phase. In the case of the aluminum profile, the aluminum concentration in the reaction zone phase adjacent to the aluminide drops below detection limits for both energy dispersive spectroscopy

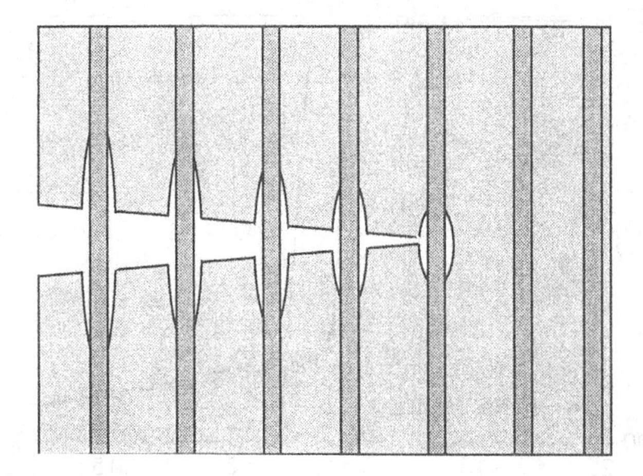

Figure 10. Schematic of crack bridging and fiber/matrix debonding in CMCs.

(EDS) and scanning Auger microscopy (SAM). This phase has been identified as $FeBe_5$. The other phase for these systems have yet to be identified.

Based on these diffusion profiles, an *in situ* diffusion barrier mechanism is believed to have been formed in these Fe-40% Al systems. Although no ternary data exists for the Fe-Be-Al system, apparently $FeBe_5$ has no, or at most trace, solubility for aluminum, and thus may be a barrier to its further diffusion. The aluminum present on the beryllide side of the $FeBe_5$ is believed to have diffused there early on in the fabrication process, prior to the formation of the phase. It is similarly possible that the longer term, slow growth may be a result of the reaction zone phases picking up their needed elements from the relatively small amounts that diffused into the terminal phases adjacent to them early on in the fabrication process.

III.3 Diffusion Barrier Layers

Efforts to directly address interdiffusional incompatibility have included attempts to develop diffusion barrier layers. One method that has been investigated has focussed on the feasibility of ion implanting to form diffusion barriers. The Pauling rules dictate that materials with considerable differences in atomic radii, valence, or electronegativity will have little solubility in one another, or will have insufficient driving force for significant interdiffusion. Based on this principle it might be expected that interdiffusion in composites may be impeded by creating a layer at the fiber/matrix interface that is very different from either of the two components.

The simple binary W/Ni system has been studied.[17] Implanted ions included Na, Ba, Ca and K. For each couple the substrate received 6.4×10^{11} ions/sq cm implanted with an accelerating potential of 190 kV. The substrate temperature was estimated to be about 700 K over the subsequent 70 hours of deposition time required to deposit about 100 microns of nickel. These diffusion couples, along with an unimplanted control couple were annealed at 1500 K for 50 hours.

The profiles for Na and Ca implanted barriers are shown with the as-received and control profiles in Figs. 9a and 9b. The evident "saturation" of W through the Ni

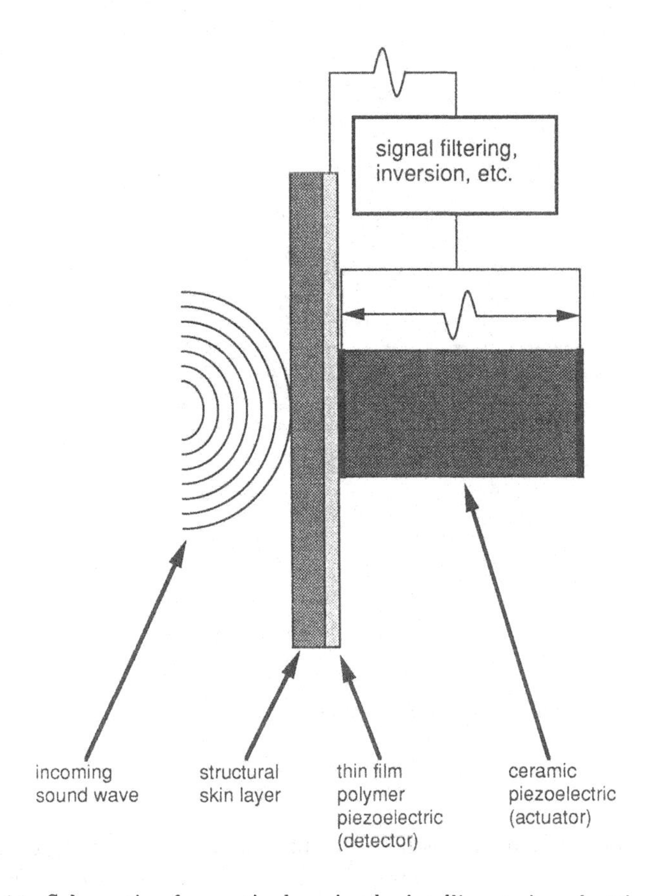

Figure 11. Schematic of acoustic damping by intelligent piezoelectric material.

matrix can be attributed to the very significant grain boundary diffusion that took place during the Ni deposition. Although this problem makes quantitative analysis impossible, qualitative analysis indicates that the Ca barrier appears to have indeed slowed the interdiffusion process. Na showed little effect. Both K and Ba showed increased interdiffusion.

A possible explanation of the acceleration effect is that the distortion of the larger ions may have caused some sort of dynamic recovery effect. This might be eliminated if lower ion doses were used. Similarly, the effect of the Ca barrier might be optimized by variation of the implant dose and accelerating potential.

IV. Ceramics and Ceramic Matrix Composites

Toughening of ceramics has been at the forefront of ceramic research for years. Measurable success has been achieved through fiber reinforcement and transformation toughening. Fiber reinforcement affects crack growth in ceramics primarily by bridging cracks and thereby resisting further crack propagation. This mechanism has two major components: fiber/matrix debonding and interface sliding friction. As a crack

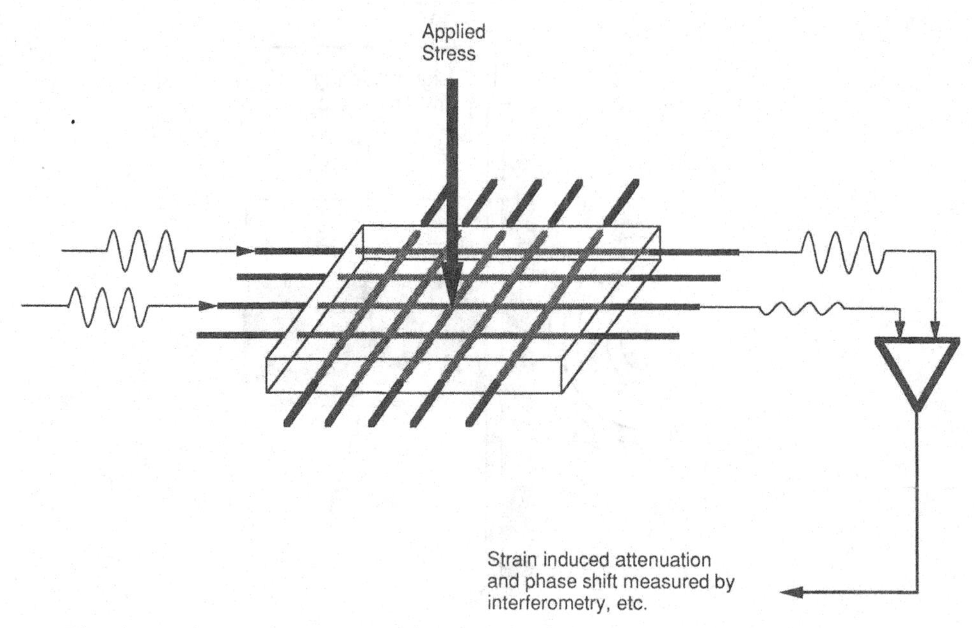

Figure 12. Schematic of damage/deformation localization by optical fiber network.

propagates through the matrix debonding occurs when the crack impacts a fiber. This serves to both absorb energy and blunt the crack[18] (see Figure 10). Further, as the crack tip passes the fiber, the crack is bridged and further debond along with interface sliding is required for the crack to widen. The net result is a vast increase in the energy and in the crack opening displacement required for the crack to incrementally grow. Fiber reinforced glasses have been reported with energy absorption values (surface energies) well above 1 kJ/m^2. Typical values for the same unreinforced glasses are in the J/m^2 range.

A variety of investigators have published quantitative formalisms for describing the behavior of "ductillized" or fiber toughened ceramics.[19–20] The form these solutions generally take relate mechanical properties to a frictional sliding resistance term and a fiber/matrix debonding energy term, in addition to other specific fiber and matrix properties. These treatments all show that a net toughening effect of the fiber reinforcement is critically related to careful engineering of the fiber matrix interface.

Another approach taken to toughen ceramic materials has been the use of monoliths or precipitates that exhibit stress induced transformations. The most studied of this class of materials are partially stabilized zirconia (PSZ, a.k.a. transformation toughened zirconia, TTZ) both as monoliths and as a reinforcing phase in alumina (ZTA). Zirconia is a polymorphic material with a low temperature monoclinic crystal structure. Toughening in the system is accomplished by alloying in order to partially stabilize the higher temperature tetragonal phase. When overstressed by proximity to an advancing crack front, the tetragonal zirconia transforms to the monoclinic structure, which effectively puts the crack tip in compression. In this manner significant increases in fracture toughness have been achieved.

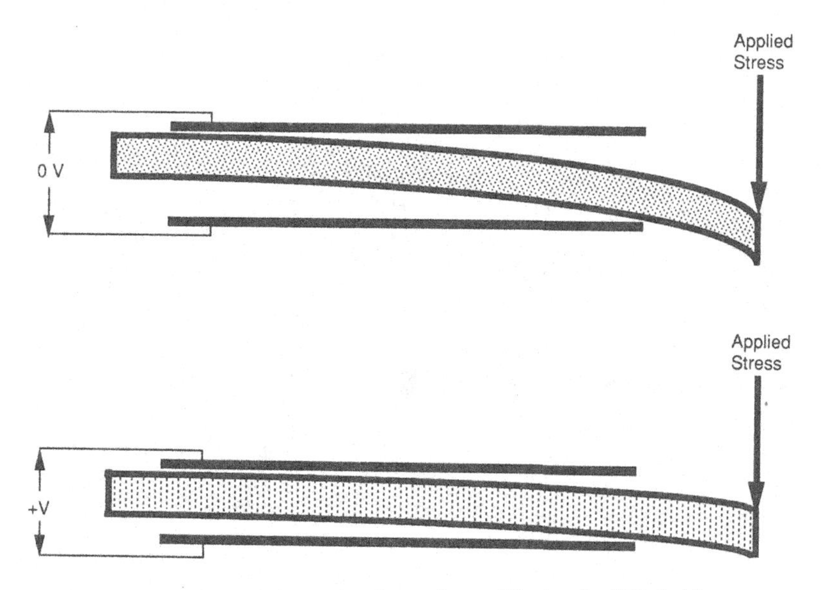

Figure 13. Schematic of member stiffening by ER fluids.

V. Intelligent Materials

The field of smart materials and structures is a newly developing multidiscplinary approach to the design of a wide range of products. The goal is to develop materials that have the inherent ability to sense their environment or internal state, and in an intelligent, or at least pre-programmed, way adapt their properties. While examples abound in the natural, living world, perhaps the only artificial example to date is photochromic glass, which homogeneously adjusts its transparency based on its exposure to ultraviolet light, without external control. Currently, research in this area is focussing on developing composite materials with discrete, embedded sensors and actuators that depend on external input or signal processing.

The most commonly studied materials for these purposes fall into four categories. Piezoelectric ceramics and polymers are being utilized as both sensors and actuators in a wide range of applications, such as robotics, acoustic transducers, etc. An example is shown in Fig. 11. Networks of optical fibers can be embedded in polymer matrices and, with considerable external assistance, can be used to sense such varied quantities as temperature, pressure, degree of polymer cure and local fracture (See Fig. 12). Electro-rheological (ER) materials are those whose properties, predominantly fluid viscosity, can be controlled by an electric field, as shown schematically in Fig. 13. Efforts to dampen vibrations in structures are currently the largest beneficiaries of ER fluids. Finally, is the widely studied class of shape memory alloys. The ability to drive these alloys to a previous shape by the application of temperature leaves them open to exploitation as actuators in many uses. The shape memory effect is shown in Fig. 14. The task of mechanical engineers, materials scientists and solid state physicists is to broaden our understanding of the sensitive or adaptive properties in the above materials and to discover new materials with these properties, thereby widening the palette from which we can choose.

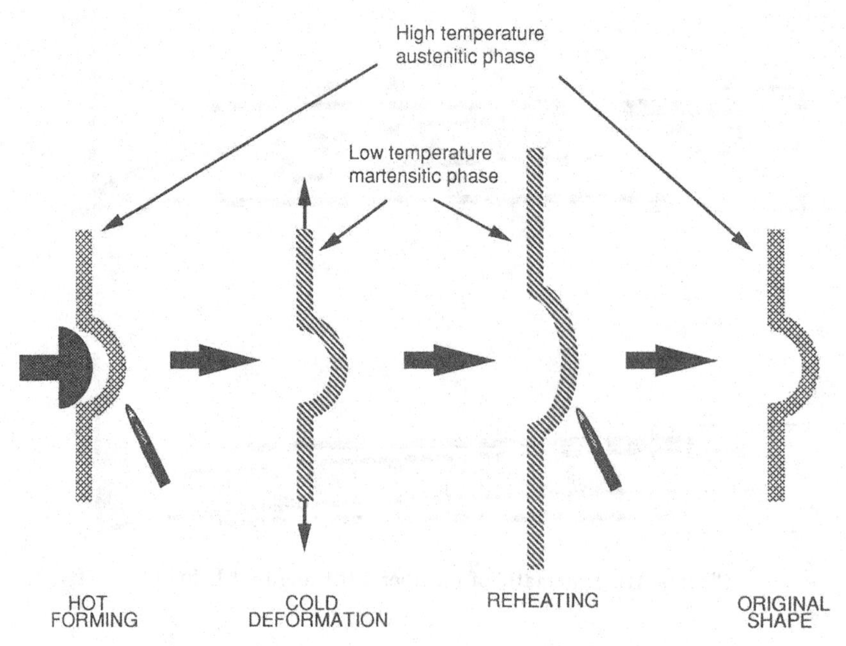

Figure 14. Schematic of the shape-memory effect.

VI. Concluding Remarks

Clearly, the challenge of developing high temperature structural materials to exceed current capabilities is being approached from a wide variety of different points of view. Although the relative maturities of the basic technologies vary for these different materials, none have yet made any real inroads into real world application. It is hoped that computational methods can be of use in advancing the technological fronts for these materials, although it is up to that community to determine what aspects of these efforts they can contribute to.

References

1. R. F. Decker and J. W. Freeman, *Trans. AIME,* **218**, 277 (1960).
2. C. Lund, J. Hockm, and M.J. Woulds, U.S. Patent 3,677,447 , (1972).
3. J. E. Doherty, B. H. Kear, and A. F. Giamei, *J. Metals,* **11**, 59 (1971).
4. F. H. Froes, *J. Metals,* **9**, 6 (1989).
5. A. B. Rodriguez and J.K. Tien, unpublished research, 1991.
6. N. Engel, *Powder Metall. Bull.,* **7**, 8 (1954).
7. L. Brewer, in *Electronic Structure and Alloy Chemistry of Transition Elements,* P. A. Bock ed., Wiley-Interscience, New York, 1963, p. 221.
8. L. Brewer, in *High Strength Materials,* V.F. Zackay ed., Wiley, New York, 1965, p. 12.
9. J. K. Gibson, L. Brewer, and K. A. Gingerich, *Metall. Trans. A,* **15**, 2075 (1984).

10. G. E. Vignoul, J. M. Sanchez, and J. K. Tien, in *High Temperature Ordered Intermetallic Alloys IV*, L. A. Johnson, D. P. Pope, and J. O. Steigler, eds., MRS, Boston MA, 1991, p. 739.
11. M. W. Kopp and J. K. Tien, in *Proc. 9th Int. Riso Symp. on Metall. and Mater. Sci.*, S. I. Andersen, H. Lilholt, and O. B. Pedersen, eds., Riso National Laboratory, Roskilde, Denmark, 1988, p. 427.
12. D. R. Tenney and J. Unnam, NASA TM-78636, NASA-Langley Research Center, Langley VA, 1978.
13. J. K. Tien, T. Caulfield, and Y. P. Wu, *Metall. Trans. A*, **20A**, 267 (1989).
14. J. K. Tien, K. E. Bagnoli, and M. W. Kopp, in *Recrystallization '90*, T. Chandra ed., TMS, Warrendale PA, 1990, p. 261.
15. L. O. K. Larsson: Ph.D. Dissertation, Chalmers University of Technology, Goteborg, Sweeden, 1981.
16. M. W. Kopp, A. J. Carbone, and J. K. Tien, *Mater. Sci. Eng. A*, in press.
17. M. W. Kopp and J. K. Tien, *Scripta Metall.*, **22**, 1527 (1988).
18. S. G. Fishman, in *Proc. Industry-University Advanced Materials Conf. II*, F.W. Smith ed., AMI, Golden CO, 1989, p. 59.
19. P. F. Becher, *J. Amer. Cer. Soc.*, **74**, 255 (1991).
20. J. Aveston, G. A. Cooper, and A. Kelly, *The Properties of Fiber Composites*, IPC Science Technology, Guildford, U.K.,1971, p. 15.

Effect of Long Range Ordering on the Magnetic and Electronic Properties of Some Transition Metal Based Alloys

M. C. Cadeville[1], J. M. Sanchez[2], V. Pierron-Bohnes[1], and J. L. Morán-López[3]

[1] *IPCMS-GEMME*
Université Louis Pasteur
4 rue Blaise Pascal
67070 Strasbourg, France

[2] *Center for Materials Science and Engineering*
The University of Texas
Austin, Texas 78712, U.S.A.

[3] *Instituto de Física*
Manuel Sandoval Vallarta
Universidad Autónoma de San Luis Potosí
San Luis Potosí, S.L.P., Mexico

Abstract

This paper presents a short overview of the effect of long-range ordering on the magnetic and electronic properties of *fcc* Ni-Pt and Co-Pt and of *bcc* Fe-Al. The first part is a summary of experimental data. The second part is devoted to their discussion in the frame either of purely statistical models developed to simulate the phase diagrams, or of models which take into account both electronic structure considerations and results of statistical calculations.

I. Introduction

Many binary transition metal-based alloys order at low temperatures on simple *fcc* or *bcc* lattices, giving rise to the commonly called $L1_2$ or $L1_0$ structures in the case

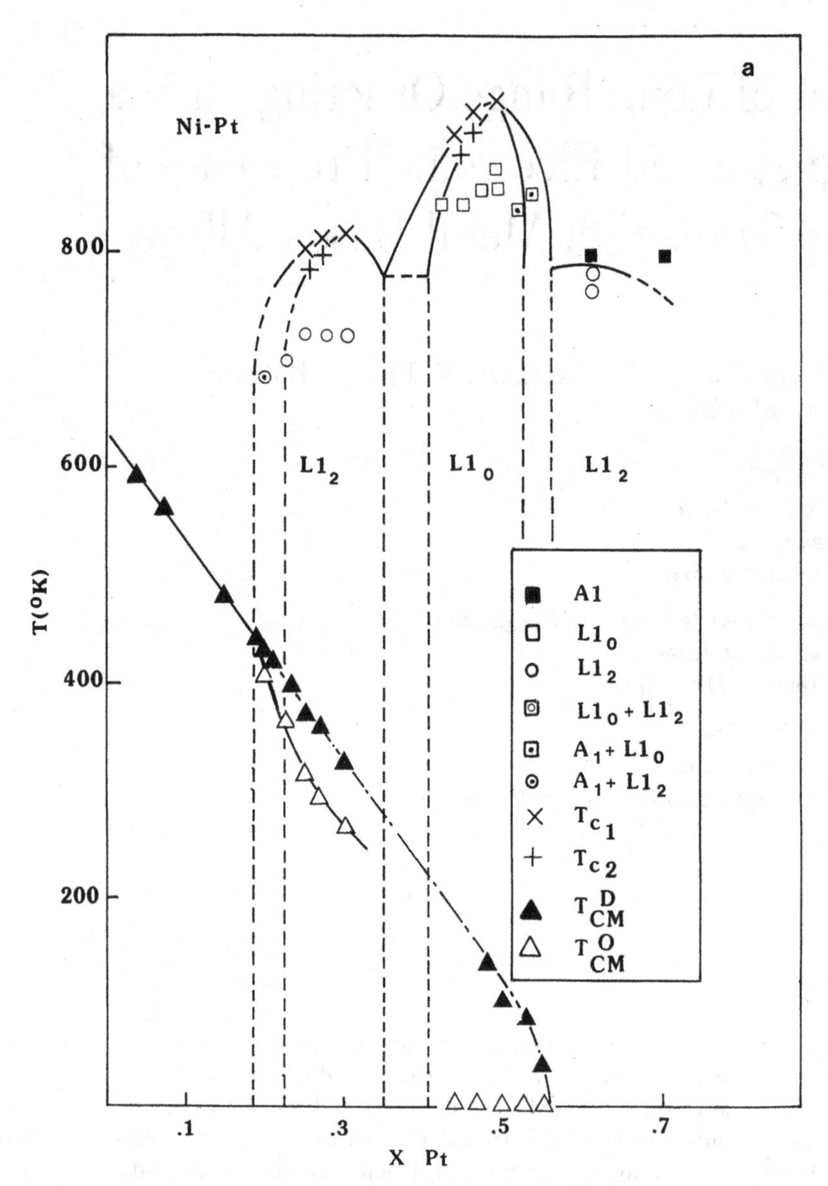

Figure 1. a) Experimental Ni-Pt phase diagram.

of the *fcc* lattice, DO_3 or B_2 structures in the case of the *bcc* one, for the A_3B-AB_3 ($L1_2$, DO_3) and AB ($L1_0$, B_2) compositions. The formation of such long-range ordered (LRO) structures strongly modifies the physical properties through the changes in the atomic and magnetic correlation functions, as well as in the local and average electronic structures, both changes being correlated.

The first part of the present contribution aims in presenting a lot of experimental results illustrating the sensitivity of magnetic, transport and electronic properties to the presence of LRO and to the value of the LRO parameter (η) when the temperature and the concentration vary. Examples are chosen among the data got by the Strasbourg's group in Co-Pt, Ni-Pt and Fe-Al systems.

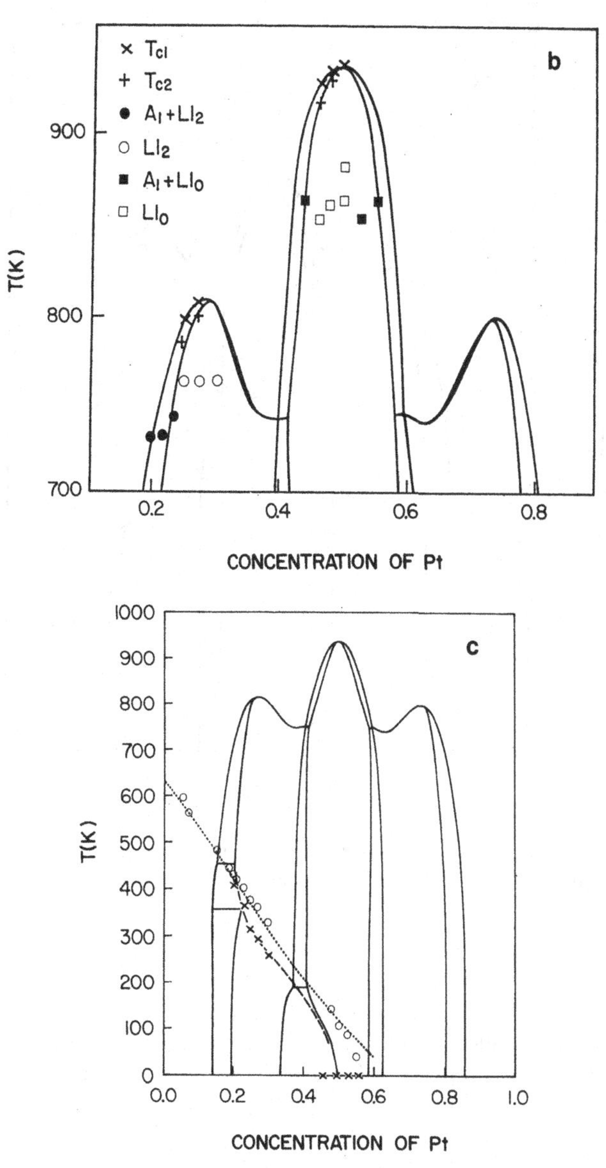

Figure 1. b), c) Calculated Ni-Pt phase diagram as obtained by means of the cluster variation method.[3a]

The second part is devoted to the discussion of these data in the frame of various approaches that take into account either statistical descriptions, or electronic structure considerations, or both of them simultaneously. It will be shown that, in Ni-Pt and Co-Pt systems, simple statistical models (CVM-type), developed within international collaborations (CNRS-CONACYT contract), that take into account the sensitivity of the magnetic moments to their chemical environment, succeed in reproducing the phase diagrams and some average magnetic properties (Curie temperatures, magnetizations, magnetic susceptibilities) as well as the contributions of the chemical and magnetic disorders to the resistivity. However, it appears that electronic structure considerations

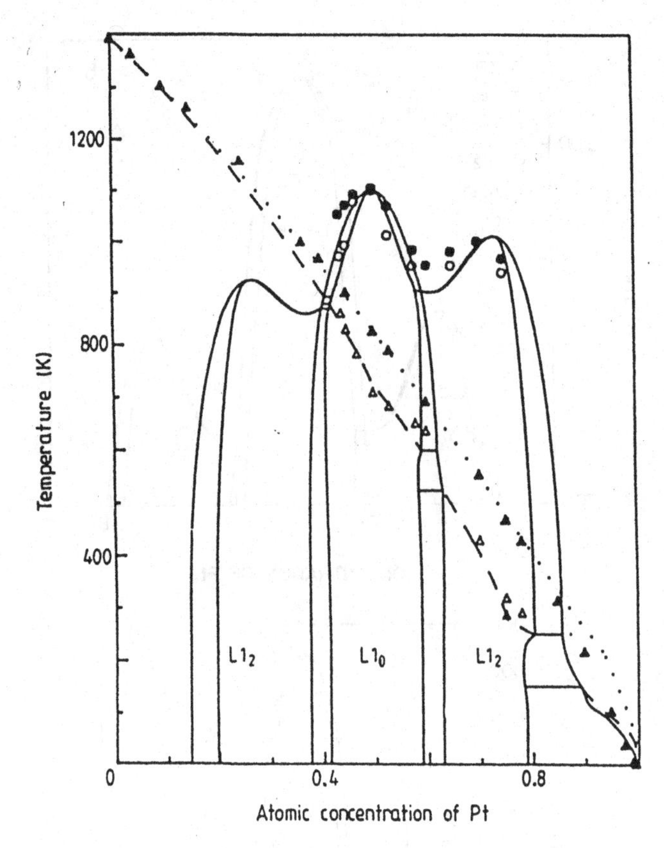

Figure 2. The Co-Pt phase diagram. •, o experimental phase limits. ▲, △ Curie temperatures in disordered and ordered phases. Full lines, dotted and chained lines are results of CVM calculations.[3b–3d]

added or not to a statistical description of alloys, have to be taken into account to explain other experimental data, such as the transport properties of non-magnetic Fe-Al alloys, or the hyperfine field distributions at Co and Pt nuclei in Co-Pt.

II. Effect of Long Range Order on Magnetic and Electronic Properties

Let us recall that all experimental data presented here correspond to well-defined thermodynamic states which can be either at equilibrium states *versus* T over the widest T-range allowed by diffusion rates towards low T, or high temperature states frozen in at low T under well controlled quenching rates. Informations on the atomic mobility in all the investigated systems (Ni-Pt, Co-Pt, Fe-Al) were obtained from previous investigations of SRO and LRO kinetics in these alloys (see for example Refs. 1 and 2).

The experimental magnetic and chemical phase diagrams of systems Ni-Pt, Co-Pt and Fe-Al on the iron-rich side are displayed in Figs. 1, 2, 3 together with the results of CVM calculations[3] when they exist. The Ni-Pt and Co-Pt diagrams are very similar to the prototype Cu-Au diagram, showing the succession of the $L1_2$, $L1_0$, $L1_2$ ordered

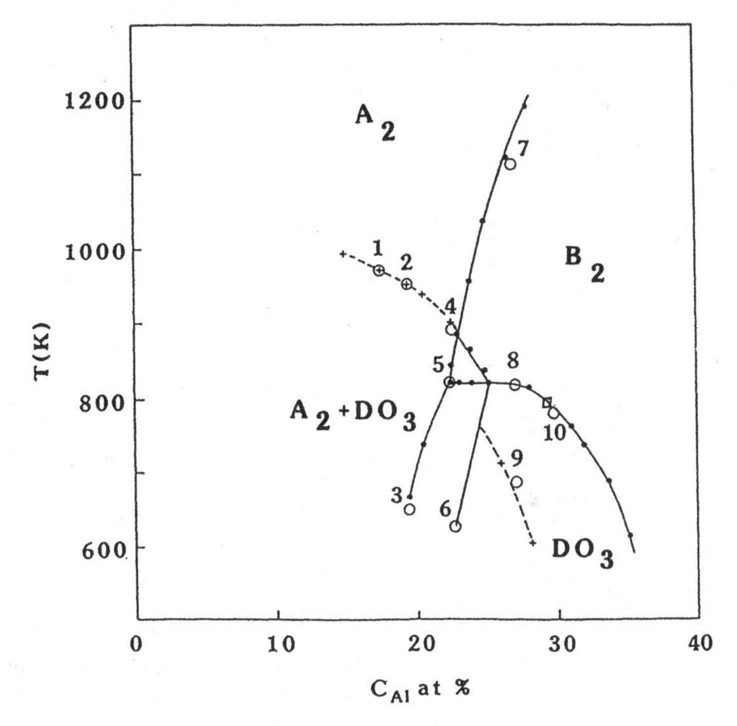

Figure 3. The experimental Fe-Al phase diagram[4] (o). The chained line represents T_{CM}.

structures respectively for the A_3B, AB and AB_3 compositions. The Fe-Al diagram on the iron-rich side[4] is characterized by the existence of the DO_3 and B_2 structures at low T, and of a single B_2 phase at high T over a broad concentration range.

In the following we summarize some results of average magnetic and transport properties of the Ni-Pt and Co-Pt systems. Then in Section I.2 we present results of local properties such as hyperfine field distributions in the Co-Pt system which are not yet published.

II.1. Average Properties

Magnetic properties

The magnetic properties (Curie temperature, average magnetization, magnetic susceptibility) of Ni-Pt and Co-Pt system have been determined in both ordered and disordered states.[5,6]

In both systems the long-range ordering decreases the Curie temperatures (T_{CM}) (Figs. 1 and 2), this effect being more drastic in Ni-Pt in which the long-range ferromagnetic order disappears with the formation of the $L1_0$ structure. A similar trend is observed in Fe-Al (Fig. 3) for which the curve $T_{CM}(x)$ displays a sudden decrease with the formation of the DO_3 phase.

The average magnetizations (σ) of ordered and disordered Ni-Pt and Co-Pt phases are displayed in Fig. 4. The behaviour of the Ni-Pt magnetization is comparable so

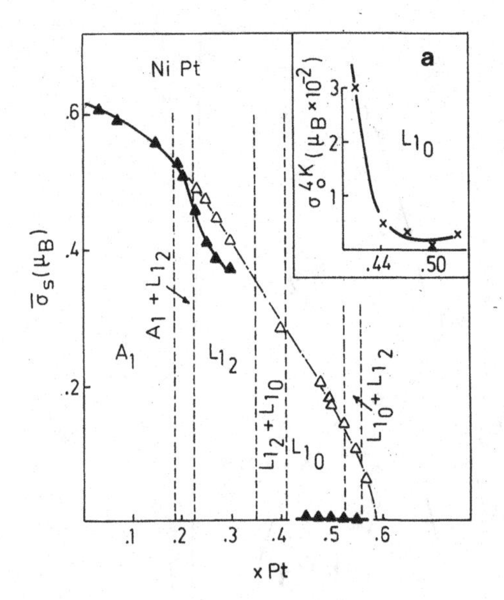

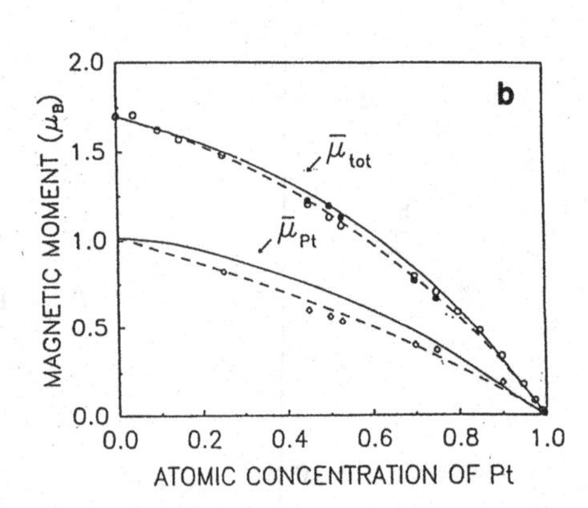

Figure 4. Concentration dependences of the average magnetizations in Ni-Pt (4a) and Co-Pt (4b) alloys. Open symbols correspond to alloys quenched from the disordered state, dark symbols to annealed (or ordered) phases. In Co-Pt, the average moment on Pt atom is deduced from $\bar{\mu}$ assuming $\mu_{Co} = 1.7\mu_B$; the full (broken) curve corresponds to results of CVM calculation[3] for the equilibrium (random) case at 1200K.[3]

that of their Curie temperatures, whereas in Co-Pt the average magnetization is nearly unsensitive to the formation of the LRO structures.

Most of investigated Ni-Pt and Co-Pt ordered phases display their order-disorder transitions above the Curie temperatures, *i.e.* in the paramagnetic range, as seen from the phase diagram. Thus the variation of the magnetic susceptibility through the order-disorder transition has been measured for Ni_3Pt, NiPt, CoPt and $CoPt_3$. All compounds display an increase of χ at T_{OD} as illustrated in Fig. 5 for CoPt and $CoPt_3$.

Thus the investigation of average magnetic properties of Ni-Pt, Co-Pt and at a least extent Fe-Al systems in both ordered and disordered states display the same trend, *i.e.* a decrease of magnetic interactions with the formation of long-range ordered structures.

Transport properties

Previous investigations[6-8] of the electrical resistivity in some intermetallic Ni-Pt and Co-Pt compounds showed that, within an extension of the Matthiessen's rule to magnetic alloys, the electrical resistivity of these concentrated alloys can be written, in a first approximation, as the sum of three contributions in both ordered (O) and disordered (D) states:

$$\rho_{\text{tot}}^{\text{O(D)}}(T) = \rho_0^{\text{O(D)}} + \mathcal{A}^{\text{O(D)}}T + \rho_m^{\text{O(D)}} \tag{1}$$

$\rho_0^{\text{O(D)}}$ is the residual resistivity or atomic disorder term which depends on temperature in the LRO state, through the variation of the LRO parameter (η), and it is nearly

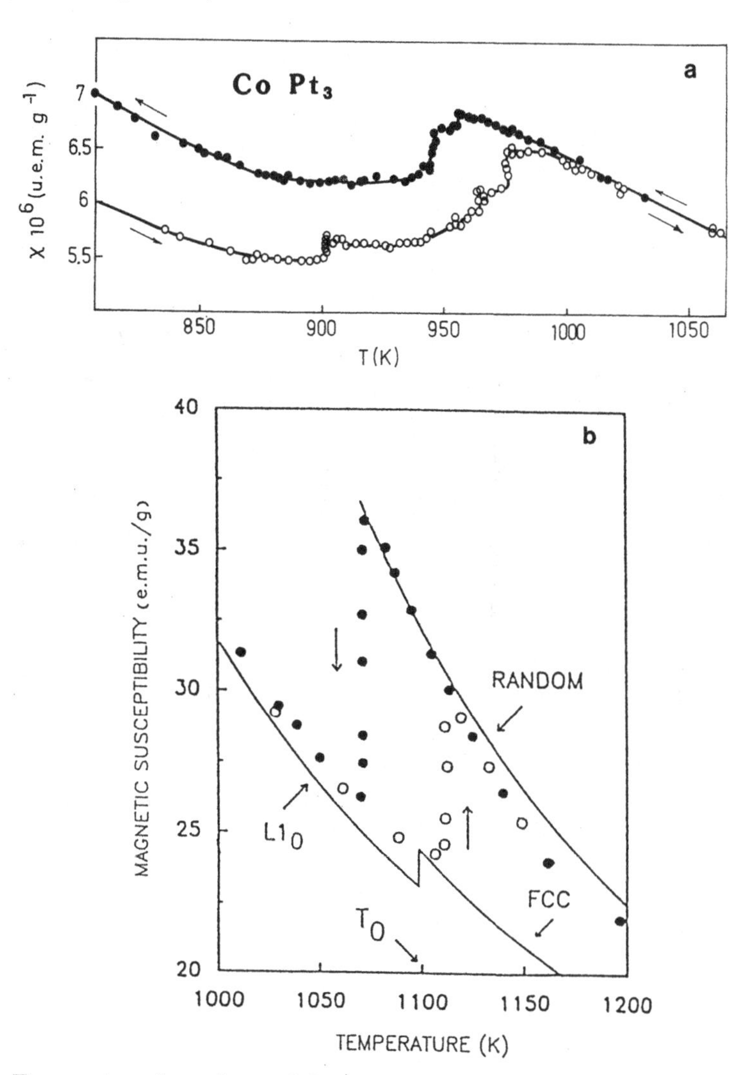

Figure 5. Temperature dependence of the magnetic susceptibility around the order-disorder transition in $CoPt_3$ (5a) and CoPt (5b). In 5b calculated results ($\times$ 1.35) are compared to experimental data for cooling ($\bullet$) and heating ($\circ$) respectively.

constant in the disordered state. $\mathcal{A}^{O(D)}T$ is the phonon term and its coefficient $\mathcal{A}$ is also dependent of temperature through η. The third term $\rho_m^{O(D)}$ is the spin disorder scattering term which is zero at low temperature in a ferromagnetic phase, and at any temperature in non-magnetic phases. In a simple localized spin model as described in Refs. 6 and 7, ρ_m is expected to increase from zero at low T to its maximum value at the Curie temperature T_{CM}, and to remain constant in the paramagnetic state.

The electrical resistivity has been investigated in non-magnetic $Fe_{1-x}Al_x$ alloys ($x \geq 0.28$) and $Ni_{1-x}Pt_x$ alloys ($x = 0.5$ and 0.7) and in ferromagnetic $Co_{1-x}Pt_x$ alloys ($x = 0.5$ and 0.7). Details on experimental procedure and analysis of data will be found in Refs. 6, 7, and 8.

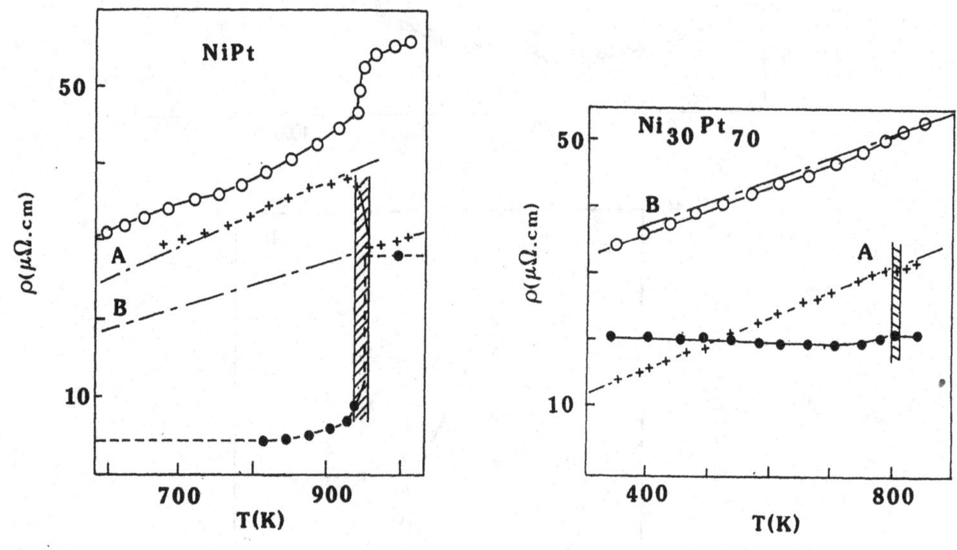

Figure 6. Resistivity data from Refs. 7 and 8 in NiPt and $Ni_{0.3}Pt_{0.7}$. Total (o), residual (•) and phonon (+) resistivities. The chain lines are least-squares fits of $\rho^{O}_{ph} = \mathcal{A}^{O}T$ (line A) and $\rho^{D}_{ph} = \rho^{D}_{0} + \mathcal{A}^{D}T$ (line B).

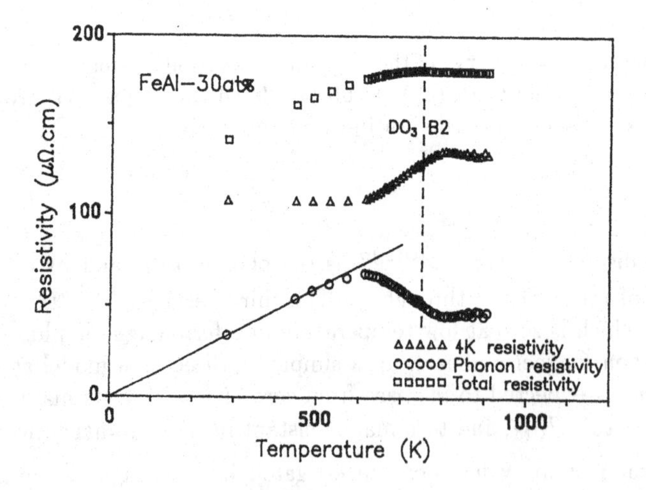

Figure 7.I.

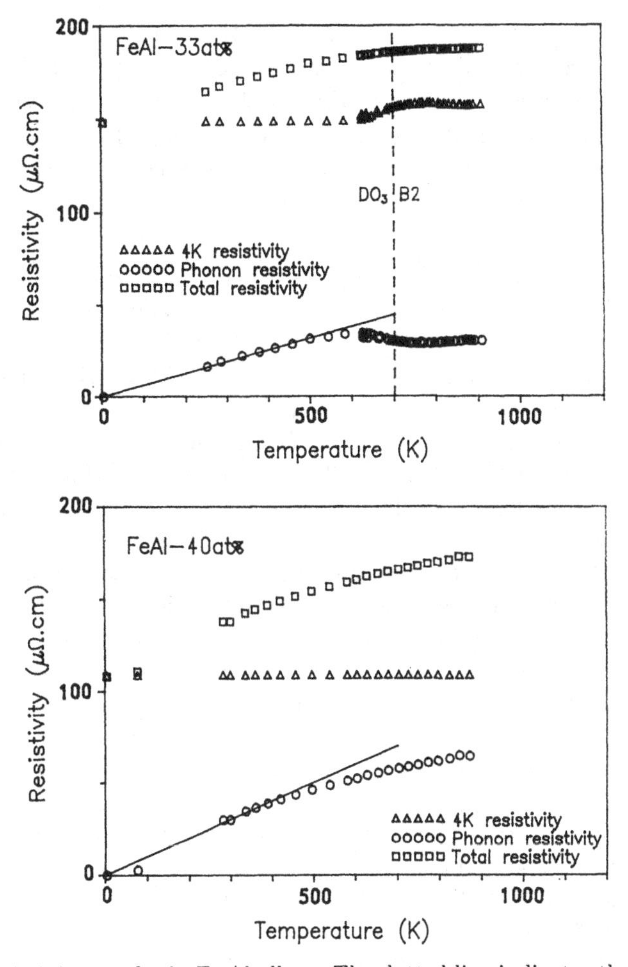

Figure 7.I. Resistivity results in Fe-Al alloys. The dotted line indicates the DO_3-B_2 transition, the full line: the slope of the phonon term when η is constant.

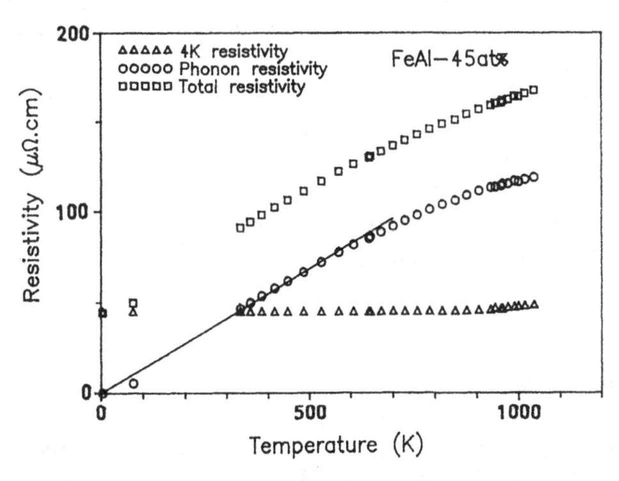

Figure 7.II. Resistivity results in Fe-Al alloys. The dotted line indicates the DO_3-B_2 transition, the full line: the slope of the phonon term when η is constant.

a) *Non-magnetic alloys.* In Ni-Pt and Fe-Al alloys, the magnetic contribution ρ_m being zero, it is simple to separate the contribution of the chemical disorder from that of the electron-phonon scattering. In addition to the total resistivity, we measure the residual resistivity due to the chemical disorder ($\rho_0^{\mathrm{O(D)}}$) by quenching the sample rapidly down to 4K, from an equilibrium state at high temperature characterized by a LRO parameter η. The phonon contribution is deduced by difference:

$$\rho_{\mathrm{ph}}^{\mathrm{O(D)}} = \mathcal{A}^{\mathrm{O(D)}}T = \rho^{\mathrm{tot}}(T) - \rho_0^{\mathrm{O(D)}}[\eta(T)].$$

The results obtained for Ni-Pt or for some $Fe_{1-x}Al_x$ alloys having either the DO_3 or the B_2 structure are shown in Figs. 6 and 7. In NiPt one observes the expected increase of ρ_0 when η decreases, whereas the phonon coefficient $\mathcal{A}$ remains constant as long as η is constant and then decreases at T_{OD}.

In $Fe_{1-x}Al_x$ alloys it is not possible to attain the disordered state because the $B_2 \rightarrow A_2$ transition temperature is superior to 900 K (our superior limit of measurement range) in the investigated x–range. In the alloys which have the DO_3 structure at low T ($x = 0.3$ and 0.33) one observes an increase of ρ_0 and correlatively a decrease of ρ_{ph} until the $DO_3 \rightarrow B_2$ transition temperature, followed by a plateau. Above about 850 K, the quenching rate is not rapid enough to frozen in the high-T equilibrium states, and the corresponding values of ρ_0, and consequently those of ρ_{ph} are not very significant. But it is clear from Fig. 7 that the phonon contribution is also very sensitive to the order state. In alloys which have the B_2 structure at low T ($x = 0.4$ and 0.45) one does not observe any significant change of both ρ_0 and ρ_{ph} because the T-measurement range is too far from the order-disorder $B_2 \rightarrow A_2$ transition to induce any important change of η.

However, by measuring the values of the phonon coefficient $\mathcal{A}$ for various concentrations over the low temperature range where the diffusionless effects frozen η in a constant value ($\eta_{T\rightarrow 0} = \eta_{\mathrm{max}}$), it is possible to observe the great sensitivity of $\mathcal{A}(T \rightarrow 0)$ to the disorder due to off-stoichiometry defects, as shown in Fig. 8. It is clear from Figs. 7 and 8 that $\mathcal{A}$ is maximum at low T for the stoichiometric compositions and is minimum either in the disordered state (Fig. 7) or for the largest off-stoichiometry (Fig. 8).

b) *Ferromagnetic alloys.* As we have three unknown quantities and only two experimental determinations, it is necessary to make some hypotheses in order to separate the various terms in (1). In view of the similarities between the vibrational properties of Co-Pt and Ni-Pt phases,[7,8] we assume that the phonon coefficients $\mathcal{A}$ are the same in both Co-Pt and Ni-Pt phases for a given composition and structure. So, using reduced T/T_{OD} temperature scales, it is possible to transfer the values of $\mathcal{A}(\eta(T/T_{\mathrm{OD}}))$ determined in $Ni_{1-x}Pt_x$ compounds to the $Co_{1-x}Pt_x$ compounds, and then to get $\rho_m(T)$ by difference from the relation in Eq. (1).

The values obtained for ρ_m in CoPt and $Co_{0.3}Pt_{0.7}$ are compared to the measured residual resistivity values in Fig. 9. One observes the expected increase of ρ_m until T_{CM} in the two phases, and an unexpected decrease in the paramagnetic range over the T-range where η decreases.

II.2. Local Properties

The hyperfine field distributions have been set up in three disordered $Co_{1-x}Pt_x$ alloys ($x = 0.25$, 0.5, and 0.75) at both ^{59}Co and ^{195}Pt nuclei, and in two ordered phases ($x = 0.5$ and 0.75) at both kinds of nuclei for $x = 0.75$ and at ^{59}Co nuclei only for

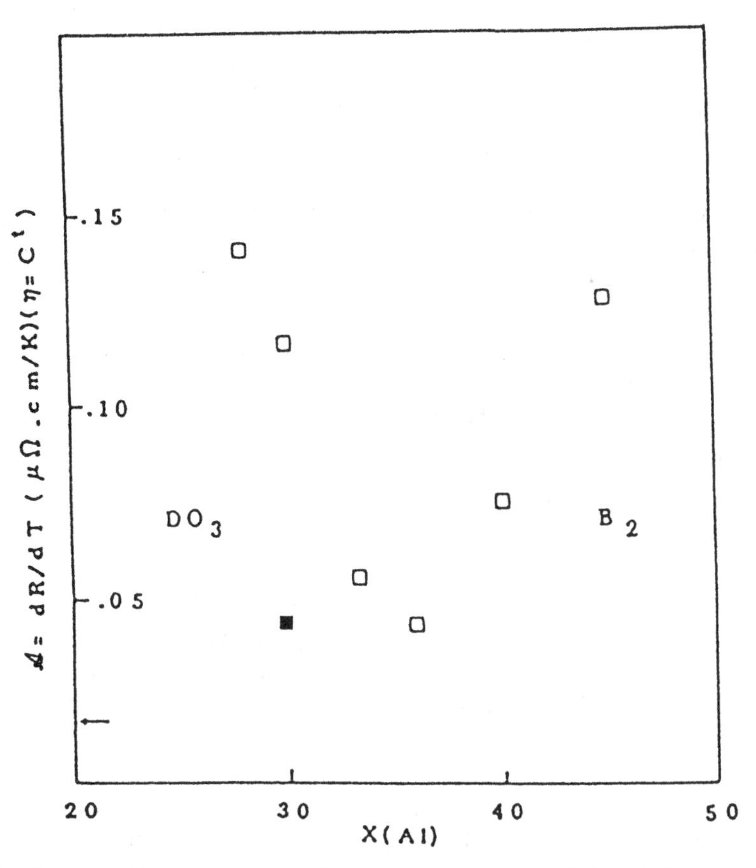

Figure 8. Concentration dependence of the linear phonon coefficient ($\mathcal{A}$) as given by the slope of straight lines in Fig. 7. The full square is the value in the disordered state as deduced from the simulation of data. The arrow indicates the values in the disordered alloys for $x \leq 0.2$.

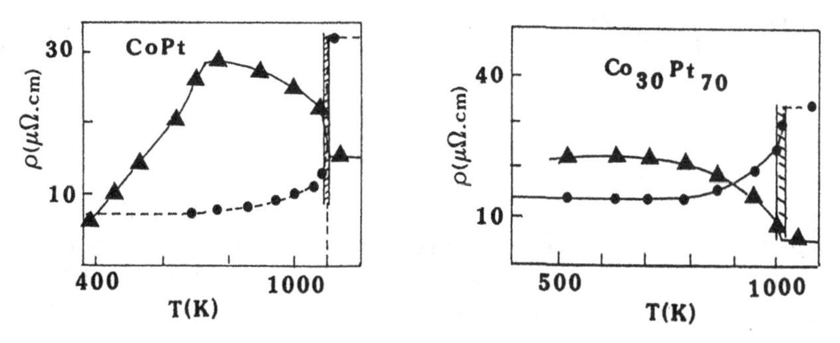

Figure 9. Resistivity data from Refs. 7 and 8 in CoPt and $Co_{0.3}Pt_{0.7}$: • measured residual resistivity, ▲ estimated magnetic contribution as described in the text.

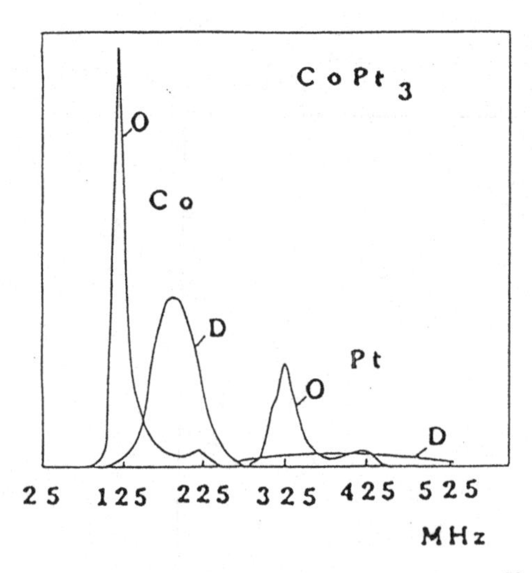

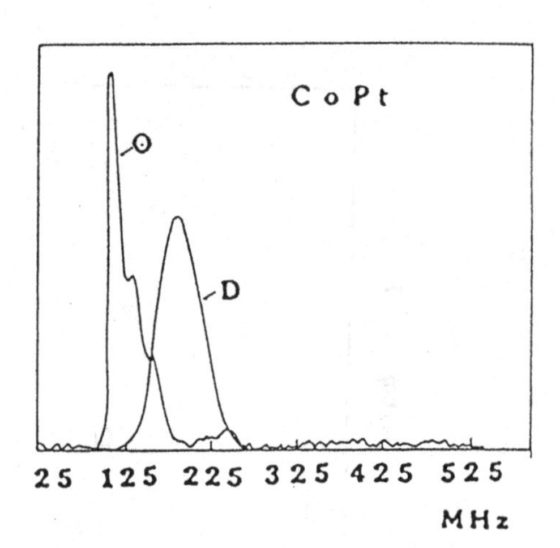

Figure 10. Hyperfine field distribution at ^{59}Co and ^{195}Pt nuclei in ordered and disordered states of $CoPt_3$ and at ^{59}Co nuclei of CoPt.

$x = 0.5$. Due to the complexity in modeling the hyperfine fields, a satisfying description of the whole of data is not yet on hand. Results in the disordered phase will be soon published. We would like only to illustrate here the important effect of ordering on the hyperfine field distribution and to discuss it qualitatively.

Figure 10a shows the "4K" hyperfine field distributions at both Co and Pt nuclei in the ordered and disordered states of $CoPt_3$ and Fig. 10b shows the same for CoPt at Co nuclei only. Besides the expected narrowing of the line with the formation of the LRO structure, one observes a big shift of the whole distribution towards lower frequencies in both compounds and at both nuclei in $CoPt_3$.

The asymmetry of the Co line in the ordered $CoPt_3$ could be assigned to Co antisite defects. The investigation of the CoPt $L1_0$ phase is not forward enough to establish whether the satellites observed are due either to antisite defects or to the magnetic asymmetry of the $L1_0$ structure.

Considering a phenomenological description of the hyperfine field at a given nucleus "i", in terms of a local contribution proportional to the local moment μ_i and of a non local contribution due to the polarization of the conduction electrons by the average moment of the alloy ($\bar{\mu}$), one has:

$$H_f^i = \alpha_i \mu_i + \beta_i \bar{\mu}.$$

As we have seen in I.1 that the average moments of CoPt and $CoPt_3$ are very slightly sensitive to the order state, and as previous neutron magnetic scattering experiments[9,10] indicate that the magnetic moments of cobalt in ordered phases are very close to its value in pure cobalt, the large decrease of H_f^{Co} with ordering can be qualitatively attributed in first approximation to an important change in the hyperfine field coupling constants (α_{Co}, β_{Co}), *i.e.* to electronic structure effects.

This simple qualitative discussion shows the important changes in the electronic structure that occur with the formation of the LRO structures.

III. Discussion

III.1. Statistical Approach

Magnetic statistical models have been developed more or less recently to reproduce the chemical and magnetic Ni-Pt and Co-Pt phase diagrams.[3] As in these models, magnetic environment effects are considered, one can say that, in that sense, such approaches, although being essentially of statistical type, take also into account the local electronic structure changes associated to various local environments. In spite of their simplicity and of some crude approximations therein, they are able to reproduce a lot of experimental trends such as the magnetic properties (T_{CM}, σ) of ordered and disordered alloys, as well as the contributions of the chemical and magnetic disorders to the resistivity.

Phase diagrams and magnetic properties

Details on the statistical method will be found in Ref. 3. Let us recall that the Hamiltonian containing both chemical and magnetic interactions was treated in the tetrahedron approximation of the Cluster Variation Method (CVM).

In the Ni-Pt system, a magnetic moment is assigned only to Ni atoms. To simulate the well-known environmental effects on the Ni magnetic moment,[11] it is assumed that the exchange interactions between two NN Ni-atoms vanishes unless a third Ni atom is also present as a common nearest to the other two, that leads for the exchange integral between Ni atoms:

$$J_{\mathrm{Ni(Pt\,Pt)Ni}} = 0,$$

and

$$J_{\mathrm{Ni(Ni\,Ni)Ni}} \neq J_{\mathrm{Ni(Ni\,Pt)Ni}} \neq 0.$$

Such a model succeeds in reproducing the high temperature paramagnetic phase diagram (Fig. 1b) as well as the Curie temperatures of both disordered (states frozen in atomic order at 2000 K) and ordered phases (states frozen at 730 K) (Fig. 1c). The model is in good agreement with experiment. The average magnetizations have not been simulated.

The Co-Pt chemical phase diagram is very similar to that of Ni-Pt. The magnetic phase diagram is different since the Co-Pt alloys are ferromagnetic at low temperature over the whole concentration range. This is due to the fact that, even on the Pt-rich side, the Co atoms wear a magnetic moment.[12] An important question which is still debated is to know whether the magnetic moments on Co and/or Pt atoms are strongly dependent or not on their environment. The Co-Pt magnetic and chemical phase diagram was fitted to experimental values in the same tetrahedron approximation of the CVM as for the Ni-Pt system. The magnetic moment on Co atoms was assumed to be constant and equal to 1.7 μ_B, its value in pure cobalt. The magnetic moments on Pt atoms were deduced from the measured average magnetizations of ordered compounds assuming $\mu_{\mathrm{Co}} = 1.7\mu_B$ and taken respectively equal to 0.6 μ_{Co}, 0.45 μ_{Co}, and 0.3 μ_{Co} in Co_3Pt, CoPt and $CoPt_3$. A satisfying description of the chemical phase diagram is obtained (Fig. 2). The calculated equilibrium Curie temperatures are in good agreement with experiment, as well as those of the disordered (quenched) alloys. The calculated average total magnetic moments and the average magnetic moments on Pt atoms in disordered alloys are correctly reproduced (Fig. 4b).

Nevertheless it seems clear today that such a model enhances the values of Pt moments and their concentration dependence. Very recent electronic structure calculations of magnetic moments in the disordered Co-Pt alloys by Ebert[13] indicate an increase of Co moment with x_{Pt} from 1.6 μ_B ($x_{Pt} = 0$) to about 2.4 μ_B ($x_{Pt} = 1$), whereas the Pt moment remains constant around 0.25±0.05 μ_B for x_{Pt} between 0 and 0.8, then decreasing rapidly towards zero in pure Pt.

Moreover, spin polarized band structure calculations of ordered Co-Pt compounds by Kootte *et al.*[14] yield very few concentration dependent moments on both Co and Pt atoms. Their calculated values of the magnetic moments in CoPt and $CoPt_3$ are in excellent agreement with the values for the ordered compounds as determined by neutron diffraction[9,10], *i.e.* $\mu_{Co} \simeq 1.69\,\mu_B$ and $\mu_{Pt} = 0.37\,\mu_B$ in CoPt and $0.27\,\mu_B$ in $CoPt_3$. However, when recalculating the average magnetizations of compounds from these values, one obtains clearly smaller average moments than the measured values. This probably means that other contributions that spin magnetism, such as orbital magnetism, are included in the experimental data. Our hypothesis of a Co moment nearly constant through the series is reasonable at least in ordered compounds. Probably a non negligible contribution of orbital moment is included in our values of Pt moments.

Transport properties

In addition to the free energies and phase diagram, the CVM yields a detailed description of the state of order in terms of temperature-dependent tetrahedron probabilities $\chi(\sigma, s)$ where σ is the chemical operator and s is the spin operator. We have shown[15] that in this approximation, all functions of the configuration of the system, such as the expectation value of resistivities can be written in terms of the probability distribution of the occupation variables σ_i (i for the site) and s_i or, equivalently, in terms of their correlation functions, up to the maximum tetrahedron cluster, giving:

$$\rho^\gamma = \sum_{(\sigma,s)} \Delta\rho^\gamma(\sigma, s)\chi(\sigma, s) \tag{2}$$

$\Delta\rho^\gamma(\sigma, s)$ gives the effective contribution to the resistivity arising from the chemical and magnetic disorder due to a tetrahedron cluster in the configuration (σ,s) in phase γ. If the configuration (σ,s) matches that of the cluster characteristic of phase γ, e.g. (σ^γ,s^γ), we expect the corresponding $\Delta\rho^\gamma(\sigma, s)$ to vanish.

We can further simplify Eq. (2) if we assume that, for clusters that are not chemically matched to the ordered phase γ, *i.e.* $\sigma \neq \sigma^\gamma$, their contribution to the residual resistivity is essentially the same irrespective of the spin disorder, and equal to $\Delta\rho_0$. On the other hand, those clusters matching the chemical environment of γ which are in the majority in the stoichiometric phases are assumed to contribute only a spin disorder term equal to $\Delta\rho_m$. With this in mind, the residual resistivity can be separated into chemical and magnetic contributions:

$$\rho^\gamma = \rho_0^\gamma + \rho_m^\gamma \tag{3}$$

with,

$$\rho_0^\gamma = \Delta\rho_0 \sum_\sigma{}' \sum_s \chi(\sigma, s) \tag{4}$$

and,

$$\rho_m^\gamma = \Delta\rho_m \sum_s{}' \chi(\sigma^\gamma, s) \tag{5}$$

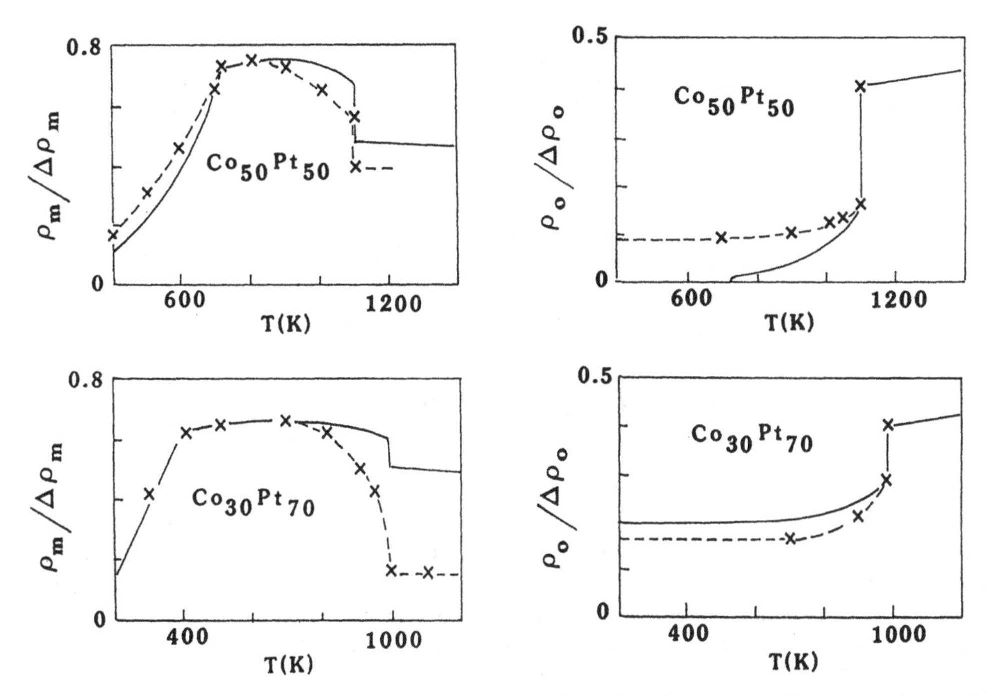

Figure 11. Comparison between calculated curves (solid lines) and normalized experimental data (×).

where in Eq. (4) the sum over the chemical occupation σ excludes those clusters chemically matched to the ordered phase γ, and in Eq. (5) the sum over s excludes tetrahedron clusters with all spins up (or down).

The calculated curves of $\rho_0/\Delta\rho_0$ and of $\rho_m/\Delta\rho_m$ obtained using Eqs. (4) and (5) are shown in Fig. 11 for CoPt and $Co_{0.3}Pt_{0.7}$, together with the experimental data normalized to one point on the theoretical curve.

In CoPt the agreement between theory and experiment is fairly good. For the chemical disorder term, the resistivity jump at T_{OD} is well reproduced. Concerning the magnetic disorder term, the changes in both ferromagnetic and paramagnetic phases at T_{OD} are correctly reproduced. In $Co_{0.3}Pt_{0.7}$ the agreement is good for the chemical disorder term. However, the experimental spin disorder term displays a much larger variation around T_{OD} than that given by the model. This discrepancy could be partially due to errors in the evaluation of the phonon term, the decrease of which around T_{OD} is expected to be more important in $Co_{0.3}Pt_{0.7}$ than in $Ni_{0.3}Pt_{0.7}$. Thus, part of the decrease in $\rho_m(T)$ at T_{OD} is due to the variation of the phonon contribution which has not been evaluated correctly.

Nevertheless, despite this discrepancy, we conclude that both the chemical and magnetic disorder contributions to the resistivity of Co-Pt intermetallic compounds are correctly described by the model, namely as sums of resistivities of individual tetrahedral clusters, with probabilities calculated in the tetrahedron approximation of a chemical and magnetic CVM.

This oversimplified description of the resistivity in terms of Eqs. (2–5) obviously has its limitations. Among the most important ones, one can mention that it can be ap-

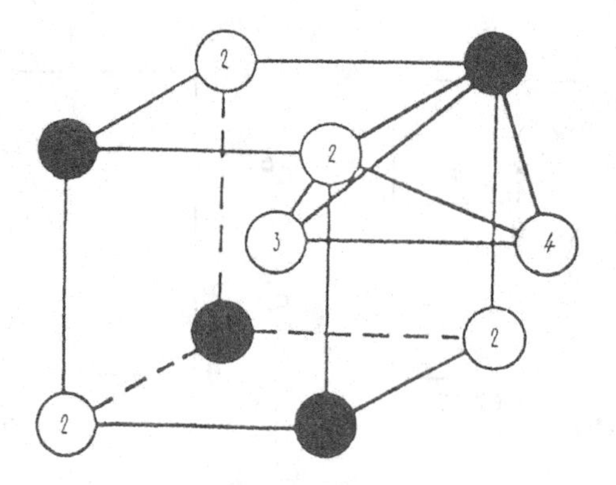

Figure 12. DO_3 (Fe_3Al) structure. • Al, o Fe.

plied only to concentrated alloys having small electron mean free paths, and to magnetic alloys having good localized moments in their paramagnetic state (static approximation). Moreover, it does not take into account the changes in the electronic structure associated with the formation of the LRO structures, as the following approach does.

III.2. Approach Lying on both Electronic Structure Considerations and Results of Statistical Calculations

A theoretical approach has been developed about ten years ago by Rossiter[16] in order to reproduce the overall dependence of the resistivity with the LRO parameter η in non-magnetic alloys. In the case of a stoichiometric composition and above Θ_D the Debye temperature, it yields:

$$\rho(\eta, T) = \rho_0^D \frac{1-\eta^2(T)}{1-A\eta^2(T)} + \frac{B}{n_0} \frac{1}{1-A\eta^2(T)} T, \tag{6}$$

where T is the measuring temperature, ρ_0^D, n_0 and B/n_0 are respectively the residual resistivity, the density of conduction electrons, and the coefficient temperature of the phonon contribution in the disordered state.

The relation (6) was obtained by considering the effect of long-range atomic ordering on the scattering of conduction electrons in an appropriate pseudo-potential model treated in the Bragg-Williams approximation, the resistivity being described within the simple relaxation time approximation:

$$\rho = \frac{m^*}{n_{\text{eff}}} \frac{1}{e^2 \tau}, \tag{7}$$

where τ is order-dependent through:

$$\tau^{-1} = \tau_0^{-1} \left[1 - \eta^2(T)\right], \tag{8}$$

τ_0 is the relaxation time corresponding to the disordered state, m^* and n_{eff} are respectively the effective mass and the effective density of the electron conductions.

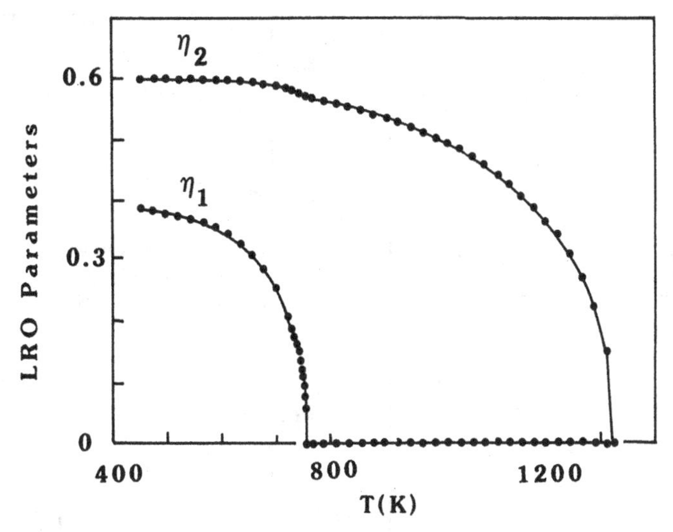

Figure 13. Calculated LRO parameters in $Fe_{0.7}Al_{0.3}$ from Sanchez.[17]

As the long-range atomic ordering may introduce new gaps into the Fermi surface at the superlattice Brillouin zone boundaries, Rossiter has shown that the value of n_{eff} and/or m^* will be order-dependent according to the following expression:

$$\frac{n_{\text{eff}}}{m^*} = \frac{n_0}{m_0^*}\left[1 - A\eta^2(T)\right]. \tag{9}$$

The coefficient A depends upon the relative positions of the Fermi surface and the superlattice Brillouin zone boundaries. Its sign determines the evolution of the electronic band structure near the Fermi level with the formation of the LRO structure. For a given concentration, the relation (6) can be separated into two contributions:

$$\rho^{\text{tot}} = \rho_0\left[\eta(T)\right] + \mathcal{A}T, \tag{10}$$

ρ_0 is the residual resistivity term which is T-dependent through $\eta(T)$, and $\mathcal{A}T$ is the phonon contribution whose coefficient $\mathcal{A} = B/n_0[1-A^2(T)]$ is also T-dependent through η.

This model which has been already applied by Rossiter[16] (1980) to Cu_3Au and Fe_3Al is *a fortiori* able to describe qualitatively the behaviours of ρ_0 and ρ_{ph} observed in non-magnetic Ni-Pt and Fe-Al alloys (Figs. 6 and 7). Here we apply it to a quantitative description of the resistivity of $Fe_{0.7}Al_{0.3}$ by using the LRO parameters calculated by Sanchez[17] and introducing them in Eq. (6). These theoretical values of η were obtained through a fitting of the non-magnetic Fe-Al phase diagram to experimental values within the tetrahedron approximation of the CVM which includes, in a *bcc* lattice, N- and NN-neighbour interactions.

Let us recall that, in the BW approximation the order state of the DO_3 phase which consists of 4 interpenetrating sublattices (Fig. 12), two of them (3 and 4), being equivalent, is determined by two LRO parameters:

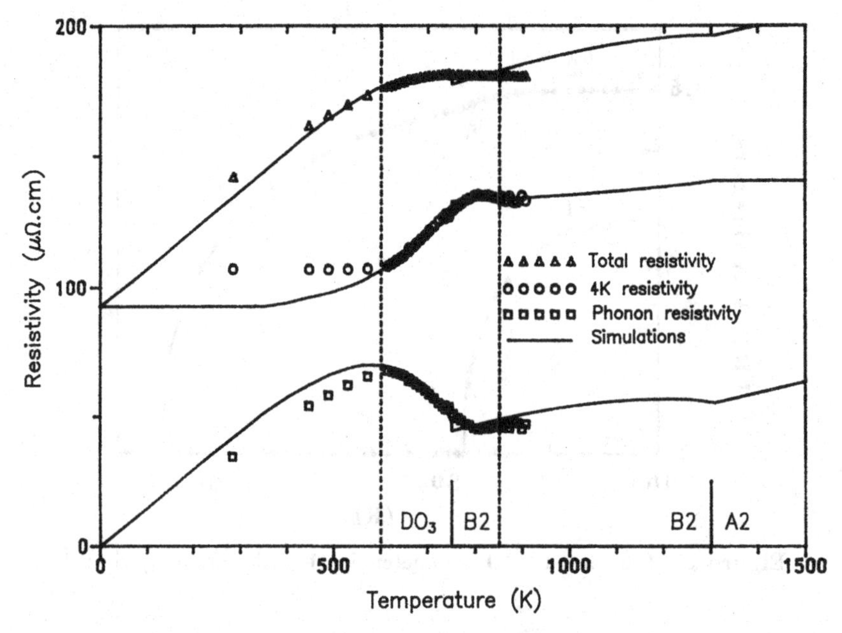

Figure 14. Resistivity data in $Fe_{0.7}Al_{0.3}$. The vertical dotted lines limit the T-range where equilibrium states can be experimentally investigated.

$$\eta_1 = \frac{1}{2}(P_A^1 - P_A^2),$$

$$\eta_2 = \frac{1}{2}(P_A^1 + P_A^2 - 2\,P_A^3),$$

where P_A^i is the occupation probability of the "i" sublattice by an Al atom.

For a stoichiometric alloy (Fe_3Al) and at $T = 0$, $P_A^1 = 1$, $P_A^2 = P_A^3 = 0$. That gives $\eta_1(T = 0) = \eta_2(T = 0) = 0.5$. For an off-stoichiometric $Fe_{3-x}Al_{1+x}$ alloy one gets at $T = 0$: $\eta_1^{\max} = (1 - x)/2$ and $\eta_2^{\max} = (1 + x)/2$. In the B_2 structure, the sublattices 1 and 2 become equivalent. The values of η_1 and η_2 obtained by Sanchez in $Fe_{0.7}Al_{0.3}$ are shown in Fig. 13. Introducing them in both terms of Eq. 6, it is possible to reproduce the experimental dependences of ρ_0 and ρ_{ph} using a unique set of A and B/n_0 parameters:

$$A = 0.8787,$$

$$B/n_0 = 0.042\,\mu\Omega\,\mathrm{cm}\,\mathrm{K}^{-1},$$

and an effective LRO parameter: $\eta_{\mathrm{eff}} = \eta_1 + 3/4\,\eta_2$.

Over the 600–850 K T-range where equilibrium states can be experimentally observed the experiments are in good agreement with the model (Fig. 14). At low T as the order state is frozen in a thermodynamic state of about 600 K, the measured residual resistivity is superior to the calculated values which correspond to equilibrium values of η. At high T, above 850 K, the quenching rate is not rapid enough to frozen in high temperature equilibrium states that explains the uncertainty in the experimental points. Nevertheless, the theoretical behaviour reproduces the experimental trend as measured until 900 K.

A small discrepancy ($\approx 30\,K$) is observed between the temperatures of the theoretical and experimental cusps corresponding to the $DO_3 \rightarrow B_2$ transition. This could have two origins: (i) rounding effects on the experimental curve due to the presence of SRO correlations, (ii) an effect due to the experimental device in which the thermocouple is not in good thermal contact with the sample.

The value of B/n_0 which corresponds to the phonon coefficient in the disordered state has a correct order of magnitude, compared for example to the value measured at low T in the most compositionally disordered alloy, $Fe_{0.64}Al_{0.36}$ (Fig. 8).

The positive value of A corresponds to an increase of n_{eff} (or n_{eff}/m^*) with increasing T, *i.e.* with increasing disorder. Let us mention that results of electronic specific heat coefficients (γ) around 30 Al at.% indicate a strong increase of γ in a quenched state[18] with respect to an annealed state.[19] But in lack of electronic structure calculations in these Fe-Al alloys, it is difficult from the above considerations to get some more reliable informations in the electronic structure rearrangement consecutive to the order variation.

It is obvious that the same formalism could be used to reproduce other resistivity data in non-magnetic Ni-Pt and Fe-Al alloys, as long as the values of LRO parameters deduced from the simulation of the phase diagrams are available.

It could even be extended to the simulation of the spin disorder scattering term in the paramagnetic state since, similarly to the other contributions, it is also dependent of n_{eff}. But the situation is more intricated in that case due to the presence of short range magnetic correlations.

IV. Conclusion

The effect of LRO on some average and local properties of Ni-Pt, Co-Pt, and Fe-Al systems has been reported and discussed.

The phenomenological statistical model that have been previously developed to simulate the Ni-Pt and Co-Pt phase diagrams are able to reproduce some general trends of the average magnetic and transport properties such as for example the decrease of magnetism with the formation of LRO structures and the temperature dependence of the chemical and magnetic disorder contributions to resistivity. However, such models that do not take into account any change in the electronic structure due to the LRO structure formation are unable to explain the important shifts in the hyperfine field distributions at ^{59}Co and ^{195}Pt nuclei consecutive to the formation of LRO structures. The anomalous T-dependence of the phonon contribution to the resistivity cannot also be explained by such a model. It is shown that an approach lying on both electronic structure considerations and calculation of LRO-parameters in a reliable statistical model fitted to the phase diagram is more satisfying. For example, the various components of the resistivity of a non-magnetic $Fe_{0.7}Al_{0.3}$ are well reproduced. Such a model will be nextly extended to other ordered phases magnetic or not.[20] Interesting information on the electronic structure dependence upon the degree of order should be obtained.

References

1. C. E. Dahmani, M. C. Cadeville, and V. Pierron-Bohnes, *Acta Metall.* **33**, 369 (1985).
2. P. Vennégues, M. C. Cadeville, V. Pierron-Bohnes, and M. Afyouni, *Acta Metall. Materialia* **38**, 2199 (1990).
3. (a) C. E. Dahmani, M. C. Cadeville, J. M. Sanchez, and J. L. Morán-López, *Phys. Rev. Lett.* **55**, 1208 (1985); (b) J. M. Sanchez, J. L. Morán-López, C. Leroux, and M. C. Cadeville, *J. Phys. Condens. Matter* **1**, 491 (1988); (c) J. M. Sanchez, J. L.

Morán-López, C. Leroux, and M. C. Cadeville, *J. Physique, Colloque C8, Suppl. N. 12,* **49**, 107 (1988); (d) J. M. Sanchez, J. L. Morán-López, and M. C. Cadeville, *Proceedings of MRS Conference*, Boston (1990).
4. (a) M. Afyouni, Thesis of Louis Pasteur University. Strasbourg, France (1989); (b) W. Köster and T. Gödecke, *Z. Metallk.* **71**, 765 (1980).
5. (a) M. C. Cadeville, C. E. Dahmani, and F. Kern, *J. Magn. Magn. Mater.* **54–57**, 1055 (1986); (b) C. E. Dahmani, Thesis of Louis Pasteur University, Strasbourg, France (1985).
6. C. Leroux, Thesis of Louis Pasteur University, Strasbourg, France, (1989).
7. C. Leroux, M. C. Cadeville, V. Pierron-Bohnes, G. Inden, and F. Hinz, *J. Phys. F: Met. Phys.* **18**, 2033 (1988).
8. C. Leroux, M. C. Cadeville, and R. Kozubski, *J. Phys. Condens. Matter* **1**, 6403 (1989).
9. B. Van Laar, *J. Physique* **25**, 600 (1964).
10. F. Menzinger and A. Paoletti, *Phys. Rev.* **143**, 365 (1966).
11. R. E. Parra, and J. W. Cable, *J. Appl. Phys.* **50**, 7522 (1979); *Phys. Rev. B* **21**, 5494 (1980).
12. B. Tissier, Doctoral dissertation, USMG Grenoble, France, unpublished (1977).
13. H. Ebert, private communication, Strasbourg (1991).
14. A. Kootte, C. Haas, and R. A. de Groot, *J. Phys. Condens. Matter* **3**, 1153 (1991).
15. J. M. Sanchez, M. C. Cadeville, and V. Pierron-Bohnes, *Proceedings of MRS Conference*, Boston (1990).
16. P. L. Rossiter, *J. Phys. F: Metal Phys.* **9**, 891 (1979); **10**, 1459 (1980); **11**, 615 (1981); P. L. Rossiter, *The electrical resistivity of metals and alloys*, Cambridge Solid State Science Series, R. W. Cahn, E. A. Davis and I. M. Ward, editors. Cambridge University Press, Cambridge (1987).
17. J. M. Sanchez, private communication (1991) and this volume Chapter 10.
18. C. H. Cheng, K. P. Gupta, C.T. Wei, and P. A. Beck, *J. Phys. Chem. Solids,* **25** 759 (1964), R. Kuentzler, *J. Physique* **44**, 1167 (1983).
19. H. Okamoto and P. A. Beck. *Monatsh. Chem.* **103**, 907 (1972).
20. M. C. Cadeville, V. Pierron-Bohnes, and J. M. Sanchez, paper in preparation.

Ferromagnetic Behavior of Pd- and Pt-Based Alloys

R. E. Parra and A. C. González

Centro de Física
Instituto Venezolano de Investigaciones Científicas
Apartado 21827, Caracas 1020 A
Venezuela

Abstract

A review of the magnetic properties of Pd- and Pt-based alloys is presented. Starting with Ni-Pt and Ni-Pd alloys, we discuss the effects of ferromagnetic clusters on the onset of ferromagnetism of these alloys, using a magnetic environment model. The model is then extended to giant moment dilute alloys, where Pd or Pt are the host atoms and Fe or Co are the magnetic impurities. In both cases, for concentrated and dilute alloys, several magnetic properties are calculated in the ferromagnetic region, and the results compared with experimental values. The agreement obtained proves the validity of this model.

I. Giant Moments and Exchange Enhancement

The term giant moment usually refers to the case of a localized, isolated, impurity moment dissolved into a nonmagnetic host, producing a much greater moment than the one corresponding to the bare impurity atom. The total moment is produced by the one corresponding to the impurity plus the induced moment on the surrounding host atoms. This total moment is usually treated as one entity called the "giant moment" or "polarization cloud". Typical examples of this are the PdFe and PtFe alloys. In the first case, an Fe atom, with a moment[1] of 3.5 μ_B produces a polarization on the surrounding Pd atoms, inducing small moments on them (about 0.07 μ_B) but affecting so many host atoms that the net result is a giant moment[2,3,4] of 10 μ_B. This means that the cloud covers more than 200 host atoms within a range of approxately 10Å. In the PtFe case, neutron measurements have given a value of 3.18 μ_B for the Fe moment[5] and according to magnetization results[2] the value of the total moment of the cloud is 4.9 μ_B.

For binary transition metal alloys, two elements are normally present in the formation of giant moments; first, a ferromagnetic impurity (although Mn at dilute concentrations acts also as a ferromagnet) and second, an exchange enhanced host such as Pd or Pt. In this paper, we will limit ourselves only to alloys where the host is one of these metals. We will start, however, with two systems which are not really classified as giant moment alloys, in the way we have defined it here, but that produce large polarization clouds and that have been of vast interest for a number of years. Also, their analysis will conduce ourselves to the study of dilute giant moment alloys. These systems are Ni-Pd and Ni-Pt, in which Pd and Pt acquire induced moments that behave magnetically like Ni.

II.1 Ferromagnetic Clusters in Concentrated Alloys

In 1970, Cable and Child[6] published their results of polarized neutron measurements on Ni-Pd near the critical concentration. The average moment of the alloys were known to decrease gradually to about 60 at.% Pd, and afterwards to decrease more rapidly toward the dilute limit of Ni. The neutron measurements showed that the Ni and Pd moments increased with Pd content to a concentration of about 90 at.% Pd, and then both moments dropped rapidly. In 1980, Parra and Cable[7] published their neutron scattering results on Ni-Pt alloys. In this case, both the Ni and the Pt moment decreased with concentration. These alloys also showed inhomogeneous moment distributions in the form of ferromagnetic clusters and indicated the importance of chemical order in the moment distribution of these systems.

In the same year Parra and Medina[8] published a magnetic environment model for Ni-Pt and Ni-Pd alloys. Here we present a review of this model, the details of which can be found in Ref. 8. Assuming a random alloy, we take the moment at each magnetic site to be a function of the number of its impurity neighbors and of an exchange field produced by its nearest neighbors. The moment at site $\mathbf{n}$ is:

$$\mu_{\mathbf{n}} = (1 - p_{\mathbf{n}})F_i(H^h_{\mathbf{n}}, \nu_{\mathbf{n}}) + p_{\mathbf{n}}F_h(H^i_{\mathbf{n}}, \nu_{\mathbf{n}}), \tag{1}$$

where h and i denote host and impurity atoms, respectively, $\nu_{\mathbf{n}}$ is the number of impurity neighbors, $H_{\mathbf{n}}$ is the exchange field, and $p_{\mathbf{n}}$ is a site occupation operator which is unity if there is an impurity atom at $\mathbf{n}$ and is zero otherwise. In this equation:

$$H^h_{\mathbf{n}} = \sum_{\mathbf{k}}[J_{hh}(1 - p_{\mathbf{n+k}}) + J_{ih}\,p_{\mathbf{n+k}}]\mu_{\mathbf{n+k}}, \tag{2}$$

$$H^i_{\mathbf{n}} = \sum_{\mathbf{k}}[J_{ii}p_{\mathbf{n+k}} + J_{ih}(1 - p_{\mathbf{n+k}})]\mu_{\mathbf{n+k}}, \tag{3}$$

and

$$\nu_{\mathbf{n}} = \sum_{\mathbf{k}} \mu_{\mathbf{n+k}}, \tag{4}$$

where the sums are over the $\mathbf{k}$ nearest neighbors. We do not determine the absolute values of the exchange constants J_{mn} but only their relative values, therefore we define the following normalized fields:

$$h^h_{\mathbf{n}} = (Z_1\,J_{hh})^{-1}H^h_{\mathbf{n}} = \frac{1}{Z_1}\sum_{\mathbf{k}} \mu_{\mathbf{n+k}}(1 - p_{\mathbf{n+k}} + \alpha p_{\mathbf{n+k}}) \tag{5}$$

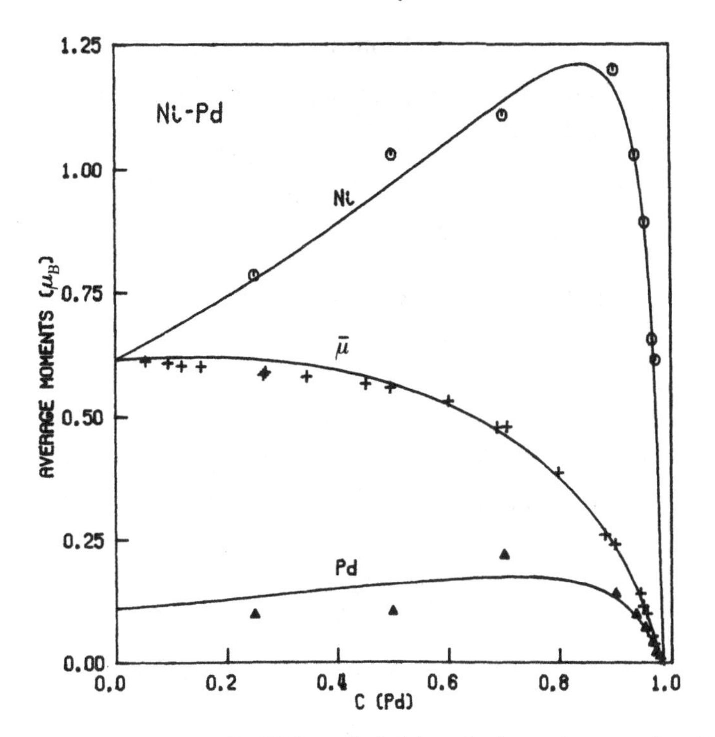

Figure 1. Average moments of Ni-Pd from Ref. (8) and other references therein. The curve represents the fitted values obtained with the magnetic environment model.

and

$$h_{\mathbf{n}}^{i} = (Z_1\, J_{ii})^{-1} H_{\mathbf{n}}^{i} = \frac{1}{Z_1} \sum_{\mathbf{k}} \mu_{\mathbf{n+k}} \left[p_{\mathbf{n+k}} + \beta(1 - p_{\mathbf{n+k}})\right], \tag{6}$$

where Z_1 is the coordination number and α and β are given by

$$\alpha = \frac{J_{ih}}{J_{hh}}, \tag{7}$$

and

$$\beta = \frac{J_{ih}}{J_{ii}}. \tag{8}$$

The moment of an atom, therefore, depends on its chemical environment and, through an exchange field, on its magnetic environment.

In order to solve the infinite set of coupled non-linear equations given by Eq. (1) we make some approximations. The first one is the linearization of the equations. With small moment fluctuations, we expand the response functions about an effective field h_{eff} and the average number of nearest impurity neighbors Z_1c. The effective fields h_{eff} are defined by the following equations:

$$\bar{\mu}_h = F_h(h_{\text{eff}}^h, Z_1c), \tag{9}$$

and

$$\bar{\mu}_i = F_i(h_{\text{eff}}^i, Z_1c), \tag{10}$$

Table I

Parameters from the fit of $\bar{\mu}_i$, $\bar{\mu}_h$, and $\bar{\mu}$ *vs.* c for Ni-Pd and Ni-Pt.

	Ni-Pd	Ni-Pt
Γ_i	0.947(9)	0.875(227)
α	5.6(2.7)	3.10(47)
β	0.182(19)	0.65(23)
Q_1	0.0072(369)	−0.163(25)
Q_2	0.0726(66)	0.056(28)

where μ_h and μ_i are the host and impurity moments. The moment at site $\mathbf{n}$ is then a function of the following parameters:

$$\mu_{\mathbf{n}} = f(\bar{\mu}_h, \bar{\mu}_i, \Gamma_h, \Gamma_i, \rho_h, \rho_i), \tag{11}$$

where

$$\Gamma_h = \frac{\partial F_h}{\partial h}, \tag{12}$$

$$\Gamma_i = \frac{\partial F_i}{\partial h}, \tag{13}$$

$$\rho_h = \frac{\left[\frac{\partial F_h}{\partial \nu}\right]}{\left[\frac{\partial F_h}{\partial h}\right]}, \tag{14}$$

and

$$\rho_i = \frac{\left[\frac{\partial F_i}{\partial \nu}\right]}{\left[\frac{\partial F_i}{\partial h}\right]}. \tag{15}$$

To make the second approximation, we use Marshall's expansion[9]

$$\mu_{\mathbf{n}} = (1 - p_{\mathbf{n}})[\bar{\mu}_h + \sum_{\mathbf{k}} g_{\mathbf{k}}(p_{\mathbf{n+k}} - c) + \cdots] + p_{\mathbf{n}}[\bar{\mu}_i + \sum_{\mathbf{k}} h_{\mathbf{k}}(p_{\mathbf{n+k}} - c) + \cdots]. \tag{16}$$

Here $g_{\mathbf{m}}$ and $h_{\mathbf{m}}$ are the impurity-induced moment disturbances at host and impurity sites, respectively, and the dots indicate nonlinear terms. The approximation consists of neglecting the nonlinear terms. Using Eqs. (5), (6) and (16) we obtain the effective fields

$$h^h_{\text{eff}} = (1 - c)(\bar{\mu}_h - cg_1) + \alpha c(\bar{\mu}_i - ch_1), \tag{17}$$

$$h^i_{\text{eff}} = c[\bar{\mu}_i + (1 - c)h_1] + \beta(1 - c)[\bar{\mu}_h + (1 - c)g_1]. \tag{18}$$

The moment disturbance function $M(\mathbf{K})$ measured by neutron scattering is obtained in a similar way. The resulting equation is Fourier transformed giving the following expression:

$$M(\mathbf{K}) = \Delta\bar{\mu} + \frac{U + VF_1(\mathbf{K}) + WF_1(\mathbf{K})^2}{1 - \Gamma F_1(\mathbf{K}) + DF_1(\mathbf{K})^2}, \tag{19}$$

where

$$F_1(\mathbf{K}) = \frac{1}{Z_1}\sum_{\mathbf{m}} e^{i\mathbf{K}\cdot\mathbf{m}}, \tag{20}$$

$$\Gamma = (1 - c)\Gamma_h + c\Gamma_i, \tag{21}$$

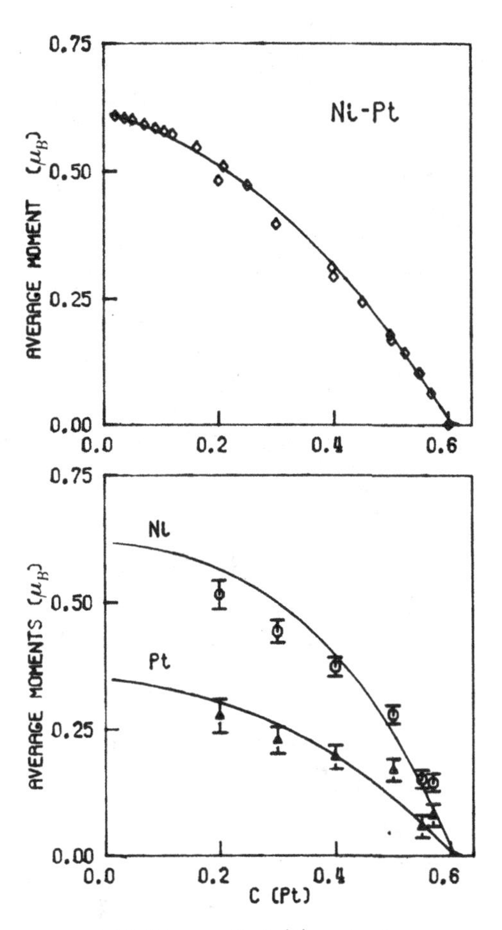

Figure 2. Average moments of Ni-Pt from Ref. (8) and other references therein, fitted with the magnetic environment model.

and U, V and W are functions of α, β, Γ_i, Γ_h, ρ_i, ρ_h, μ_i, μ_h and c. The average moments are functions of the effective fields which are themselves functions of the average moments. It is therefore possible to obtain the average moments for each concentration if we know the response functions F_h and F_i.

It is important to consider another important condition for these alloys. It has been shown[10] that $M(0)$ is related to the change of moment with concentration by the following equation:

$$M(0) = \frac{d\bar{\mu}}{dc}. \tag{22}$$

Due to the approximations of the model, this condition is not necessarily met, but it can be used as a test of the goodness of the method.

For Ni-Pd and Ni-Pt alloys the response functions F_h and F_i are not known, but since the neutron scattering values[6,7] of the moments of Pd and Pt are small we make the approximation

$$F_i(h, \nu) = \Gamma_i h. \tag{23}$$

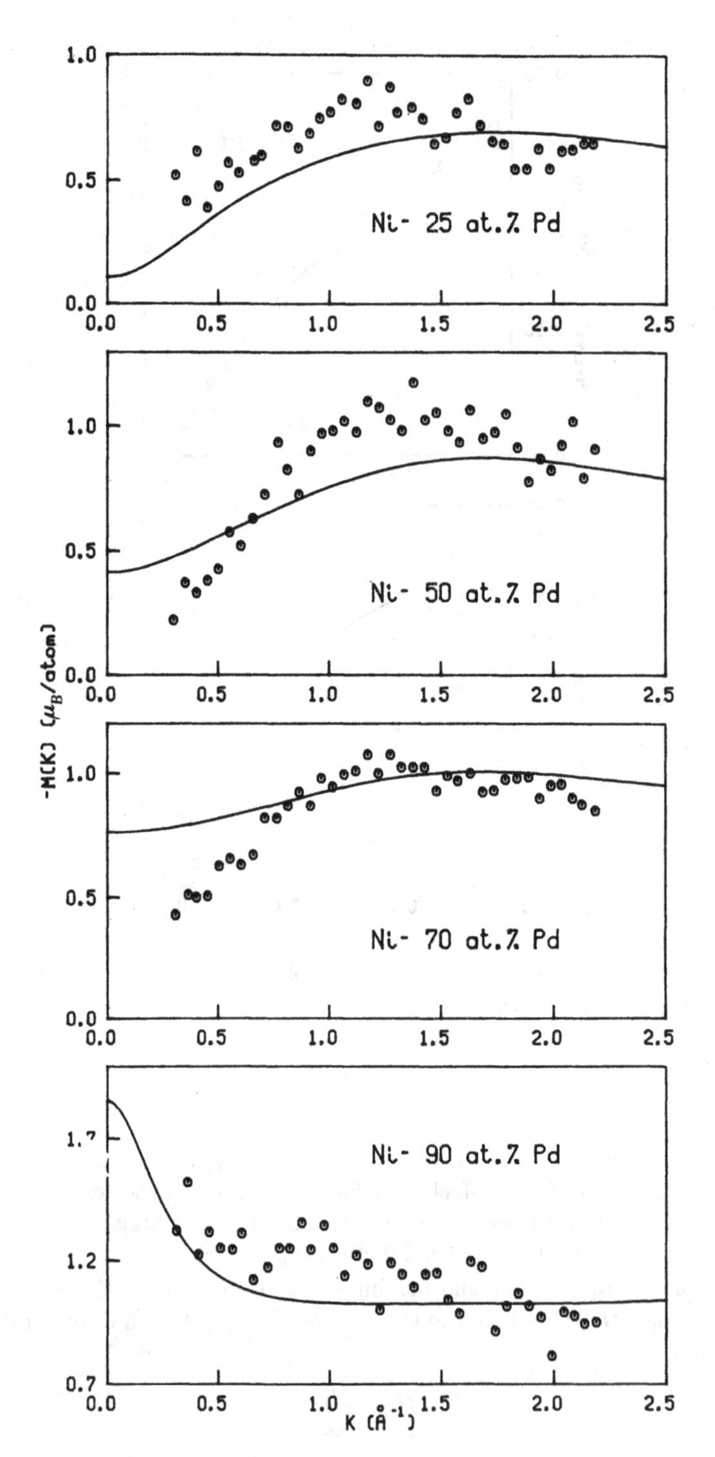

Figure 3. Moment disturbances of Ni-Pd alloys, from Ref. (6), determined with polarized neutrons. The continuous curve corresponds to the values calculated with the parameters of the magnetic environment model.

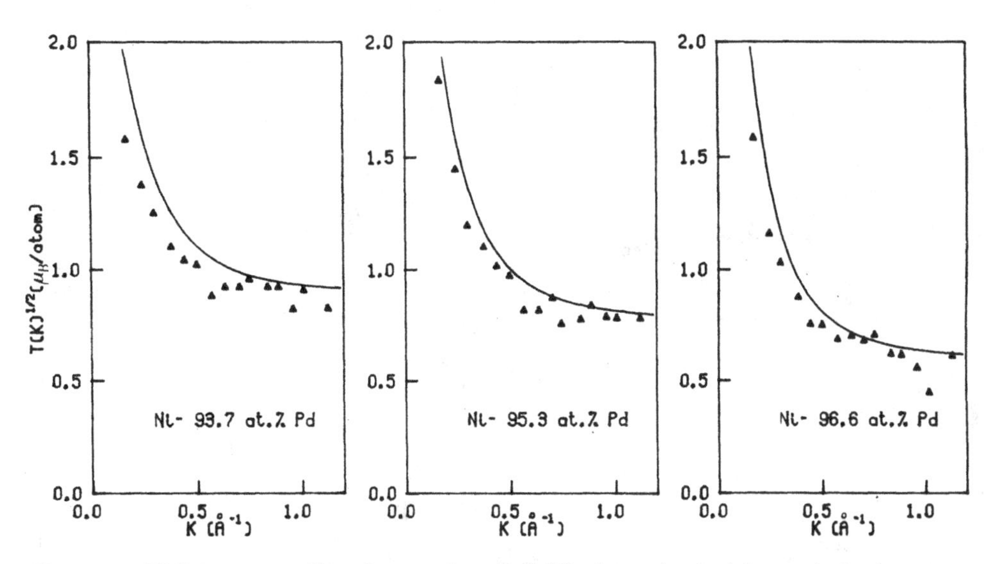

Figure 4. Ni-Pd moment disturbances, from Ref. (8), determined with unpolarized neutrons (dots), and values calculated with the model (continuous curve).

The same approximations cannot be made for Ni, therefore we use an analytical expression, similar to the one used by Hicks[11] and by Medina and Cable[12] for Ni-Cu alloys:

$$F_h(h,\nu) = \frac{A(\nu)h}{1+B(\nu)h}\,. \tag{24}$$

The parameter $A(\nu)$ is the initial susceptibility while $A(\nu)/B(\nu)$ is the high field moment. We do not know the form of the functions $A(\nu)$ and $B(\nu)$ but we approximate them with the following expressions:

$$A(\nu) = A\Phi_1(\nu)\,, \tag{25}$$

and

$$\frac{A(\nu)}{B(\nu)} = (\frac{A}{B})\Phi_2(\nu)\,, \tag{26}$$

with Φ_1 and Φ_2 expressed as simple exponentials:

$$\Phi_1(\nu) = e^{Q_1\nu}\,, \tag{27}$$

and

$$\Phi_2(\nu) = e^{Q_2\nu}\,. \tag{28}$$

One is then able to determine the average moments μ_i and μ_h.

The constants A and B were determined by Parra and Medina[8] for Ni-Cu alloys, obtaining $A = 2.58 \pm 0.12$ and $B = 2.56 \pm 0.19$. We used these values to treat the cases of Ni-Pd and Ni-Pt. The average alloy moment of Ni-Pd and of Ni-Pt and their average host and impurity moments were fitted simultaneously in order to obtain the parameters Γ_i, α, β, Q_1 and Q_2. The function Γ_h, considered unknown, was to be determined by condition (22).

The results of the fitting are shown in Figs. 1 and 2, and the parameters are given in Table I. The values of Γ_i are close to one, in agreement with the high magnetic

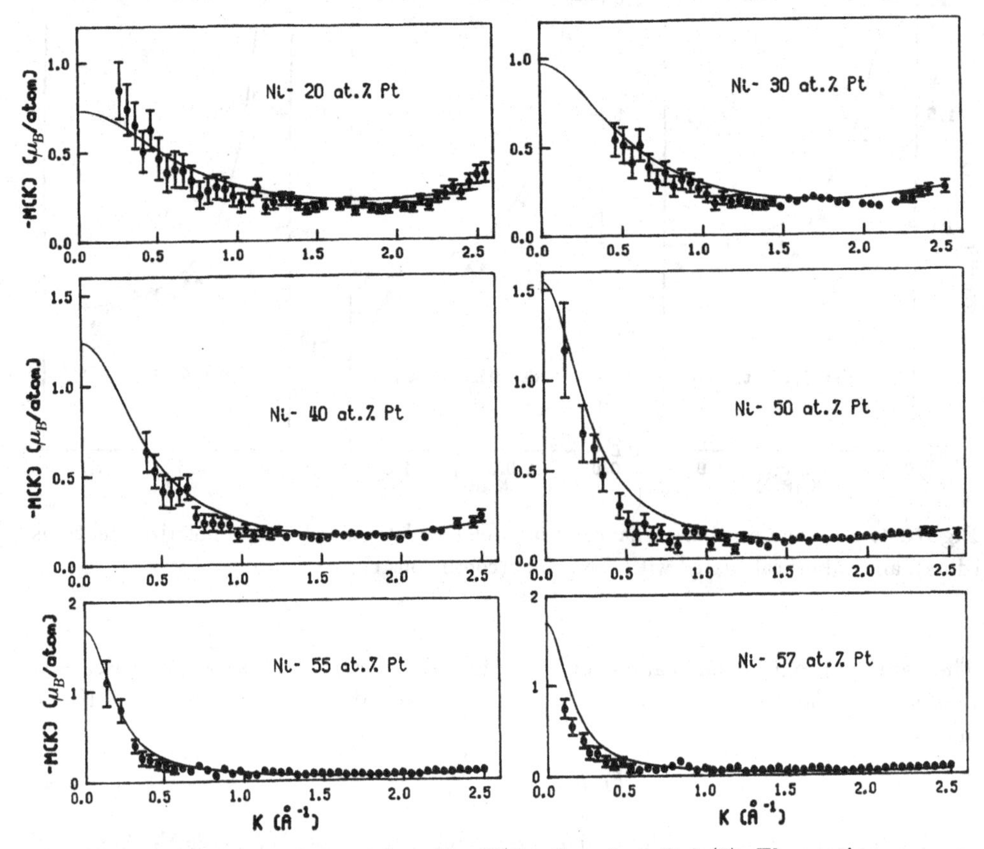

Figure 5. Moment disturbances of random Ni-Pt alloys from Ref. (8). The continuous curve corresponds to the values obtained with the model.

polarizability of Pd and Pt. The values of α and β indicate that a Pd atom produces an exchange field about five times bigger than the one produced by a Ni atom with the same moment, independently of the kind of atom that the field acts upon. A Pt atom produces an exchange field on a Ni atom about three times bigger than the one produced by a Ni atom with the same moment and about one and a half times bigger if it acts upon another Pt atom. In both cases the presence of an impurity atom increases the saturation moment of nearby Ni atoms as expressed by the values of Q_2. This probably indicates that Ni transfers electrons to impurities nearby.

With the parameters obtained, we calculated the values of the spherical average of $M(K)$ with Eq. (19) and they are compared with the neutron data in Figs. 3, 4 and 5. Considering the approximations and assumptions made in the model, the agreement is excellent for Ni-Pt at all concentrations and good in the case of Ni-Pd near the critical concentration. There are, however, differences between data and calculated values for Ni-Pd at smaller concentrations. This is probably due to the fact that the $M(K)$ data of Ni-Pt were corrected for chemical short-range order (SRO), but they were not for Ni-Pd. SRO effects are minimal as the concentration approaches the extreme values 0 and 1; this explains the good agreement obtained near the critical concentration of Ni-Pd.

The importance of SRO for the Ni-Pd data is evident when one compares the concentration derivative of the average moment with the $M(0)$ value obtained by the extrapolation of the data at the same concentration. The two values should coincide for random alloys. On the other hand, the concentration derivative for the 70 at.% Pd alloy is 0.72, and the value calculated with the model is 0.76.

Some comments must be made about the approximations used in this model. The response and the chemical effect functions used are arbitrary, but simple, and they give a good description of the effects. Although contributions to the exchange field due to neighbors beyond the first ones cannot be excluded, the agreement of the calculated $M(K)$ with the neutron scattering data shows that only first neighbor interactions are necessary to explain the observations. The moment perturbations produced by first neighbors are propagated to first neighbors of those neighbors and so on. This would explain the long-range moment disturbances measured with neutrons.

III. Giant Moment Ferromagnetism in Dilute Alloys

Alloys of Pd and Pt with magnetic impurities at low concentration have long been considered as typical examples of giant moment ferromagnetic systems, as has been mentioned in Section I. The model[13] we apply to these systems is an extension of the one utilized for Ni-Pt and Ni-Pd. As in Sec. II, here we only present a summary of the model for dilute alloys, with further details found in Ref. (13). In this model the impurity moments are not dependent on their magnetic environment while the host moment at site $\mathbf{n}$ is given by

$$\mu_{\mathbf{n}} = \chi_0 h_{\mathbf{n}} , \tag{29}$$

where χ_0 is the nonenhanced susceptibility and $h_{\mathbf{n}}$ is a local field given by

$$h_{\mathbf{n}} = b_{\mathbf{n}} + J_h \sum_{\mathbf{k}} \mu_{\mathbf{n+k}}[1 + (\alpha - 1)p_{\mathbf{n+k}}] . \tag{30}$$

Here we are summing over the $\mathbf{k}$ nearest neighbors, J_h is an exchange constant between host atoms, αJ_h is the exchange constant between host and impurity atoms, $p_{\mathbf{n}}$ is a site occupation operator and $b_{\mathbf{n}}$ is an external field.

When the number of impurities tends to zero, we obtain from Eqs. (29) and (30) the following expression for the enhanced susceptibility:

$$\chi = \frac{\chi_0}{1 - \Gamma} , \tag{31}$$

where $\Gamma = J_h Z_1 \chi_0$, with Z_1 being the number of first neighbors. The parameter Γ was determined in Sec. II using neutron scattering data for Ni-Pd and Ni-Pt alloys. The reported enhanced susceptibilities[14,15] of Pd and Pt are: $\chi_{\mathrm{Pd}} = 6.85 \times 10^{-6}$ emu/g, $\chi_{\mathrm{Pt}} = 1.08 \times 10^{-6}$ emu/g, from which we can estimate the values of the nonenhanced susceptibilities of Pt, $\chi_0^{\mathrm{Pt}} = 1.35 \times 10^{-7}$ emu/g, and of Pd, $\chi_0^{\mathrm{Pd}} = 3.63 \times 10^{-7}$ emu/g, and the exchange constants, $J_{\mathrm{Pt}} = 1039\ K$ and $J_{\mathrm{Pd}} = 766\ K$.

Considering a magnetic impurity atom at $r = 0$ with a moment μ_i, we get the magnetization cloud by Fourier transforming Eq. (29), obtaining

$$M(\mathbf{K}) = \mu_i f_i(\mathbf{K}) \left[\frac{\alpha}{\Phi_0[1 - \Gamma F_1(\mathbf{K})]} + (1 - \alpha) \right] , \tag{32}$$

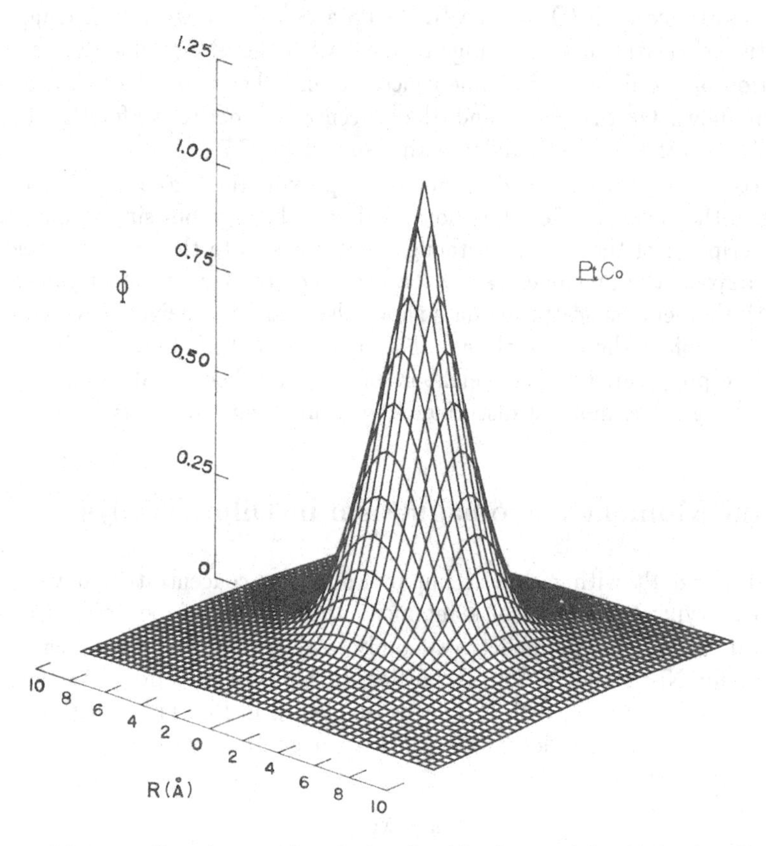

Figure 6. Moment disturbances Φ calculated with the model. R is the distance (in ångstroms) from the impurity site to the disturbed moment.

Table II

Exchange constants and energy of magnetization cloud calculated with the model.

	$J_{ik}10^{-3}$ eV	E_C (K)
PdFe	8.23(1.11)	-149(34)
PtFe	7.97(2.06)	-75(20)
PtCo	8.40(2.23)	-33(19)

where $F_1(K)$ is the first shell structure factor as defined in Sec. II and

$$\Phi_{\mathrm{n}} = \frac{1}{V_c^*} \int \frac{e^{i\mathbf{K}\cdot\mathbf{n}}\, d^3\mathbf{K}}{1 - \Gamma F_1(\mathbf{K})}, \tag{33}$$

where the integrals are over a reciprocal lattice unit cell. We take $M(0) = d\bar{\mu}/dc$ as the total moment of the cloud, where c is the impurity concentration and $\bar{\mu}$ is the average magnetic moment of the alloy, that can be obtained from magnetization measurements. The impurity moment can be obtained from neutron scattering measurements, and therefore α and J_{ih} can be calculated.

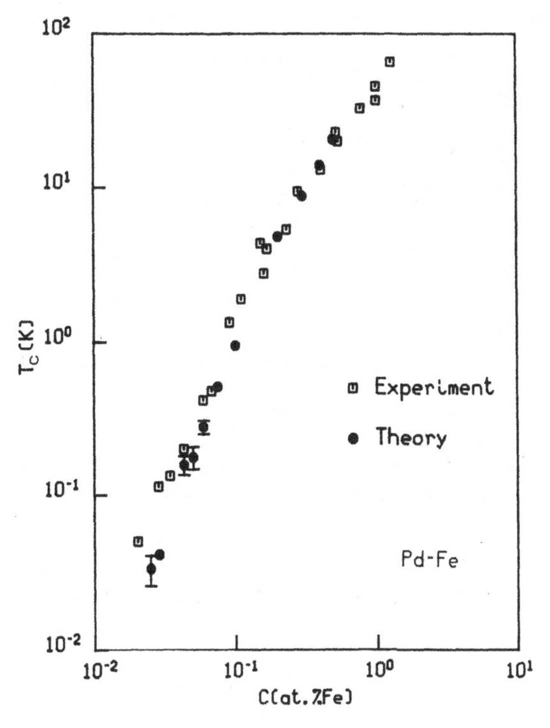

Figure 7. Critical temperatures for the onset of ferromagnetism of PdFe, as obtained with our model by Monte Carlo calculations, are compared with experimental data from Ref. (3) and other references therein.

Introducing the variable $m_{\mathbf{n}}$ defined to be equal to $\mu_{\mathbf{n}}$ for host atoms and to $\alpha\mu_i$ for the impurity atoms at site 0, the K dependent part of Eq. (29) becomes

$$m(\mathbf{K}) = \sum_{\mathbf{n}} m_{\mathbf{n}} e^{i\mathbf{K}\cdot\mathbf{n}}\,, \tag{34}$$

with

$$m_{\mathbf{n}} = \frac{\alpha\mu_i}{\Phi_0}\Phi_{\mathbf{n}}\,, \tag{35}$$

and where the Φ's are given by Eq. (33). From this equation it can be seen that the Φ's are the moment disturbances of the host moments due to the magnetic impurity.

We are assuming that around each impurity atom there is a magnetization cloud that follows the orientation of the impurity moment and that the interaction energy can be expressed as a sum of interactions between pairs of magnetization clouds.

We find that the interaction energy between two giant moments at positions **i** and **j**, and with directions **S** is,

$$E = -\frac{1}{2}\sum_{\mathbf{i},\mathbf{j}} p_{\mathbf{i}} p_{\mathbf{j}} J_{\mathbf{ij}}(\mathbf{S}_{\mathbf{i}} \cdot \mathbf{S}_{\mathbf{j}})\,, \tag{36}$$

where the $J_{\mathbf{ij}}$'s are site-dependent exchange constants that are calculated using the values of Φ. The $J_{\mathbf{ii}}$ value corresponds to the self energy of a magnetization cloud. In

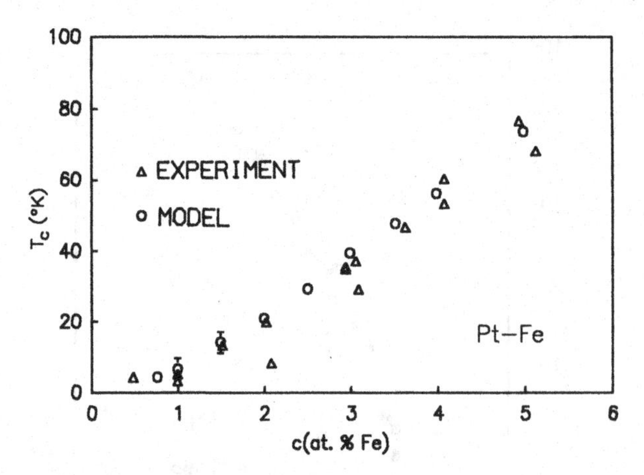

Figure 8. Critical temperatures of PtFe determined with our model by Monte Carlo calculations are compared with experimental results from Ref. (19) and other references therein.

the case of an isolated impurity atom with its cloud this energy is given by

$$E_c = -\frac{1}{2} J(\alpha\mu_i)^2 Z_1 \frac{\Phi_1}{\Phi_0} \,. \tag{37}$$

We used the Monte Carlo method for a Heisenberg system with classical spins to obtain the critical temperatures for the onset of ferromagnetism. Samples of 100 impurity spins were used with 2000 Monte Carlo steps per spin. The critical temperatures T_c were obtained from the maximum of the magnetic susceptibility against temperature, where we have used a susceptibility that is given by the fluctuation of the magnitude of the total moment M:

$$\tilde{\chi} = \frac{\langle (M - \langle M \rangle)^2 \rangle}{NkT} \,, \tag{38}$$

where N is the number of spins (further details of the method used are found in Ref. (13)).

We have applied this model to dilute PdFe, PtFe and PtCo alloys.[13,16,17] The exchange constants J_{ih} were obtained using Eqs. (31) and (32), and the energy of the cloud is calculated utilizing Eq. (37). The results are shown in Table II. The values of the moment disturbances were calculated by numerical evaluation of Eq. (33) and fitted with a spline function. The results are shown in Fig. 6, giving us a graphic description of the magnetization cloud. Critical temperatures were obtained by Monte Carlo simulation using Eq. (38), and the values were compared with experimental data. The results are shown in Figs. 7, 8 and 9. The model allows also the calculation of other magnetic quantities such as the susceptibilities above T_c, the spin wave stiffness coefficients and the calculation of unpolarized neutron diffuse scattering cross sections, all of which have been compared with experimental values or have become a prediction of experimental results to be obtained. The model has also been applied recently to PdMn alloys. These calculations, however, are not shown here but can be found in Refs. (13), (18), (19) and (20).

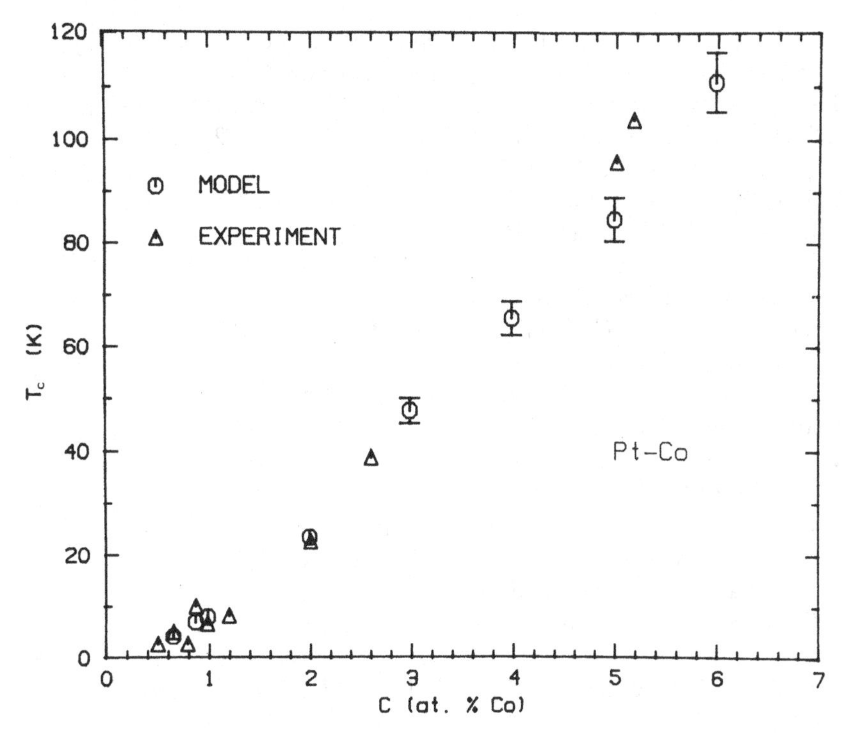

Figure 9. Critical temperatures of PtCo determined with the model by Monte Carlo calculations, are compared with the experimental results from Ref. (17) and other references therein.

IV. Conclusions

The model we have presented describes well several magnetic properties of the alloys considered. We find that we can analyze the onset of ferromagnetism of these alloys based on the following points: (a) Only first-neighbor interactions (with propagation to first neighbors of those neighbors, etc.) are needed to explain the observations, (b) for concentrated alloys, the moment on a Ni atom is a function not only of its chemical environment but also of the magnetic moments of the surrounding atoms, (c) for dilute alloys, no direct interactions were considered between the magnetic impurities, (d) ferromagnetism is a consequence of interactions between magnetization clouds, and (e) these clouds are due to the exchange enhancement of the host metal.

The models presented here are phenomenological, therefore a fundamental theory is needed. However, they describe with few parameters the magnetic behavior of Pd and Pt based alloys and the job of a fundamental theory would be reduced to an explanation of those parameters.

References

1. G. G. Low and T. M. Holden, *Proc. Phys. Soc.* London **89**, 119 (1966).
2. J. Crangle and W. R. Scott, *J. Appl. Phys.* **36**, 921 (1965).
3. G. Chouteau and R. Tournier, *J. Phys. (Paris), Colloq.* **32**, C1-1002 (1971).
4. T. J. Hicks, T. M. Holden, and G.G. Low, *J. Phys.* *C***1**, 528 (1968).
5. J. C. Ododo, *J. Phys F* **12**, 1821 (1982).
6. J. W. Cable and H. R. Child, *Phys. Rev.* *B***1**, 3809 (1970).
7. R. E. Parra and J. W. Cable, *Phys. Rev.* *B***21**, 5494 (1980).
8. R. E. Parra and R. Medina, *Phys. Rev.* *B***22**, 5460 (1980).
9. W. Marshall, *J. Phys.* *C***1**, 88 (1968).
10. R. Medina and J. W. Garland, *Phys. Rev.* *B***14**, 5060 (1976).
11. T. J. Hicks, *J. Phys F***7**, 481 (1977).
12. R. Medina and J. W. Cable, *Phys. Rev.* *B***15**, 1539 (1977).
13. R. Medina and R. E. Parra, *Phys. Rev.* *B***26**, 5187 (1982).
14. B. H. Verbeek, G. J. Nieuwenhuys, J. A. Mydosh, C. van Dijk, and B. D. Rainford, *Phys. Rev.* *B***22**, 5426 (1980).
15. O. Krogh Andersen, *Phys. Rev.* *B***2**, 883 (1970).
16. A. C. González and R. E. Parra, *J. Appl. Phys.* **55**, 2045 (1984).
17. R. E. Parra and R. A. López, *J. Appl. Phys.* **61**, 3989 (1987).
18. R. E. Parra and R. A. López, *J. Appl. Phys.* **63**, 3617 (1988).
19. R. Medina, R. E. Parra, G. Mora, and A. C. González, *Phys. Rev.* *B***32**, 1628 (1985).
20. R. E. Parra, A. C. González, and R. A. López, *J. Phys.: Condens. Matter* **2**, 7309 (1990).

Monte Carlo Simulation of Order-Disorder Phenomena in Binary Alloys

Werner Schweika

Institut für Festkörperforschung
Forschungszentrum Jülich
5170 Jülich
Federal Republic of Germany

Abstract

An introduction is given into the Monte Carlo technique of phase diagram calculations. Beside the standard approaches used to determine first and second order phase transitions, a most promising development is discussed, namely the analysis of the probability distributions by histograms. Applications to real alloys are considered, such as the decomposition in Cu-Co and Cu-Ni, the short-range order and critical scattering in Cu-Au type alloys, the phenomenon of surface induced order in Cu-Au models having a free (100) surface, and the interplay of magnetism and chemical ordering in Fe-Al.

I. Introduction

Monte Carlo (MC) simulations have been established as a useful tool to study order-disorder phenomena in alloys. In view of the number of excellent reviews and textbooks,[1–6] the intention of this paper is to give a brief tutorial introduction into the Monte Carlo simulation-technique and the phase diagram calculation of alloys.

In comparison to analytic approximations, such as the simple mean-field type or more sophisticated ones used by the Cluster Variation method, the Monte Carlo method can provide highly accurate results, with manageable statistical errors for many-body system with a large number of degrees of freedom. Another advantage is the flexibility of this method which allows one to observe quite different physical phenomena occuring during the equilibration process almost as seen in an experiment.

Structural and Phase Stability of Alloys
Edited by J.L. Morán-López *et al.*, Plenum Press, New York, 1992

The configurational energy of a binary alloy on a given lattice can be described by the Ising model:

$$H = -\sum_{i \neq j} J_{ij} s_i s_j - h \sum_i s_i \qquad s_i = \begin{cases} +1, & \text{if atom A is at site } i; \\ -1, & \text{if atom B is at site } i; \end{cases}$$

where J_{ij} denotes the effective interaction between the sites i and j. In the grand canonic ensemble, one uses the intensive thermodynamic variable, the chemical potential difference h, as the independent variable, the corresponding conjugate extensive variable, the concentration $c_{(A)} = \frac{1}{2}(1 + \langle s \rangle)$, is then observed and averaged in a straightforward manner.

The interactions J_{ij} are sometimes fitted to experimental phase diagrams, but with more predictive power they can be independently determined either from *ab-initio* theoretical calculations or from the measured short-range order (SRO) using the Inverse Monte Carlo method.[7–9]

II. Monte Carlo Method and Simulation-Technique

Except for very small systems, including a few atoms only, and some trivial cases (i.e. for $T = 0$ or ∞), exact analytical solutions do not exist for the configurational equilibrium properties of binary alloys. With increasing number of atoms N, the possible configurations $\vec{x}$ increase as 2^N, while the occupation probability $P(\vec{x}) = \frac{1}{Z} e^{-\beta H(\vec{x})}$ (Boltzmann distribution) becomes sharply peaked close to equilibrium configurations $\vec{x}_{eq}$.

Metropolis *et al.*[10] have introduced the MC method with the idea of importance sampling: in thermodynamic equilibrium the principle of detailed balance must hold $P(\vec{x})W(\vec{x} \to \vec{x}') = P(\vec{x}')W(\vec{x}' \to \vec{x})$. A transition probability W, which satisfies this equation, ensures that the most likely states are created, and as ergodicity can be presumed, the equilibrium will be reached from any (and even unlikely) starting configuration.

In practice we proceed as follows : construct a lattice of linear dimensions L_1, L_2, L_3 with periodic boundary conditions. Visit a site i and calculate the local energy H_i. An exchange of atoms ("spin-flip") $s_i \to -s_i$ should be performed if the transition probability, for instance

$$W = e^{-\beta(H_i^{final} - H_i^{initial})},$$

is larger than a random number $\eta \in (0,1)$. Alternatively one may use $W' = W/(1+W)$ (kinetic Ising model).[11,12] In the case of more than two different states —there are three for a ternary alloy or a binary alloy with one magnetic component ($\to FeAl$)— the so-called heat-bath algorithm[13,14] is advantageous. Here only the final energies are regarded for the transition probability W_i, which has to be normalized by the sum of the transitions probabilities to all possible final states.

In the canonic ensemble exchanges of $s_i \leftrightarrow s_j$ (often between nearest neighbours) are performed. To compete with critical slowing down one has tried "flipping" of whole clusters has been tried[15] and other MC steps can be imagined. Such a MC step needs to be often repeated for all sites to relax the whole system towards equilibrium, and then further repeated to obtain many configurations to determine averages of composition, energy and other quantities for each T and h.

High flexibility and performance of a program can be achieved, if the geometry is treated separately. Therefore, the lattice is mapped onto a 1-dim chain, as defined by a regular walk along all sites. Neighbouring, cartesian coordinates $\vec{R} = (l, m, n)$ are transformed by a matrix $\mathbf{T}$ to those of the primitive cell orientations and projected onto the distance i in the chain by $\mathbf{T}\vec{R} \cdot \vec{L} = i$, with $\vec{L} = (1, L_1, L_1 \cdot L_2)$;
i.e.
for the 3-dim simple cubic lattice $i = l + mL_1 + nL_1L_2$
with $L_1 \| [100]$, $L_2 \| [010]$, $L_3 \| [001]$;
for the 3-dim *fcc* lattice $i = \frac{1}{2}(l + m - n) + \frac{1}{2}(l - m - n)L_1 + nL_1L_2$
with $L_1 \| [110]$, $L_2 \| [1\bar{1}0]$, $L_3 \| [101]$.

This procedure leads to helical periodic boundary conditions, which are intrinsically fulfilled within the chain, while at its ends an identical copy has to be added. It also facilitates our ability to vectorize a program on supercomputers. A considerable further efficiency (paid by some loss in flexibility) is gained by using multi-spin coding techniques, where the occupation variables are stored in the bits of an integer word, and logical operations allow for parallel computations.[16,17] Thus, 335 million MC steps per second have been achieved on single processor of a CRAY YMP.[18]

Before starting the first MC simulation particular attention has to be paid to the quality of the pseudo random numbers. For discussions of appropriate random number generators we refer to the literature.[19,20]

The inverse MC method is used to determine effective interactions from experimental data for the equilibrium SRO. The method has been described in detail elsewhere.[7,8] It should be noted that the regular MC method can be applied as well. Therefore the interactions have to be varied until the experimental SRO is reproduced.[21]

III. Applications to phase transitions and phase diagram calculations of alloys

The MC method itself is very simple and with increasing effort the results may be as accurate as desired. However, the quality of the results depends on a careful analysis of the statistical and systematic errors.[22] In particular, finite size effects may have to be considered to locate a phase transition precisely.

III.1 Phase diagrams of decomposing alloys

The simplest type of phase diagrams are found for alloys showing a miscibility gap at lower temperatures. Positive interactions $J_{ij} > 0$ (ij may be nearest neighbours, then we shall use also J_1 and for next nearest neighbours J_2, etc.) express that different types of atoms do not like each other. The phase diagram is easily calculated within the grand canonic ensemble with $h = 0$. For a given temperature the composition will relax quickly towards $\langle s \rangle = 0$ for $T > T_c$ and to the miscibility boundary below T_c. Using the canonic ensemble may appear more natural for the simulation of ordering in alloys. Indeed it is very suitable to study the process itself, for instance the kinetics of spinodal decomposition and nucleation and growth. However, this ensemble is much less efficient for the determination of the phase boundaries, because of its slower relaxation.

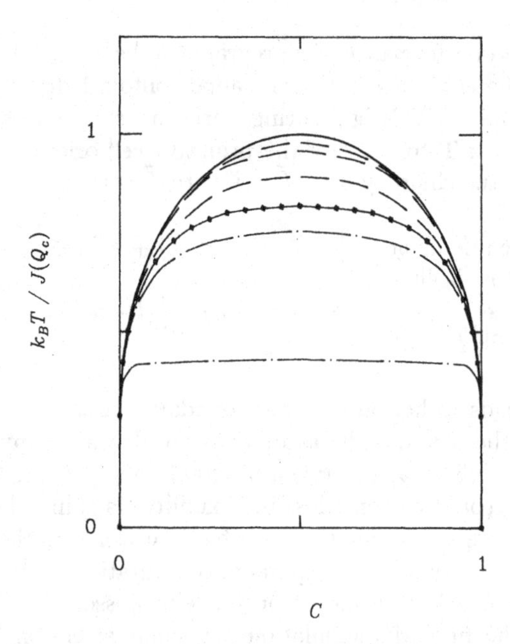

Figure 1. Phase Diagrams calculated for decomposing alloys: MF, — MC $J_1 > 0$; ......increasing range of interaction $J_1 - J_2$, $J_1 - J_5$, $J_1 - J_{10}$; .-.-. competitive interactions $J_2 = -\frac{1}{2}J_1$ and $J_2 = -J_1$.

III.1.1 Various interaction models and a comparison to the MF result

How good is the mean-field (MF) solution (Bragg-Williams)? Figure 1 reveals that the MF- and MC-results coincide asymptotically for low T and c. Correlation effects and finite size effects are almost negligible, i. e. long MC runs on small systems yield comparable results to shorter runs on large systems. Larger discrepancies are observed around the critical temperatures T_c. Compared to the MC simulation of a simple *fcc* nearest neighbour model the MF result overestimates T_c by about 25%.

However, considering the critical temperatures a clear dependence of the MC-results is found depending on the way how the total interaction is distributed over neighbouring atoms. If the interaction energy is distributed monotonicly onto further and further neighbours, the MC- and MF-result should finally coincide as expected. The contrary is found for oscillating interactions. A competing (negative) next nearest neighbour interaction lowers T_c (although the sum of interaction energies is still kept constant). A qualitatively different behaviour is also found in MC simulations of nucleation and phase seperation. While monotonous interactions favour nuclei of a more spherical shape, competing, oscillating interactions tend to form flat boundaries and even cube-like nuclei.

One practical application is to estimate the interactions from experimental phase diagrams using the curvature of a solubility boundary line. For example, the boundaries of solubilities of Co in Cu are reproduced within 1% deviations by a nearest neighbour interaction $J_1 = 16.35 \pm 0.12$ meV. Since the deviations seem to be still of systematic nature it appears to be justified to also include interactions J_2 to second nearest neighbours. This yields perfect agreement with the experimental solubilities,[23] but demonstrates also the limits of such a method to determine interactions: $J_1 = 21.4$ meV, $J_2 = -9.8$ meV for **CuCo**. Taking even further interactions into account would yield arbitrarily many different sets of solutions. On the other hand, if the interactions are not determined from a macroscopic quantity, but from the atomic SRO using the inverse MC method, the solution is in most cases stable with respect to the number of interaction parameters. Furthermore this allows one to predict phase transitions at lower temperatures in alloys, where due to the low mobility an experimental observation is difficult. For CuNi the interactions $J_1 = (6.1 \pm 0.3)$ meV and $J_2 = (-2.8 \pm 0.3)$ meV have been determined from SRO, and a coherent miscibility gap is found by MC simulations[9] below $T = 495K$.

III.1.2 Second order phase transitions

Locating precisely the second-order transition at T_c turns out to be difficult. Two essential aspects have to be kept in mind: the correlation length and the correlation time diverge at T_c. Therefore as $\xi \gg L$ the finite size effects must be analysed and long MC runs are required.

According to finite size scaling theory (Fisher 71)[24-30] we expect that instead of a real singularity, the susceptibility $k_B T \chi(T) = N(\langle s^2 \rangle - \langle s \rangle^2)$ in a finite system exhibits rather a rounded maximum at $T_c(L)$ scaling with the linear dimensions L as $\propto L^{\gamma/\nu}$. A plot of $T_c(L)$ *versus* $L^{-1/\nu}$ yields the T_c for an infinite system. Difficulties may arise if the exponent ν is not known.

Analysing the finite size dependence of the 4th order cumulant

$$\tilde{U} = 1 - \frac{\langle s^4 \rangle}{3\langle s^2 \rangle^2}$$

of the order parameter $\langle s \rangle$ is a quite elegant method to locate a second-order transition. Binder (1981)[28] has shown that the 4th order cumulant of the order parameter depends on T and L but crosses at T_c for different values, where the fixpoint $\tilde{U}^*(T_c)$ is an universal quantity. At $T = 0$, $\tilde{U} = 2/3$, for $T > T_c$ the cumulant tends to 0, the value for a gaussian distribution. The exponent ν is determined independently from the slope

$$\frac{\partial \tilde{U}_{L'}}{\partial \tilde{U}_L}|_{\tilde{U}^*} = (\frac{L'}{L})^{1/\nu}.$$

The most promising advance to improve the accuracy of the MC results for the study of critical properties utilizes the analysis of distribution functions.[31,32] From a single long MC run a distribution of energies E and compositions $m = \langle s \rangle$ is obtained as a histogram $P_T(E,m)$. One can extrapolate from this distribution to the one of another at T' —T_c is of course of most interest—, by reweighting the possible states according to their occupation probability at T' :

$$P_{T'}(E,m) = \frac{P_T(E,m)e^{-(\beta'-\beta)H}}{\sum_{E,m} P_T(E,m)e^{-(\beta'-\beta)H}}$$

Then the averages of quantities of E or m are determined at the transition temperature T_c. Finite size scaling, as already discussed, should be applied of course again. Thus the most accurate determination of a critical transition temperature ($\beta_c = 0.2216595 \pm 0.0000027$, 3-dim sc) and of critical exponents ($\nu = 0.627 \pm 0.002$) has been obtained by the MC method recently,[33,34] while before exact series expansions and renormalization group techniques have been superior. For small systems, a single MC run is sufficient to determine the equilibrium properties at all temperatures.[35] However, with increasing system size the distribution functions become narrower, and then the estimates for different temperatures remain reliable only in the vicinity of T where the distribution has been sampled.

III.2 MC simulations of alloys with CuAu-type of ordering

Since the *fcc* lattice geometry implies nearest neighbour triangles, a simple antiferromagnetic type of order is not possible without frustration. Several peculiar aspects of ordering may be related to this fact. No continuous but only first-order phase transitions are usually observed, long period superstructures may occur due to small anti-phase boundary energies, the ordering tendency at a surface may be stronger than in the bulk, and the SRO depends unusually upon the temperature.

III.2.1 First-order phase transitions

At first order phase transitions, sufficiently large systems behave practically non-ergodic, and the configurations sampled depend on the starting configuration. To locate the phase transition in the presence of a more or less pronounced hysteresis one must determine the phase of lower free energy. An accurate estimate of the free energy F is possible by calculating F of the ordered and disordered phase by integration according to the standard thermodynamic relations $S = -(\partial F/\partial T)_h$, further with $U = -T^2(\partial(F/T)\partial T)_h$, $(\partial S/\partial T)_h$ one obtains the free energy beginning from reference points at $T = 0$ and $T = \infty$, where the entropy is known:

$$F = U - TS(0,h) - k_B T \int_0^T (\partial U/\partial T)_h dT/k_b T,$$

$$\beta F = -S(\infty, h)/k_B + \int_0^\beta U d\beta.$$

From the other relation $\langle s \rangle = m = -(\partial F/\partial h)_T$ one may continue the path integral for F in the plane T, h:

$$F(T, h_2) = F(T, h_1) - \int_{h_1}^{h_2} m\, dh\,.$$

As an example, MC results for the *fcc* model with $J_1 < 0$ are given. The first-order transition temperature for $L1_0 \leftrightarrow fcc$ at $h = 0$ is found to be $T = 1.7386 \pm 0.0002$, where 30 steps and a parabolic interpolation were used for the numerical integration of the ordered and disordered branches. Different lattice sizes, $8000 < N < 216000$, have been considered. Since the correlations decay exponentially, finite size effects are very small apart from statistical accuracy. However, in sufficiently small systems the transition would appear to be practically continuous.

Calculations by the Cluster Variation Method (CVM) as well as MC simulations have indicated that the character of the transition $L1_0 \leftrightarrow fcc$ should change to second-order, if a further next-nearest neighbour interaction $J_2 > -0.3J_1$ is introduced.[36] At first sight, it is not easy to distinguish by MC simulations a weakly first-order from

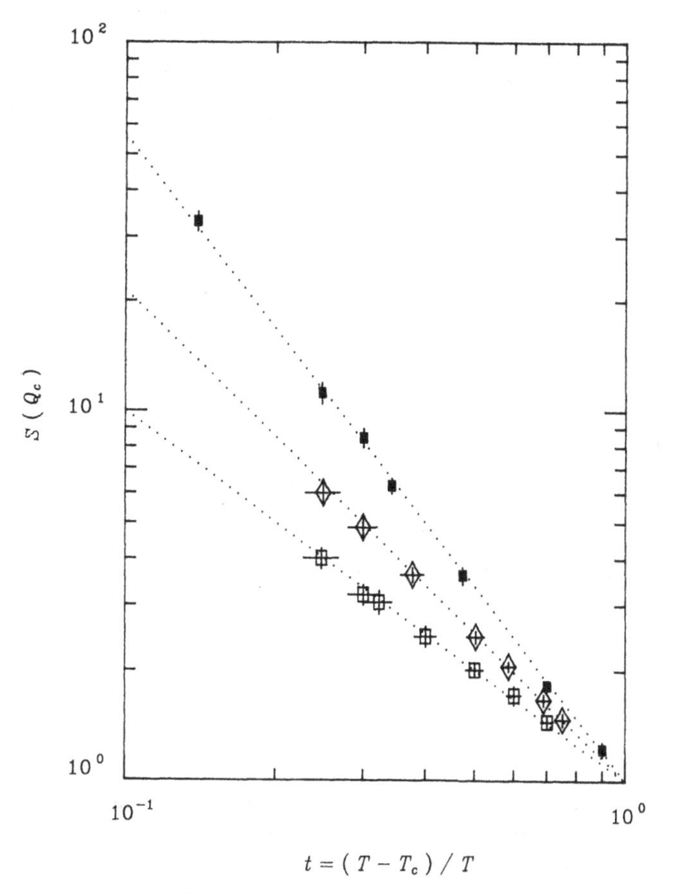

Figure 2. Log–log plot of the high temperature susceptibility (=SRO scattering $S(\vec{Q}_c)$) from MC simulations: 2 dim square lattice $J_1 > 0$ (filled squares): $\gamma = 1.751 \pm 0.005$ "CuAu" model $J_1 = J_2$ (diamonds), $\gamma = 1.33$; "Cu_3Au model $J_1 = -J_2$ (open squares), $\gamma = 1.0$.

a second-order phase transition. Recently,[37,38] therefore, the distribution function of energy was analysed close to the transition temperature. Even for $J_2 = -J_1$ the energy distribution function remained doubly peaked for the states of the ordered and disordered phase respectively. This indication of a first-order transition becomes more pronounced as the system size is increased. Possibly, this transition might change to second-order only for $J_2/J_1 \to \infty$.

III.2.2 Short-range order and critical behaviour

Critical exponents are usually defined only in the asymptotic limit $T \to T_c$. However, it is interesting to study, for instance, how far from a second-order transition the susceptibility can be well described by a power law[39] $\chi \propto (1 - \frac{T}{T_c})^{-\gamma}$, for $T > T_c$. May this also apply further to the susceptibility above a first-order transition? We note that the susceptibility is related to the SRO scattering at the ordering wave vector $\vec{Q}_c$ by

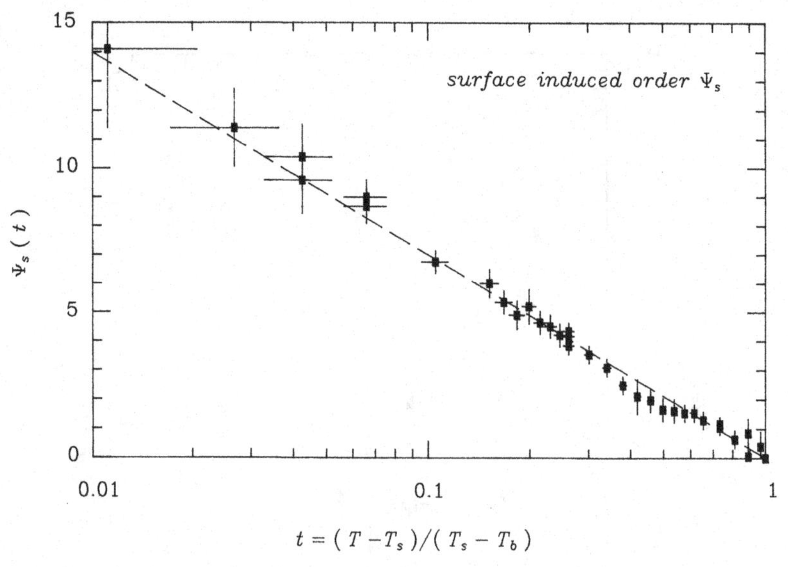

Figure 3. Semi-log plot of the surface induced order.

$$k_B T \chi = S(\vec{Q}_c) = \sum_j \alpha_j e^{i\vec{Q}_c \vec{R}_j}, \quad \text{with} \quad \alpha_j = \frac{\langle s_i s_{i+j} \rangle - \langle s \rangle^2}{1 - \langle s \rangle^2}.$$

Although not very well known, theoretical investigations[40,41] and experiments[42] on β-CuZn have shown that for several systems this power law is a very good approximation for all $T > T_c$. By MC simulations of large systems, $L >> \xi$, for the 2-dim square lattice (fig. 2), the exponent $\gamma = 1.751 \pm 0.005$ was calculated for the high temperature limit, whereas the exact asymptotic value for $T \to T_c$ is known to be 7/4.

The MC results for the SRO above the transition temperatures of CuAu and Cu_3Au models may be surprising. Again a single power law is found for all $T > T_c$, where T_c is now the spinodal ordering temperature which is just below the first-order transition temperature. Instead of $\gamma = 1.24$ as expected for the universality class of the 3-dim Ising model, the simulations yield $\gamma \approx 1.33$ for "CuAu" and $\gamma \approx 1.0$ for "Cu_3Au" (fig.2). Both systems belong not to the Ising universality class due to the higher dimensionality of their order parameters having necessarily 3 and 4 components respectively,[43] but to the same class as the 3 and 4 states Potts model.[44]

A MF description (Curie-Weiss law)[45–47] with $\gamma = 1$ for the SRO at high temperatures is therefore only valid for a few exceptional cases, like for "Cu_3Au". However, one may approximate $(1 - T_c/T)^{-\gamma}$ by $(1 - \gamma T_c/T)^{-1}$, which explains the typical overestimation of $T_c^{MF} \approx \gamma T_c$.

III.2.3 Surface induced (dis-) order

Typically the degree of order tends to decrease close to free surfaces. This phenomenon of surface induced disorder (SID), present for instance in Cu_3Au, has attracted much interest.[48–61]

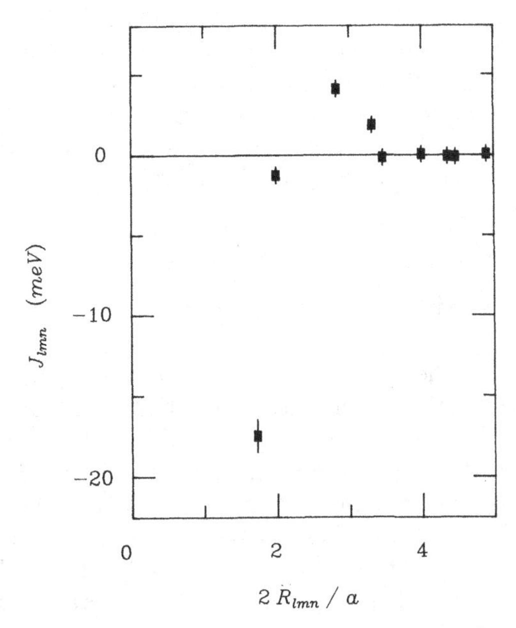

Figure 4. Effective interactions determined by the inverse MC from SRO measured at 1013 K with a single crystal of $Fe_{0.8}Al_{0.2}$.

Recent MC simulations[62] have shown that also surface induced ordering (SIO) can occur in alloys showing a CuAu-type of order ($L1_0$). At a free (100)-surface, a c(2×2) superstructure is more stable than the bulk order for models with $J_1 < 0$ and a small $J_2 \geq 0$.[63] The reason is that the interactions in the surface are not frustrated in contrast to those in the bulk. The phase transition at the surface at T_s is continuous although the bulk transition at a lower temperature T_b is of first-order. Approaching T_b from T_s (fig.3) the ordered surface wets the disordered bulk, while the thickness of the ordered film diverges logarithmically as predicted (Lipowsky).[48–51]

Surface induced phase transitions, like SID and SIO and also surface induced melting, are physically very interesting with new qualitatively different behaviour between 2 and 3 dimensions.

III.3 Ordering in models of *bcc* magnetic FeAl alloys

III.3.1 Phase diagrams and interactions

In the iron-rich side of the FeAl phase diagram, the chemical ordering is interfered by a further ferromagnetic phase transition. Several attempts have been made by CVM[64–66] and by the MC method[66] to simulate this phase diagram.

The agreement with the experimental diagram is fairly satisfactory. In particular the two phase region of a ferromagnetic A2 (disordered *bcc*) and a paramagnetic B2 was not found in the simulations so far. On the other hand, the SRO has been investigated by X-ray [67] and neutron scattering experiments.[67–69] Using the inverse CVM and MC method, values for the interaction parameters are available (Fig. 4).

However, so far the magnetic SRO is not known, and the chemical SRO is expected to be independent of the magnetic interaction only far above the ferromagnetic phase transition. Nonetheless using these interaction parameters, a recent MC calculation[70] yields a qualitative correct picture of the phase diagram including the two-phase field A2-B2. One essential ingredient for this success was probably that the magnetic coupling was treated by the Heisenberg model allowing for a continuous degree of freedom for the Fe-spins.

III.3.2 Interplay of magnetism and chemical order

Looking a bit closer to the chemical and magnetic interactions, one notices that there is a competition between both. The chemical interaction $J_1 < 0$ prefers unlike nearest neighbours, whereas the ferromagnetic interaction $J_1^m > 0$ would indirectly favour Fe-Fe neighbour pairs with aligned magnetic moments. Indeed, a new phase has been found by neutron scattering experiments[69] in the $Fe_{0.8}Al_{0.2}$ alloy at temperatures below 380° C. This phase, B32, prefers unlike next nearest neighbours rather than unlike nearest neighbours, as is typical for the phase B2 of FeAl. Hence, it is apparently formed because of the influence of the magnetic interactions on the chemical ordering.

IV. Final remarks

Monte Carlo simulations have contributed considerably to the understanding of ordering phenomena in alloys. Techniques as well as computer power are rapidly improving. In view of such a delicate aim, as the prediction of the phase stability of real alloys, further progress in our understanding will be achieved by incorporating to a greater extent all of the relevant physics. Therefore it is not a best fit of a phase diagram which is needed, but rather a better control of the real, microscopic mechanism and its parameters. For instance, further important advances can probably be expected by including the local relaxations of atoms.

Ab-initio theories and scattering experiments may provide this information. MC simulations can solve a lot of problems along this way, and since the burden is left to the computer it can be fun.

References

1. K. Binder, *J. Comput. Phys.* **59**, 1 (1985).
2. K. Binder in Festkörperprobleme (*Advances in Solid State Physics*), edited by P. Grosse, Vieweg, Braunschweig, 1986. Vol. 26, p.133.
3. K. Binder in *International Meeting on Advances on Phase Transitions and Disorder Phenomena* in Amalfi June 1986, edited by G. Busiello, L. De Cesare, F. Manchini, and M. Marinaro, World Scientific, Singapore, 1987. p. 1.
4. *The Monte Carlo Method in Statistical Physics,* 2nd ed, edited by K. Binder, Springer–Verlag, Berlin, 1986).
5. *Applications of the Monte Carlo Method Method in Statistical Physics*, 2nd ed, edited by K. Binder, Springer–Verlag,Berlin,1987.

6. *The Monte Carlo Method in Condensed Matter Physics,* edited by K. Binder, Springer–Verlag, Berlin, 1991.
7. V. Gerold and J. Kern, *Acta Metall.* **35**, 393 (1987).
8. W. Schweika and H. G. Haubold, Phys. Rev. B37, (1988) 9240
9. W. Schweika, in *Alloy Phase Stability*, edited by G. M. Stocks and A. Gonis, Kluwer Acad. Publ., Dordrecht, 1989. Proc. NATO ASI, Maleme, 1987.
10. N. Metropolis, A. W. Rosenbluth, M. N. Rosenbluth, A. H. Teller and E. Teller, *J. Chem. Phys.* **21**, 108 (1953).
11. R. J. Glauber, *J. Math. Phys.* **4**, 294 (1963).
12. K. Kawasaki in *Phase Transitions and Critical Phenomena*, edited by C. Domb and M. S. Green, Academic Press, New York, 1972. Vol. 2, p.443.
13. M. Creutz, *Phys. Rev.* **D36**, 515 (1987).
14. S. L. Adler, *Phys. Rev.* **D38**, 1349 (1988).
15. R. H. Swendsen and J.–S. Wang, *Phys. Rev. Lett.* **58**, 86 (1987).
16. S. Wansleben, *Comp. Phys. Comm.* **43**, 315 (1987).
17. N. Ito and Y. Kanada, *Supercomputer 25,* **3**, 31 (1988).
18. H. O. Heuer, *Europhys. Lett.* **12**, 551 (1990); to appear in *Comput. Phys. Comm.*
19. S. Kirkpatrick and E. Stoll, *J. Comp. Phys.* **40**, 517 (1981).
20. F. James, *Repts. Prog. Phys.* **43**, 1145 (1980).
21. F. Livet and M. Bessiere, *J. Phys.* (Paris) **48**, 1703 (1987).
22. A. M. Ferrenberg, D. P. Landau, and K. Binder *J. Stat. Phys.* **63**, 867 (1991).
23. G. Tammann and W. Oelsen, *Z. Anorg. Chem.* **186**, 260 (1930).
24. M. E. Fisher, *Proceedings of the International Summer School Enrico Fermi*, Course 51, Academic Press, New York, 1971.
25. M. E. Fisher, in *Critical Phenomena*, edited by M. S. Green, Academic Press, New York, 1971.
26. M. N. Barber, in *Phase Transitions and Critical Phenomena*, eds. C. Domb and J. L. Lebowitz, Academic Press, New York, 1983, Vol. 8, p. 145.
27. V. Privman and M. E. Fisher, *J. Stat. Phys.* **33**, 385 (1983); *Phys. Rev.* **B32**, 447 (1985).
28. K. Binder, *Z. Phys.* **B43**, 119 (1981).
29. K. Binder, D. P. Landau, *Phys. Rev.* **B30**, 1477 (1984).
30. K. Binder, *Rep. Prog. Phys.* **50**, 783 (1987).
31. A. M. Ferrenberg and R. H. Swendsen, *Phys. Rev. Lett.* **61**, (1988).
32. R. H. Swendsen, J.-S. Wang and A. M. Ferrenberg, *The Monte Carlo Method in Condensed Matter Physics*, edited by K. Binder, Springer-Verlag, Berlin, 1991.
33. A. M. Ferrenberg and D. P. Landau, *Phys. Rev. B44*, 5081 (1991).
34. A. M. Ferrenberg, D. P. Landau, and P. Peczak, *J. Appl. Phys.* **69**, 6153 (1991).
35. J. Lee and J. M. Kosterlitz, *Phys. Rev. Lett.* **65**, 137 (1990).
36. J. L. Lebowitz, M. K. Phani, and D. F. Styer, *J. Stat. Phys.* **38**, 413 (1985).
37. J. F. Fernández and J. Oitmaa, *J. Phys.* **C8**, 1549 (1988).
38. J. Oitmaa and J. F. Fernández, *Phys. Rev.* **B39**, 11920 (1989).
39. M. E. Fisher and R. J. Burford, *Phys. Rev.* **156**, 583 (1967).
40. A. Arrot, *Phys. Rev.* **B31**, 2951 (1985).
41. M. Fähnle and J. Souletie, *J. Phys.* **C17**, L469 (1984); *Phys. Rev.* **B32**, 3328 (1985); *Phys. Stat. Sol.* **138**, 181 (1986).
42. C. Lamers and W. Schweika, ICNS'91 Oxford, to be published in *Physica B.*
43. F. Ducastelle, ICNS'91 Oxford, to be published in *Physica B.*
44. E. Domany, Y. Shnidman and D. Mukamel, *J. Phys. C: Solid State Phys.* **15**, L495 (1982).

45. R. Brout, *Phase Transitions*, Benjamin, New York, 1965.
46. M. A. Krivoglaz and A. A. Smirnov, *The Theory of Order-Disorder in Alloys*, MacDonald, London, 1964; M. A. Krivoglaz, *Theory of X-Ray and Thermal Neutron Scattering by Real Crystals*, Plenum Press, New York, 1969.
47. S. C. Moss and P. C. Clapp, *Phys. Rev.* **171**, 764 (1968).
48. R. Lipowsky, *Phys. Rev. Lett.* **49**, 1575 (1982).
49. R. Lipowsky and W. Speth, *Phys. Rev.* **B28**, 3983 (1983).
50. R. Lipowsky, *J. Appl. Phys.* **55**, 2485 (1984).
51. R. Lipowsky, *Ferroelectrics* **73**, 69 (1987).
52. K. Binder in *Phase Transitions and Critical Phenomena*, edited by C. Domb and J. L. Lebowitz, Academic Press, New York, 1983. Vol. 8, p. 1.
53. J. M. Sanchez and J. L. Morán-López, *Phys. Rev.* **B32**, 3534 (1985).
54. J. M. Sanchez and J. L. Morán-López, *Surf. Sci. Lett.* **157**, 297 (1985).
55. H. W. Diehl, in *Phase Transitions and Critical Phenomena*, edited by C. Domb and J. L. Lebowitz, Academic Press, New York, 1986. Vol. 10, p. 75.
56. K. Binder and D. P. Landau, *Physica* (Amsterdam) **163A**, 17 (1990).
57. S. Dietrich, *Phase Transitions and Critical Phenomena*, edited by C. Domb and J. L. Lebowitz, Academic Press, New York, 1986. Vol. 12, p. 1.
58. D. M. Kroll and G. Gompper, *Phys. Rev.* **B36**, 7078 (1987).
59. G. Gompper and D. M. Kroll, *Phys. Rev.* **B38**, 459 (1988).
60. D. M. Kroll and G. Gompper, *Phys. Rev.* **B39**, 433 (1989).
61. W. Helbing, B. Dünweg, K. Binder, and D. P. Landau, *Z. Physik B* , (1990).
62. W. Schweika, K. Binder, and D. P. Landau, *Phys. Rev. Lett.* **65**, 3321 (1990).
63. Y. Teraoka, *Surface Sci.* **232**, 193 (1990).
64. D. A. Contreras-Solorio, F. Mejía-Lira, J. M. Sanchez, and J. L. Morán-López, *Phys. Rev.* **B38**, 4955 (1988).
65. D. A. Contreras-Solorio, F. Mejía-Lira, J. M. Sanchez, and J. L. Morán-López, *Phys. Rev.* **B38**, 11481 (1988).
66. B. Dünweg and K. Binder, *Phys. Rev.* **B36**, 6935 (1987).
67. V. Pierron-Bohnes, M. C. Cadeville, A. Finel, O. Schäpf, *J. Phys. Condens. Matter.* **1**, 247 (1991).
68. W. Schweika, M. Monkenbush, and A. Ackerman, *Physica* **B156& 157**, 78 (1989).
69. W. Schweika, *Mat. Res. Soc. Symp. Proc.* **166**, 249 (1990).
70. F. Schmid and K. Binder, to be published.

Compatibility of Lattice Stabilities Derived by Thermochemical and First Principles

A. P. Miodownik

Department of Materials
Science and Engineering
University of Surrey
Guilford, Surrey GU2 5XH
England

I. Introduction

Complex phase equilibria can now be routinely calculated in multicomponent commercial alloys,[1] a development that can be traced to the single minded application of the concept of lattice stabilities by Kaufman.[2] At that time such quantities were viewed with considerable suspicion, and could certainly not be calculated with any degree of confidence. Titanium was historically one of the first elements investigated, since this element exhibits an allotropic transformation and experimental data could therefore be obtained for both phases. The expressions used to define the phase stability included Debye temperature and electronic specific heat terms as well as the lattice stability at zero K.

$$\begin{aligned}\Delta G_{\mathrm{A}}^{\alpha-\beta} &= \Delta E_{\circ}^{\alpha-\beta} + \frac{9}{8}R(\theta_{\mathrm{A}}^{\beta} - \theta_{\mathrm{A}}^{\alpha}) + f(\theta_{\mathrm{A}}^{\beta}/T) - f(\theta_{\mathrm{A}}^{\alpha}/T) \\ &+ \frac{10^{-4}}{2}R\left[\left(\frac{3}{2}T - \theta_{\mathrm{A}}^{\beta}\right)^2 - \left(\frac{3}{2}T - \theta_{\mathrm{A}}^{\alpha}\right)^2\right] - \frac{T^2}{2}\left(\gamma_{\mathrm{A}}^{\beta} - \gamma_{\mathrm{A}}^{\alpha}\right),\end{aligned} \tag{1}$$

where $\Delta E_{\circ}$ is the lattice stability at zero K, θ is the Debye temperature, γ is the electronic specific heat and f is a suitable integration function.

Extension of the same principles to Fe and Mn immediately created an awareness of the importance of magnetic terms which had to be added to the equation, and which generated a very irregular temperature dependence for all free energy differences. Nonetheless it proved possible to develop suitable techniques to give an adequate

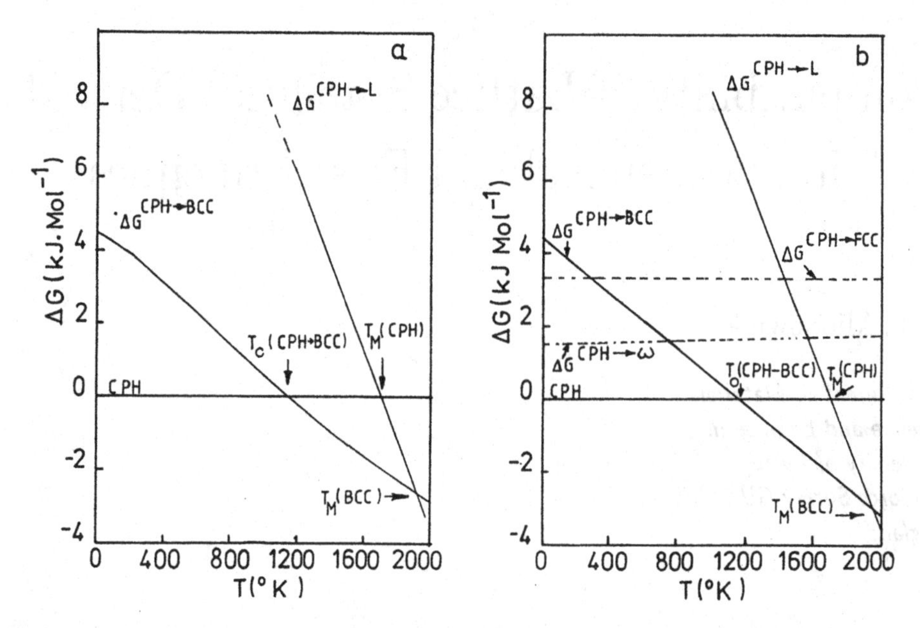

Figure 1. The relative free energies for TCΔG *cph→fcc* titanium. a) summating separate free energy contributions. b) using a linear approximation. (It also includes the relative position of the ΔG *cph→fcc* and ΔG *cph→ ω* transformation adapted from Ref. 7).

representation of these magnetic terms, and superimpose them onto a hypothetical paramagnetic reference state.

Early treatments of the apparently simpler elements Cu, Ag, and Zn generated an awareness of other problems pertinent to the current situation, namely the requirement to obtain data for the Debye temperatures and electronic specific heats of metastable phases in elements that do not exhibit allotropy in accessible regimes of temperature or pressure.

Since a determination of lattice stabilities is the cornerstone of the total free energy expression, a method had to be found which could yield an adequate approximation to the real lattice stability. It became apparent that, provided the free energy differences between different crystal structures was not too great, linear approximations could be made that differed only slightly from the more rigorous derivation (Fig. 1). This so-called Van Laar technique relies on a combination of extrapolated melting points and a knowledge of the entropy of fusion, as expressed by Eq. 2, and which for the purpose of this paper will be called the Thermochemical lattice stability TCΔG:

$$\Delta H_T^{\mathrm{B-A}} = T_f^{\mathrm{B}} \Delta_f^{\mathrm{B}} - T_f^{\mathrm{A}} \Delta_f^{\mathrm{A}} \tag{2a}$$

$$\mathrm{TC}\Delta G_{T=0}^{\mathrm{B-A}} \approx \Delta H_{T=T_m}^{\mathrm{B-A}} \tag{2b}$$

where data is not available, the TCΔG assumes that any differences in Debye temperature and electronic specific heat of the pair of phases are either small or self cancelling, so that, if magnetic terms are ignored, the lattice stability can be represented to a first approximation by the difference of the two enthalpies of melting. Estimates for the entropy of melting of metastable phases could be made on the basis of a number

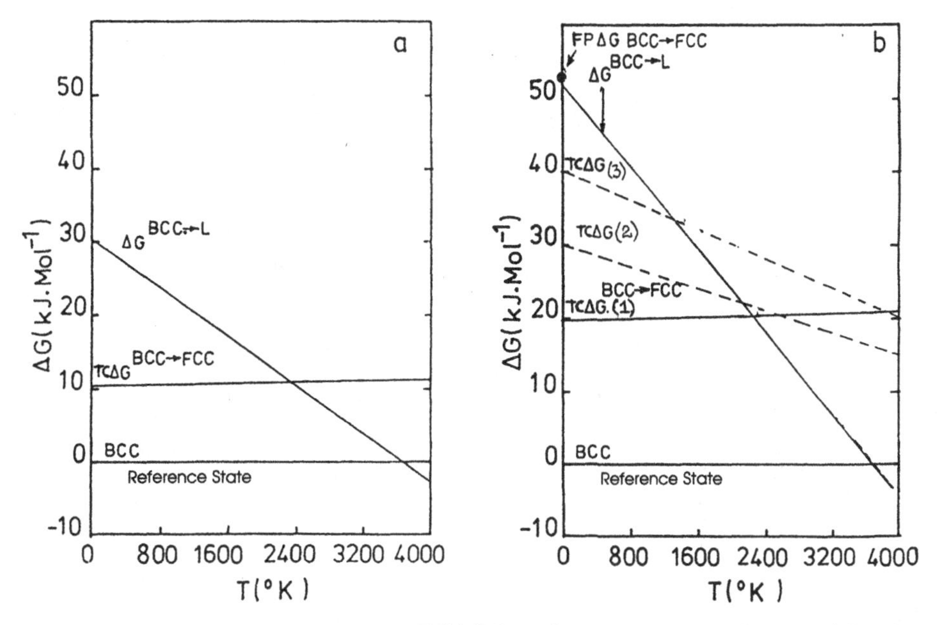

Figure 2. The relative free energies for TCΔG *bcc→fcc* tungsten according to: a) Kaufman[2] and b) Gustaffson[6] and Miodownik.[7]

of empirical observations such as developed by Ref 3. Many questions can be raised about the validity of this procedure,[4] especially when the metastable melting points become appreciably lower than the melting point of the stable phase. Despite the inherent problems in making long extrapolations, the requirement of deriving a single set of parameters for each phase that yields satisfactory calculations for many different binary combinations gave confidence in these values. It is important to realize that a large number of the lattice stabilities used by the Calphad community have perforce been derived in this way. More to the point, in many cases such values agree very closely with lattice stabilities derived from First Principles (FPΔG).[5]

A further important check on any derived lattice stabilities is the requirement that, when combined with suitable algorithms for excess free energies, the chosen parameters must not only generate the phase diagrams, but also simultaneously reproduce all other available thermochemical data. This offers an efficient way of distinguishing between alternative but self–consistent data sets, but this procedure can unfortunately not be applied when metastable phases are experimentally inaccesible.

Inconsistent lattice stabilities lead to considerable complications when data sets from different sources have to be combined for multicomponent system, and also generate doubts that the whole process is based on arbitrary fitting parameters which have no physical significance. Both of these undesirable consequences can be removed if the lattice stabilities entered into the data base can be calculated from first principles; this is why it has always been important to ensure convergence between TCΔG and FPΔG values. It was therefore disturbing to find the two methods yielding widely different results in more recent times and this paper will attempt to trace the developments that have taken place in the last five or six years to reconcile these divergent results.

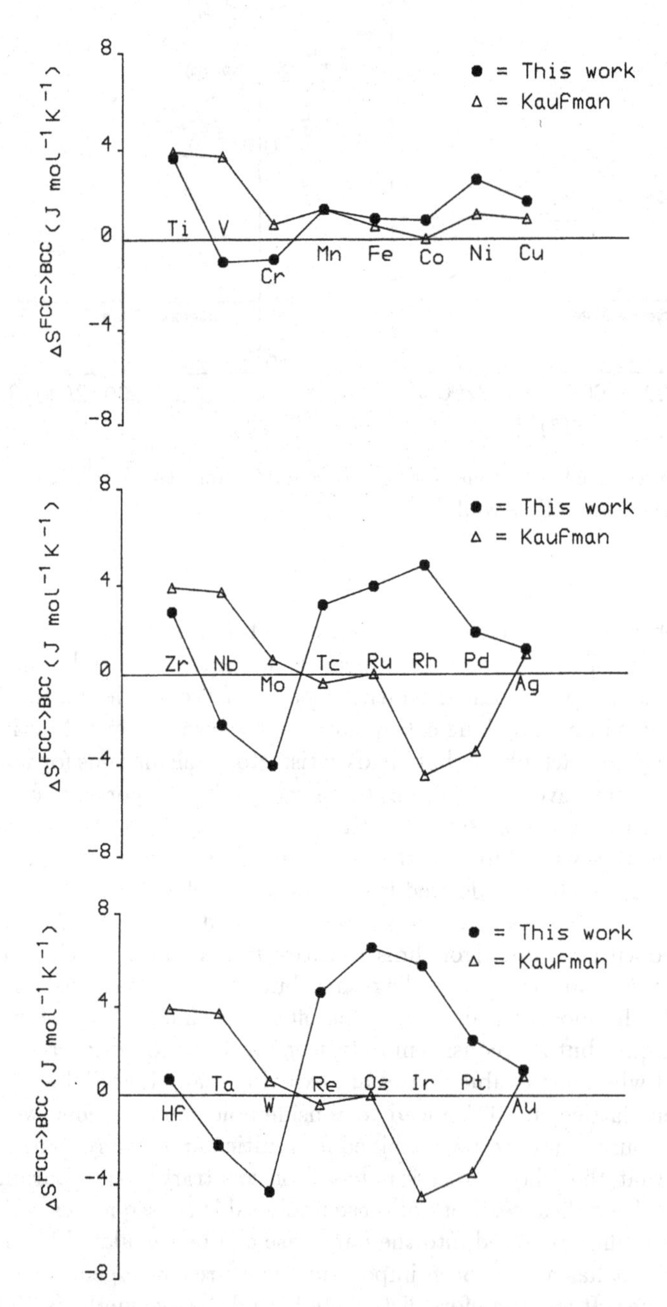

Figure 3. Comparison of ΔS of *fcc*→*bcc* for 3d, 4d and 5d elements, ● Ref. 5a, △ Ref.2.

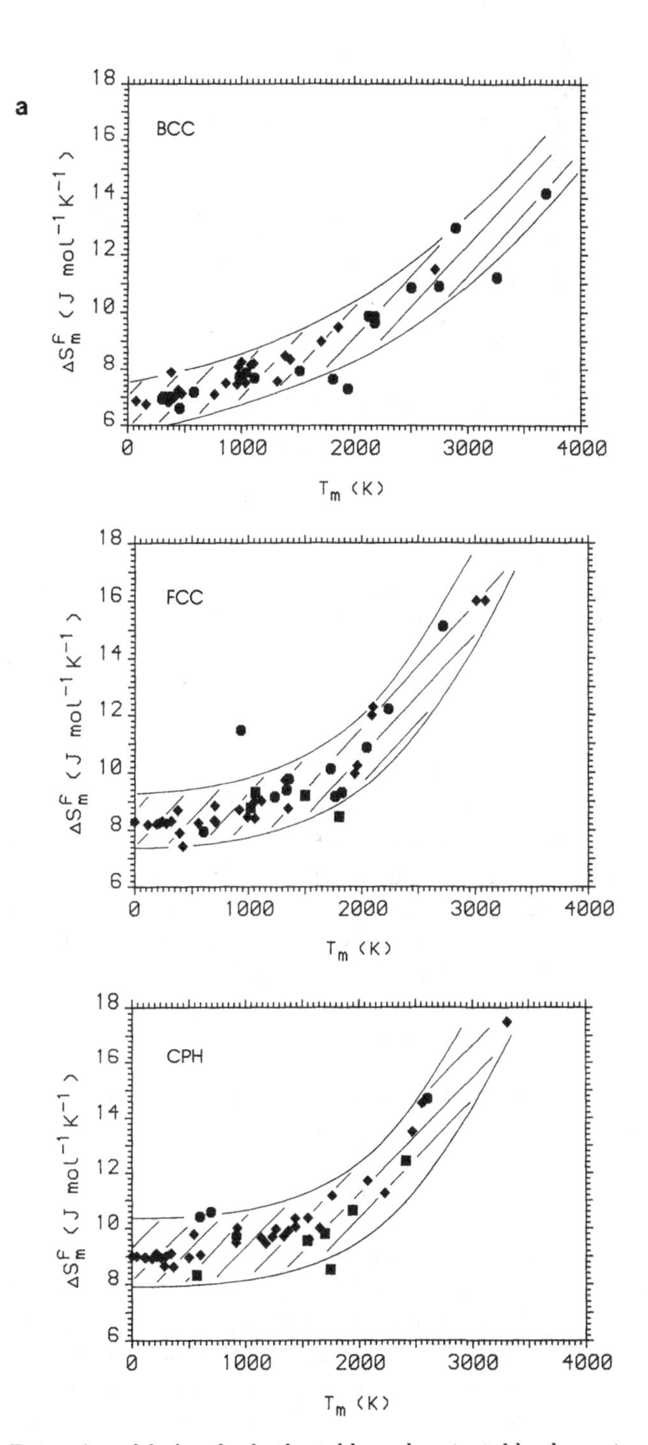

Figure 4a. Entropies of fusion for both stable and metastable elements exhibiting *bcc, fcc* and *cph* structures: ● Ref. 5a, ▲ Ref.2, ■ Ref. 41.

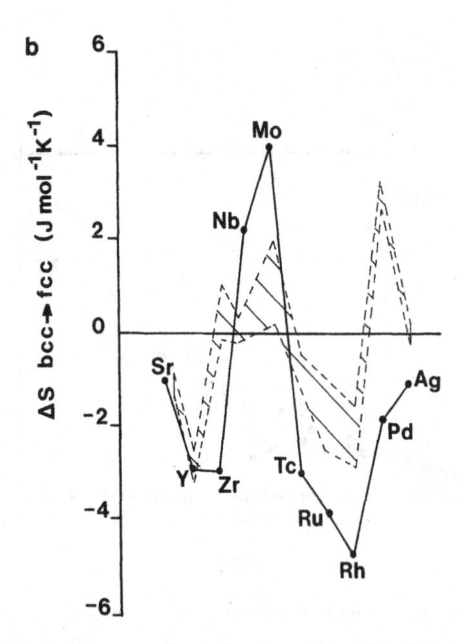

Figure 4b. Comparison between evaluated values for ΔS of *bcc*→*fcc* for the 4d elements and values predicted from band structure calculations.[39]

II. Role of Entropies of Fusion

Recent revision of the experimental values of the entropies of fusion for some of the high melting point transition elements provided the key for a correspondig revision of TCΔG values. Thus if the new entropy of fusion for *bcc* W is entered into Eq. (1) keeping all other data constant, the apparent TCΔG (*fcc–bcc*) for W is effectively doubled[6] (compare Fig. 2a and ΔG1 in Fig. 2b). This lead to the realization[7] that even lower metastable melting points for the *fcc* phase would give subtantially higher values for the TCΔG (see ΔG2 and ΔG3 in Fig. 2b). There is however a need to place some theoretical limits on such an approach, since otherwise it would be possible to arbitrarily match any proposed FPΔG calculation by invoking artificially low metastable melting points.

Because of the significant effort involved in making major revisions to comprehensive data bases, the possibility of new phase stability values was subjected to extensive review in the UK[5], USA,[8] and Sweden.[4,9,10]

The re-examination of the lattice stability of tungsten gives a good indication of the way in which such changes are being viewed by the Calphad community (Fig. 7). Interestingly these reviews came to a variety of different conclusions. The UK approach developed the approach raised by Ref 7 which led to quite substantial revisions of the previously held lattice stabilities (Figs. 3–6), while the american approach suggested only minor revisions. The swedish investigation began with analysis of the potential limitations of the Van Laar technique, and while accepting some revised values, placed severe limitations on the acceptable magnitude of any proposed change.

However all three groups are in agreement that:

(i) There remains a major discrepancy between FPΔG and TCΔG for both *fcc–bcc* and *fcc–cph* Cr.

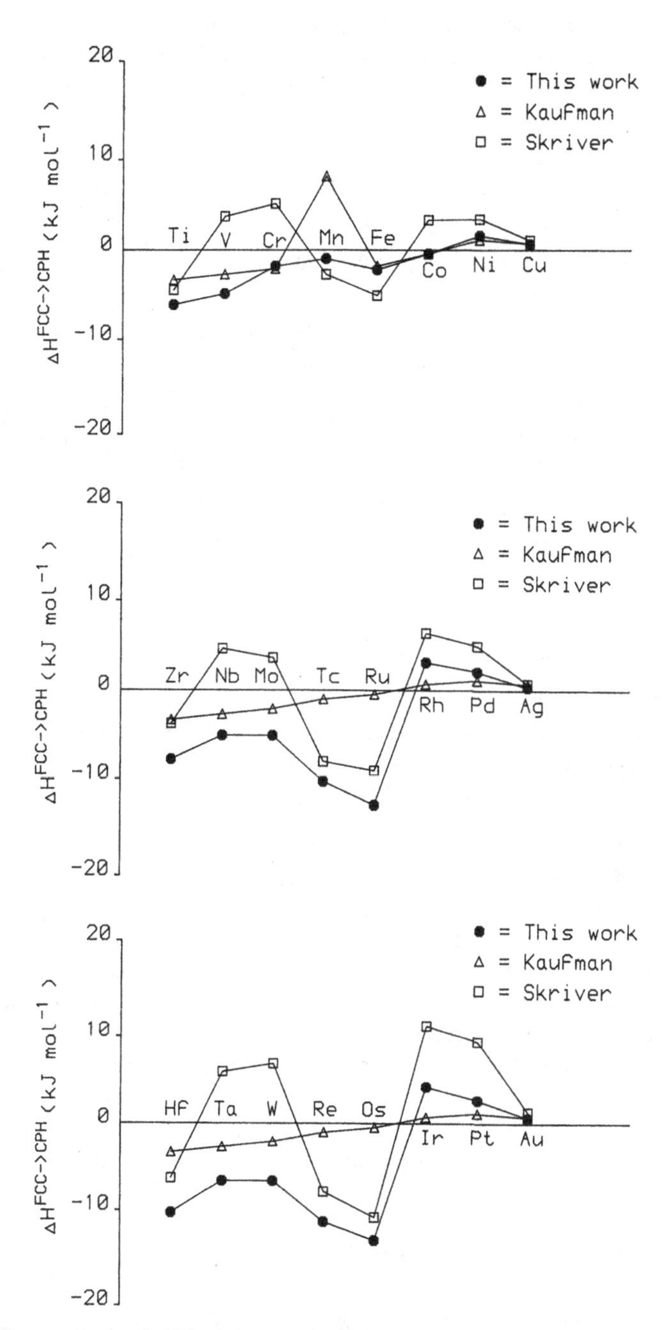

Figure 5. Comparison of ΔH of *fcc*→*cph* for 3d, 4d and 5d elements ● Ref. 5a, △ Ref.2, □ Ref. 41.

Table I

The Properties of the metastable *fcc* and *chp* modifications of Nb and Ta according to the estimates by Kaufman,[2] Saunders and Miodownik[5a] and Swedish workers.[10] The units for enthalpy, entropy and temperature are: (J/mol), (J/mol K) and (K), respectively

	fcc→*bcc* Nb				*chp*→*bcc* Nb			
Ref.	Δ H	Δ S	T_f^{fcc}	$\Delta_f S^{fcc}$	Δ H	Δ S	T_f^{chp}	$\Delta_f S^{chp}$
2	8996	−3.56	1170	11.92	6276	−3.35	1420	11.71
5a	22000	+2.2	919	8.71	17000	+1.2	1339	9.71
10	13500	−1.7	1301	12.5	10000	−2.4	1497	13.2

	fcc→*bcc* Ta				*chp*→*bcc* Ta			
Ref.	Δ H	Δ S	T_f^{fcc}	$\Delta_f S^{fcc}$	Δ H	Δ S	T_f^{hcp}	$\Delta_f S^{hcp}$
2	8996	−3.56	1540	11.92	6276	−3.35	1800	11.71
5a	26500	+2.2	1116	9.02	20000	+1.2	1653	10.02
10	16000	−1.7	1623	12.7	12000	−2.4	1828	13.96

(ii) There is concern at the very high values suggested for Osmium, even by Saunders *et al.*[5] who quote the highest TCΔG values.

(iii) Metastable *cph* phases are in general found to be more stable than the corresponding metastable *fcc* phase for group V and VI elements, in contradiction to theoretical expectations form FP calculations.

The values proposed by Ref. 5 are undoubtedly substantially larger than most of the values proposed by other groups. The cardinal assumption is that the entropy of fusion for metastable phases will fall on the same locus as the values given by stable elements of the same crystal structure.

This is already implicit in the earlier formulation[3] and unaffected by the fact that those locii are no longer linear with temperature; however the more recent values for the entropy of fusion of high melting point elements magnify the difference in entropy of fusion for a given difference in melting points (Fig. 8). Contrary to the popular belief, the object of Ref. 5 was not to approach the FP values as closely as possible, but merely to devise a consistent operational principle, which happens to result in much closer agreement with FP values than has been achieved before.

III. Consequences of Adopting New Lattice Stability Values

New lattice stability values cannot be adopted by the Calphad Community unless a number of important questions are answered:

(a) Can phase boundaries still be calculated to the same degree of accuracy with new lattice stability values?

(b) Are the corresponding changes in the excess free energy terms consistent with other available thermochemical measurements terms?

(c) Have all available experimental methods that could provide a means of checking relative lattice stabilities at low temperatures been used?

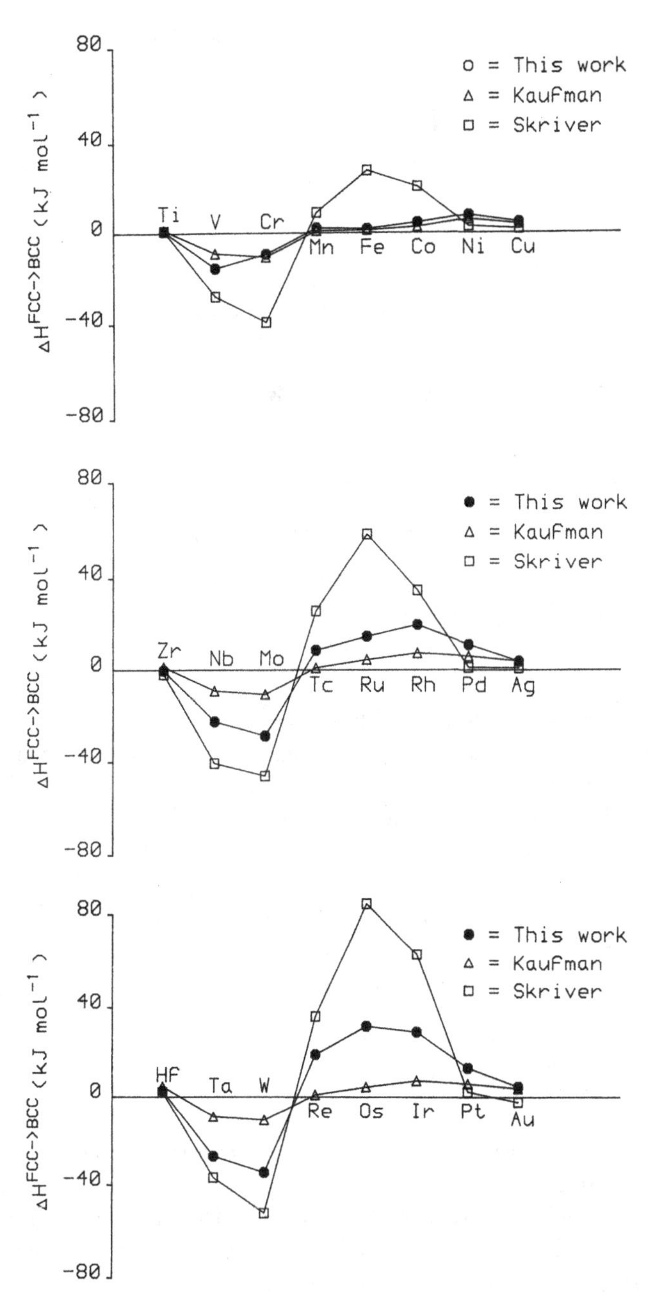

Figure 6. Comparison of ΔH of *fcc*→*bcc* for 3d, 4d and 5d elements ● Ref. 5a, △ Ref.2, □ Ref. 41.

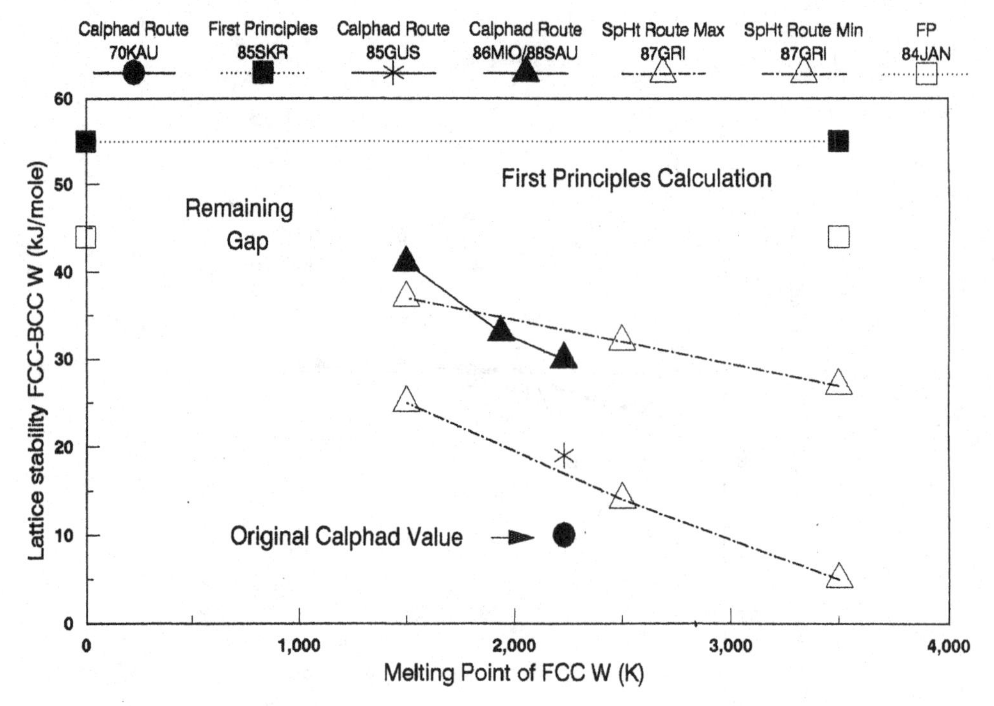

Figure 7. The lattice stability of *bcc*→*fcc* tungsten derived from various sources: ● Ref. 2, ■ Ref. 41, ∗ Ref.6, ▲ Ref. 5a, △ Ref. 38

Success in matching experimental phase boundaries with new lattice stabilities depends very much on what constraints are placed on the allowed variation of the other parameters used in the calculation. Thus Refs. 9 and 10 placed limits on the relative values of the entropy of fusion for *bcc*, *cph*, and *fcc* structures, based on the average values for this parameter, thus clearly refuting the basic assumption used by Ref. 5.

These authors also considered that the TCΔG(*fcc*–*bcc*) V could not exceed that of Cr, based on general periodic trends. However this ignores the fact that magnetic terms are substantially different[11] in Cr and V.

IV. The Gold-Vanadium System

The Au-V system was used as testing ground (Fig. 9) to establish TCΔG for Vanadium but in view of the above restrictions, it is perhaps not surprising that only limited changes were ultimately proposed for V, Nb, and Ta (and later for Cr, Mo, W) [Table I]. However the Rh-Mo diagram has been successfully reproduced using higher *fcc*-*bcc* phase stability values for Mo without departing form interaction parameters in line with predicitons by Colinet[12] (Fig. 10). By contrast difficulties have arisen when trying to fit the same diagram using Miedma's published set of lattice stabilities and heats of formation without using any further adjustable parameters. It is therefore important to appreciate how much the end result depends on the limits within which parameters are allowed to vary.[13]

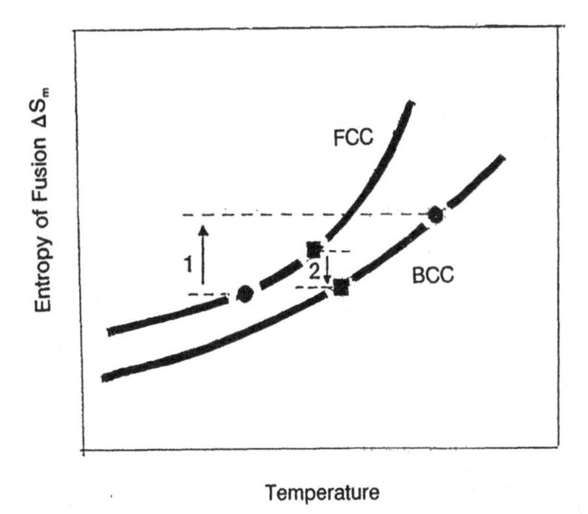

Figure 8. Schematic illustration of how the relative entropies of melting for *fcc* and *bcc* allotropes can change sign according to their relative melting points. Case 1: $T_m^{bcc} \gg T_m^{fcc}$. Case 2: $T_m^{bcc} \approx T_m^{fcc}$. See Ref. 5a and Fig. 4a.

Table II

Alternative Interaction Parameters for Fe–Ru: $\Delta G_{Fe}^{cph-fcc} = A + BT$,

$\Delta G_{Ru}^{cph-fcc} = C + DT$ and $\Delta G_{Excess}^{cph-fcc} = E$

	Ref. 17	Ref. 5a
A	−1829	−1829
B	+4.69	+4.69
C	−502	−12500
D	−3.35	+2.4
E	−8054	+800

V. The Iron-Ruthenium System

This system was sudied extensively by Blackburn *et al.*,[14] and by Stepakoff and Kaufman[15] in order to obtain the lattice stability of the *fcc-cph* transformation in Fe. The corresponding lattice stability for *fcc-cph* Ru was of secondary interest at the time, and assumed to be equal to that of Fe after substraction of the magnetic component. The system was re-assessed by Swartzenbruder and Sundman,[16] who proposed a value of 866 J/mol, which is essentially a small change in relative magnitud from the original value of 500 J/mol. These two values can be contrasted with the value[5] of 12500 kJ/mol !

In keeping with the philosophy of the TC approach, the TC assessment included martensite start and finish temperatures from which values for T_o could be deduced, experiments using the pressure variable, as well as specific heat and vapour pressure measurements. When combined with phase boundary requirements, this led to tightly prescribed values of the interaction parameters for *fcc* and *cph* solid solutions, thus

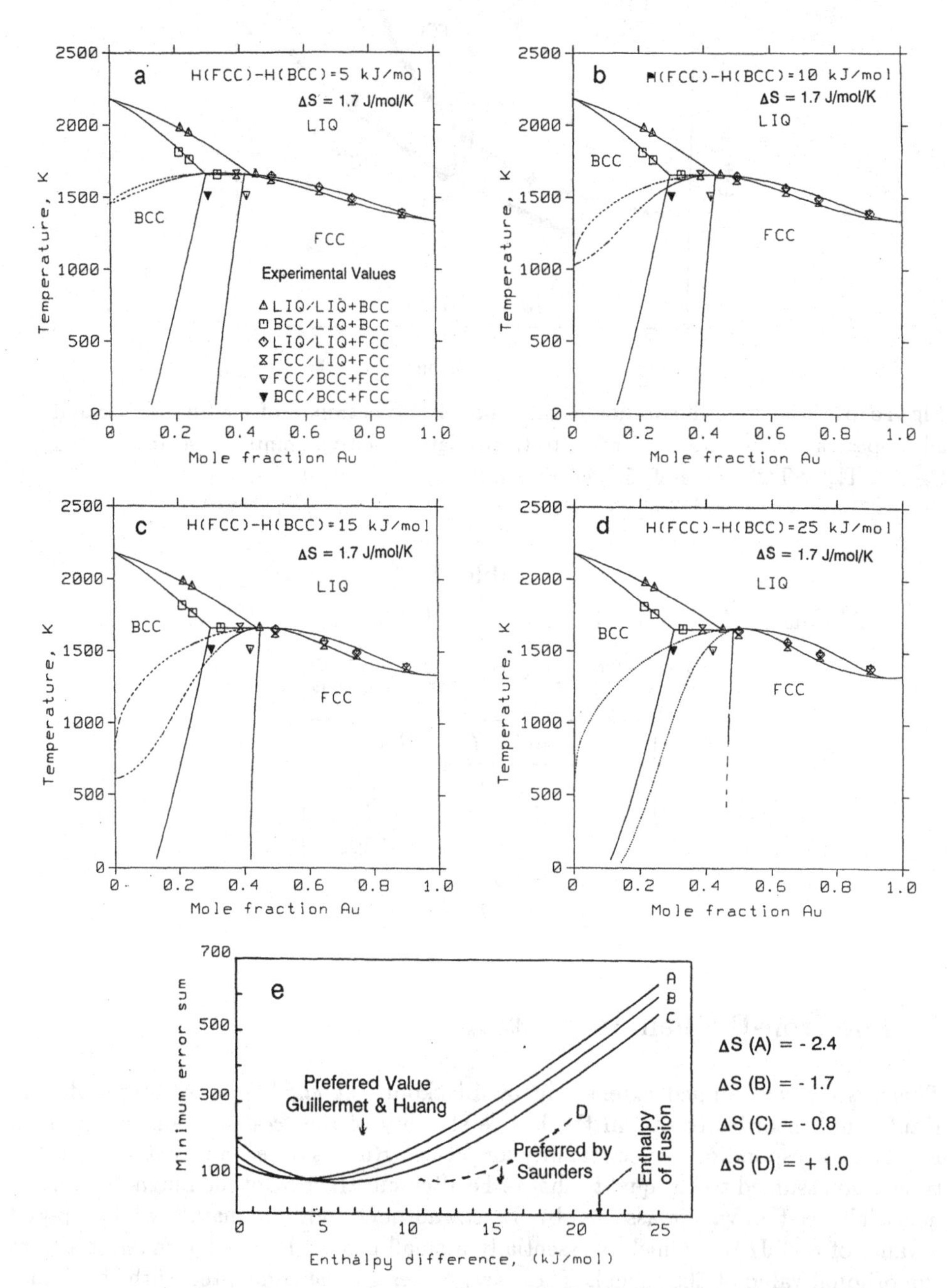

Figure 9. a)–d) The effect of varying the *fcc-bcc* vanadium phase stability on the matching of experimental and calculated phase boundaries in the Au-V system after Ref. 9 and 10. e) The effect of varying the entropy difference on the minimum error giving a preferred value of 7.5 kJ/mol if ΔS is keep negative[10] but rising towards 15 kJ/mol if positive ΔS values are allowed.[5a]

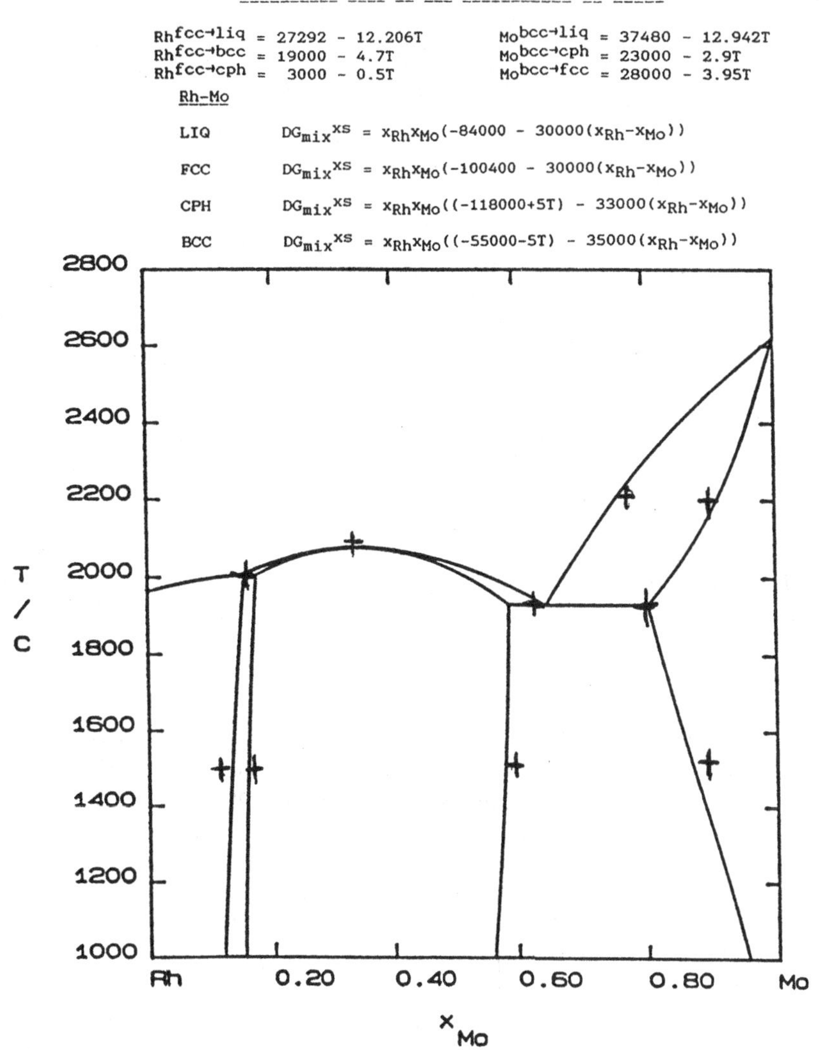

Figure 10. Calculated phase diagram for Rh-Mo using lattice stabilities from Ref. 5a and sub-regular solution model together with some critical points from the experimentally determined phase diagram.

limiting their adjustement when using a higher Ru stability (alterations in these parameters by Swartzendruber and Sundman[16] can be largely attributed to the adoption of a different formulation for the magnetic components).

This system therefore clearly provides an interesting test case, and has been re-examined in some detail by Kaufman,[17] who has also taken into account more recent estimates for the heats of solution by Watson and Bennet,[18] Miedma and his co-workers[19] and the work of Colinet,[12] all of whom suggest that these solutions should exhibit relatively small negative deviations from ideality, in keeping with interaction parameters suggested by Kaufman[2], and by Swartzenbruder and Sundman.[16]

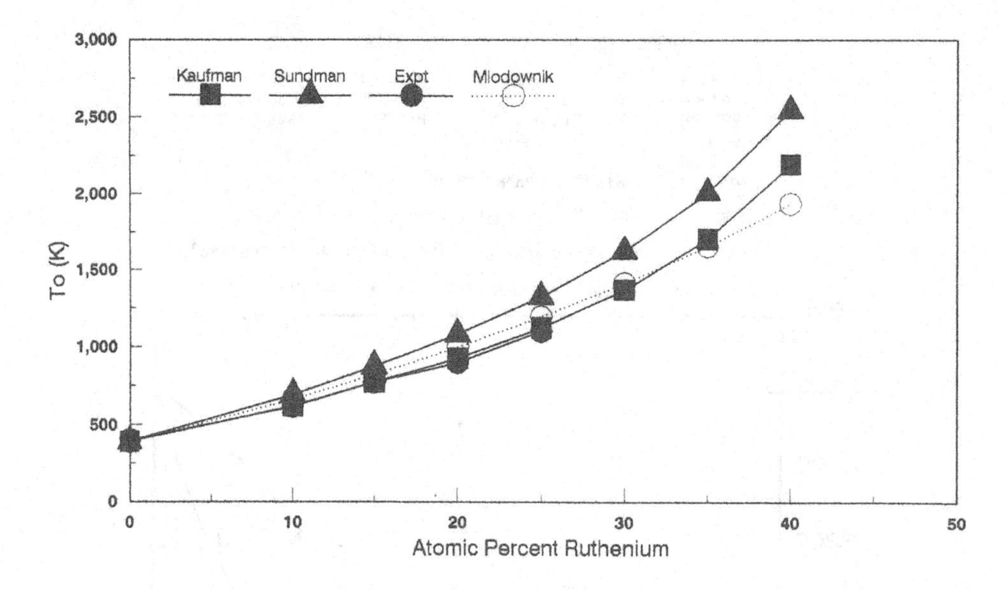

Figure 11. Comparison of T_0 values ($\Delta G^{cph-fcc} = 0$) based on different phase stabilities for ruthenium.

Table III

Comparison of SFE (mJm^{-2}) calculated from first principles[22] and the values obtained by combining the formalism in Ref. 20 with lattice stabilities derived by Ref. 22.

Element	Ref. 22	This work $\sigma = 0$
Ag	33	43
Al	124	122
Au	44	47
Cu	70	78
Ir	534	466
Ni	180	152
Pd	161	173
Rh	308	330

For a regular solution model and interaction parameters independent of temperature (as assumed by Kaufman), a simple algorithm [Eq. 3b], can be used to derive the change in the value of the interaction parameters necessary to acommodate the new value of the lattice stability, while still matching the experimentally derived values of T_0.

$$(1-x)\Delta G_{\mathrm{Fe}}^{cph-fcc} + x\Delta G_{\mathrm{Ru}}^{cph-fcc} + x(1-x)\Delta G_{\mathrm{Excess}}^{cph-fcc} = 0; \quad [T = T_0] \qquad (3a)$$

$$T_0 = \left[\frac{x^2 E - A + (A - E - C)x}{B - (B - D)x}\right] \qquad (3b)$$

($A \to E$ are defined in Table II).

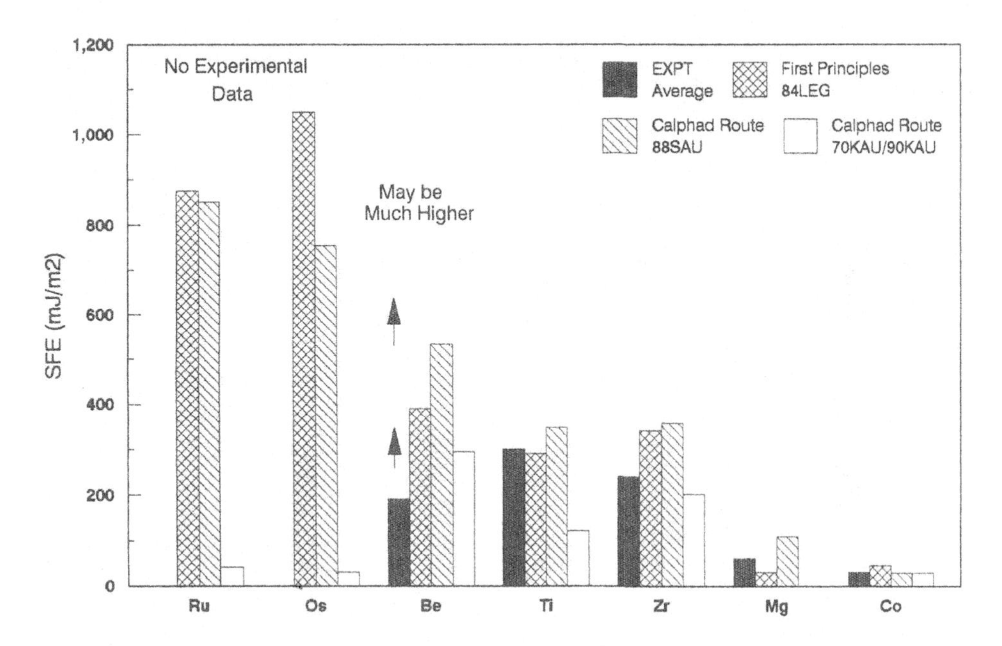

Figure 12. Comparison of experimental and calculated stacking fault energies in *fcc* metals. Data for first principle calculations from Ref. 22, and values for thermochemical calculations from Refs. 2, 5, 20 and 17.

Figure 11 shows that experimental values for T_o can be matched by various combinations of the phase stability of Ru and the difference in interaction parameters for *fcc* and *cph* solutions. A small positive sign for the difference between interaction parameters does not run contrary to the predicted and experimentally observed negative heats of formation in this system[12,18,19] (or the activity data), but a change in sign of the E (Table II) quantity was interpreted by Kaufman as support for a low value of the TC lattice stability of *cph-fcc* Ru. In fact a more serious requirement is that the T_o trajectory must meet the solidus region in the right composition range, otherwise the corresponding minimization of ΔG will produce incorrect phase boundaries. Examination of Eq. (3b) shows that the result is very sensitive to the relative values of the coefficients in the denominator, and it can be shown that a higher value for *cph-fcc* Ru can be used with a minor change in the entropy of melting for Ru (1.7 J/mol/K), but the Kaufman values do give a better fit in the iron rich region.

However when the Kaufman parameter T_o is projected to the Ru rich region additional complications are seen to occur which can again be attributed to the form of the denominator of Eq. (3) and the difference in the sign of the temperature coefficient of the lattice stabilities for Fe and Ru. One solution of this problem is to go for a sub-regular model and/or a temperature dependence for the interaction coefficients, but it is clear that a definitive solution still remains to be found and one can question the initial premise by Kaufman that the lattice stability of Ru can be derived from that of Fe by merely substracting the magnetic component in the latter. Certainly iron is not the best solute to use as a test vehicle since there are complicated magnetic interactions between iron and the precious metal group.

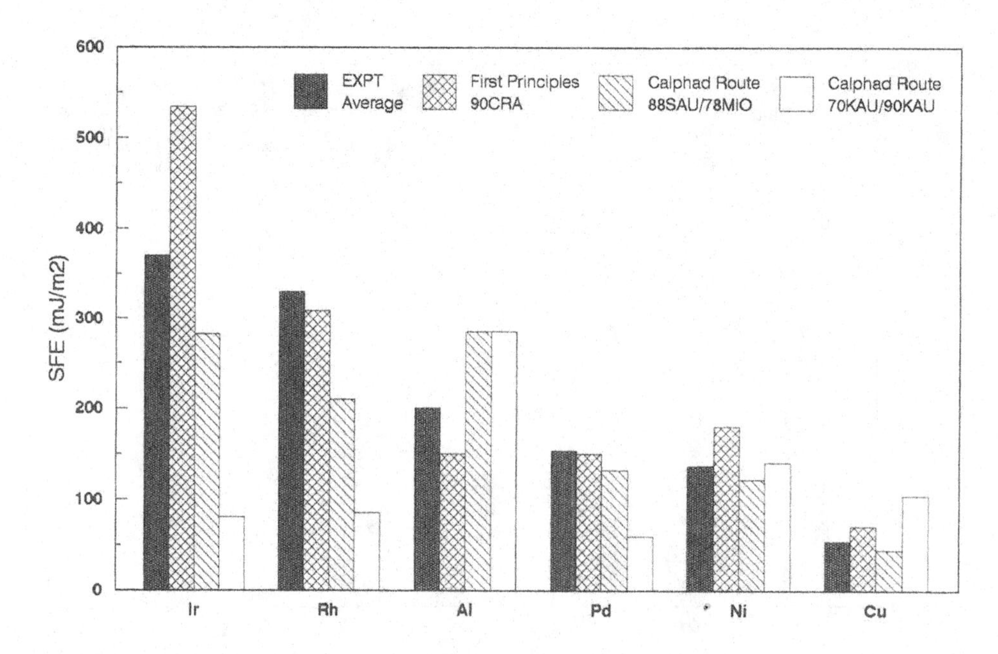

Figure 13. Comparison of experimental and calculated stacking fault energies in *cph* metals. Data for first principle calculations from Ref. 25, and values for thermochemical calculations from Refs. 2, 5 and 17.

Table IV

Comparison of lattice stabilities derived by various methods

Route	Ag	Al	Au	Cu	Ref
First Principles (FP)	840	2360	917	1210	Ref. 22
from Exptl (SFE)	256	2656	512	720	Ref. 20
Thermo Chemical (TC)	300	5000	550	600	Ref. 5a

Route	Ir	Ni	Pd	Rh	Ref
First Principles (FP)	9327	2200	3039	5607	Ref. 22
from Exptl (SFE)	7680	2000	2880	5280	Ref. 20
Thermo Chemical (TC)	4000	1500	2000	3000	Ref. 5a

VI. Utilization of Stacking Fault Energies

Since there is a well established theoretical relationship between the $\Delta G^{cph-fcc}$ and the intrinsec stacking fault energy (SFE), the latter is in principle very useful as a means of checking the relative validity of widely differing estimates of *fcc-cph* lattice stabilities.

It is first wortwhile examining how far the theoretical relationship between lattice stability and SFE is upheld in practice. A quantitative relationship between SFE and TCΔG *fcc-cph* [Eq. (4)] has been used by to predict SFE in stainless steels,[20] and in copper-tin alloys.[7] (The $[\sigma]$ term allows for non-chemical interfacial energy contributions and is relatively small.) Where it is possible to extract this from experimental results, it is expected to scale roughly with the modulus. A similar relationship has been used for nickel alloys.[21]

$$\gamma = 2(\Delta G_s^{fcc/cph} + \sigma) \tag{4a}$$

$$\Delta G_s = \frac{10^7}{N^{1/3}} \left[\frac{\rho}{M}\right]^{2/3} \Delta G_v \tag{4b}$$

$([\Delta G_s]$ in mJ/m^2; $[\Delta G]$ in J/mol; $[\rho]$ in gm/cm^3; $[M]$ in gm$)$

Intrinsic stacking energies have also been obtained via FPΔG calculations.[22,23] Using the equivalent lattice stabilities derived from a central force model listed by Crampin *et al.*,[22] it is possible to check whether Eq. (4) yields an equivalent value for the SFE given the same input parameters. The agreement shown in Table III gives some confidence for a general comparison of SFE derived by different routes.

From Figs. 12 and 13, and Table IV, where the experimental values are an average drawn from Refs. 22, 26, 27, 28 and 29, it is apparent that:

(i) There is a reasonable correlation between experimental and calculated SFE in many cases.

(ii) In view of Eq. (4) there is, as expected, a notable difference in the predicted SFE form TC lattice stabilities drawn from Refs. 2 and 5.

(iii) Disappointingly, there are no experimental values available for Ru and Os! It is however significant that high SFE values have been both observed and predicted for Rh and Ir which could be taken as confirmation of a higher amplitude for the variation of the *cph-fcc* lattice stability with filling of the d-shell than proposed by Kaufman.[2]

It would clearly be interesting to re-investigate these two elements and obtain reliable SFE values. If the SFE is as low as suggested by the value derived[17] from the lattice stability, stacking faults should be easily detected. On the other hand, with the very high SFE predicted by both FP calculations and Saunders *et al.*,[5] it is expected there might be severe experimental difficulties, consistent with the absence of any direct experimental SFE in the literature.

No FP calculations for the SFE of Cr have been traced in the literature, probably because it is assumed that this values is of little interest when the stable modification of this element is undoubtedly *bcc*. Nevertheless it is of considerable interest in relation to the value of TCΔG *fcc-cph* that has to be used in making calculations for stainless steels, which contain appreciable quantities of this element. The current consensus is that the *cph* phase is more stable than *fcc*, although the difference is so small that it is quite possible to assume a small positive value with alternative interaction parameters.[20,30] Recent calculations by Paxton *et al.*[31] confirm the general periodicity of the *fcc-cph* stability previously obtained for 3d transition elements, and, if at all, make the *cph* phase less stable. However these calculations exclude spin configurations which could easily shift the balance in the opposite direction. The current work of Moruzzi and Marcus[11] recalls some suggestions made by Weiss[32] and indicates that magnetic terms must be included and the latter certainly affect the SFE of nickel rich alloys.[20,21]

Recent work indicated that the effect of changing the c/a ratio values of *fcc-cph* lattice stabilities plays an important part in determining the SFE. The energy difference proposed by Kaufman[33] for TCΔG *hex-fcc* (Zn)(+1925 J/mol) is close to

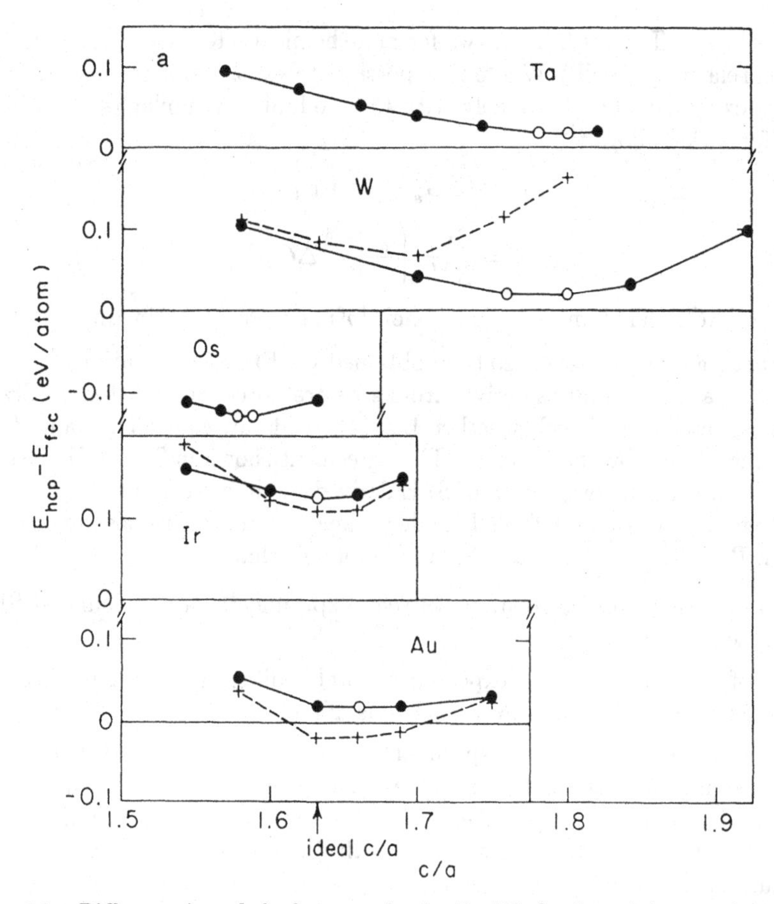

Figure 14a. Difference in *cph-fcc* free energies for Ta, W, Os, Ir and Au as a function of c/a ratio. The circles are full potential double basis set results, while crosses indicate muffin-tin single STO results.[35]

the value now proposed by Singh and Papaconstantopoulos[34] for FPΔG *hex-fcc* (Zn) of 1572 J/mol but, more importantly, they propose that the true *cph* structure is less stable than *fcc* by 1048 J/mol.

Interestingly, Fernando *et al.*[35] have indicated that the values of FPΔG *fcc-cph* for W and Ta depend appreciably on the *c/a* ratio (Fig. 14). Although these latest values still differ in sign from those obtained by Saunders and coworkers,[5] there is now clearly much closer agreement.

It should be noted that the values of Ref. 5 do in fact reproduce the sinusoidal variation expected form FP calculations, and the remaining gap might be bridged in some cases if *fcc* and *cph* lattices have a large enough difference in Debye temperatures (see next section).

VII. Additional Factors

Since there is good agreement between a large number of TC and FP lattice stabilities, one can ask whether there are some fundamental reasons for the major remaining discrepancies:

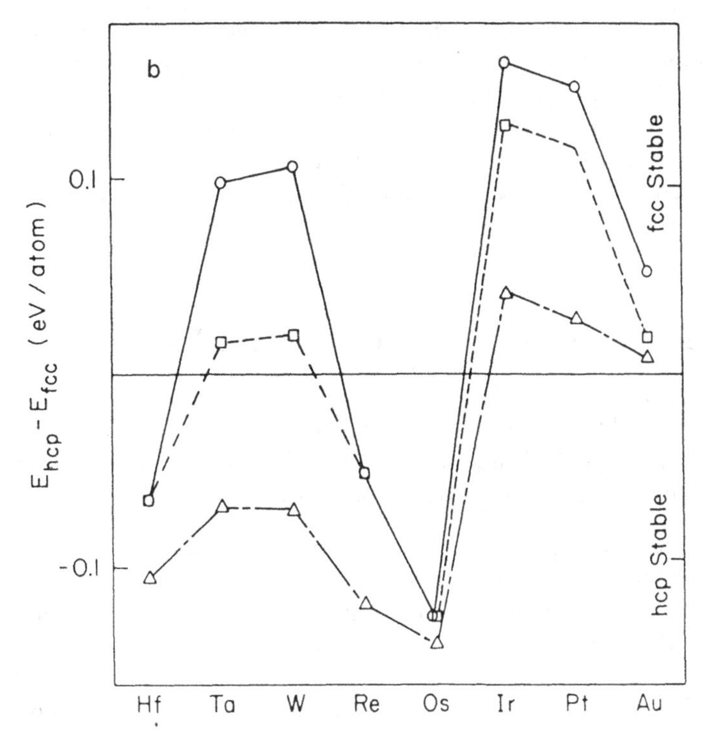

Figure 14b. Corresponding change in *cph-fcc* free energies with atomic number in the 5d transition elements[35] and a comparison with the TCΔG values of Ref. 5.

(a) Is it actually realistic to compare extrapolated TC lattice stabilities from high temperatures to FP values calculated for zero K?

(b) What are the implications of very low metastable melting points, including the extreme case of virtual melting points below zero K?

VIII. The Effect of Debye Temperatures

Historically TCΔG values have been largely obtained by simplistic extrapolation which depends on there being only small differences in the Debye temperature of the various allotropes. Metastable melting points have to be lower than stable ones and Debye temperatures generally scale with the melting point so the line approximation technique will differ from the real lattice stability by approximately $3R\Delta\theta_D$. It would be useful to see if the method used by Moruzzi[36] for the estimation of Debye temperatures from first principle calculations of the stable structures could be made to yield equivalent values for the metastable structures, to test this hypothesis. It is however unlikely that this would materially effect very large differences between TCΔG and FPΔG such as for *fcc-bcc* Tungsten and Osmium.

IX. Entropy of Fusion

If the entropy of fusion depends only on vibrational entropy terms a case can be made that $\Delta G(fcc - bcc)$ should be less that the correspondig entropy of fusion for close packed structures.[4,37] The correlation proposed by Ref. 3 supported this viewpoint, but the proposition seems to be correct only if the melting points of the alternative structures are similar, and Grimvall has made a very convincing case for a large contribution[38] from the electronic specific heat in the case of W. There is also a significant correlation between the entropies of fusion proposed by Saunders *et al.*[5] and entropies derived from band structure calculations by Watson and Weinert[39] (see Fig 4b).

X. The Effect of a Glass Transition in the Liquid Phase

It has been suggested that where extrapolation indicates a very low metastable melting point, the Van Laar method may severely overestimate the lattice stability because the liquid phase undergoes a glass transition at low temperatures.[4]

It is however clear that where extrapolations are made from experimentally observe liquidus values, these must always refer to the standard high temperature liquid; any extrapolated value is by definition using this structure as a reference state even if the real liquid subsequently undergoes a glass transition. The situation is entirely analogous to the use of a paramagnetic reference state for pure iron which in reality undergoes an insupressible ferromagnetic transition. In any event the vast majority of extrapolated metastable melting points fall above the range where a glass transition can be expected.[5]

XI. Virtual Melting Points

There are cases of extrapolations which yield a (virtual) negative melting temperature, which implies that the liquid phase is more stable than some competing metastable crystalline structures. This does not appear to violate any fundamental thermodynamics, providing there is at least one crystalline phase with a lower energy at zero K; indeed the relative lattice stabilities obtained from such virtual melting points[5b] agree favourably with later FP calculations. It should also be pointed out that when compound formation can be suppressed by rapid solidification, there is strong experimental evidence for liquid/amorphous phases being more stable than competing solid solutions, and at low temperatures their stability range can be predicted quite well from free energy curves based on extrapolation of high temperature data.[5b,40]

XII. Negative Shear Constants

There is a case to be made for examining the sign of the elastic constants for unstable crystalline modifications such as *fcc* Tungsten of *bcc* Osmium. If all the contentious cases turn out to be associated with structures exhibiting negative shear constants, this might provide a justification for using effective TC lattice stabilities extrapolated from regions where the embedded atom behaves normally. In this case Osmium may not be the best reference standard for the lattice stabilities of other 5d elements.[35]

XII. Conclusions

There is obviously no shortage of other issues that could be debated, including further evaluation of the effects of volume changes and c/a ratio (following the work in Ref. 35) but an urgent problem for accurately predicting the melting point of both stable and metastable elements directly from first principles. If this could be achieved, it would provide a direct comparison with the extrapolated values derived by the TC approach and finally provide a unified set of lattice stabilities.

In addition to the SFE measurements already mentioned, increasing use might also be made of specially targeted experimentation using rapid quenching methods to allow access to the properties of metastable phases over a wider composition range. In the end it is the experimental evidence that will finally decide which lattice stability values are to be adopted.

References

1. A. P. Miodownik, *Mater. Design* **4** 187 (1990)
2. L. Kaufman and H. Bernstein, *Computer Calculations of Phase Diagrams*, Academic Press, New York, 1970.
3. A. P. Miodownik, *Metallurgical Chemistry Symposium*, NPL–HMSO 1971, (1972). p 233–244, and p. 484–493 (discussion).
4. A. F. Guillermet and M. Hillert, *Calphad* **12**, 337 (1988).
5. (a) N. J. Saunders, A. P. Miodownik, and A. T. Dinsdale, *Calphad* **12**, 351 (1988); (b) N. J. Saunders and A. P. Miodownik, *Materials Science and Technology* **4**, 768 (1988).
6. P. Gustafson, *Int. J. Thermophys.* **6**, 395 (1985).
7. A. P. Miodownik, *J. Less Common. Metals* **114**, 81 (1985).
8. L. Kaufman, *Contract Report* FG 110–1, ASM, 1986.
9. J. O. Anderson, A. F. Guillermet, and P. Gustafson, *Calphad* **11**, 365 (1987).
10. A. F. Guillermet and W. Huang, TRITA–MAC, Materials Research Centre, Royal Inst. Technology, Stockholm, **0349** (1987).
11. V. L. Moruzzi and P. M. Marcus, *Phys. Rev.* **B42**, 8361 (1990).
12. C. Colinet, A. Pasturel, and P. Hicter, *Calphad* **9**, 71 (1985).
13. N. J. Saunders and L. Kaufman, Private Correspondence, 1987.
14. L. D. Blackburn, L. Kaufman, and M. Cohen, *Acta Metall.* **13**, 533 (1965).
15. G. L. Stepakoff and L. Kaufman, *Acta Metall.* **16**, 13 (1968).
16. L. J. Swartzendruber and B. Sundman, *Bull. Alloy Phase Diagrams* **4**, 155 (1983).
17. L. Kaufman, *The Lattice Stability of the Iron Group Elements*, Mats. Hillert Festschrift, Preprint, 1990.
18. R. E. Watson and L. H. Bennett, *Calphad* **5**, 25 (1981).
19. A. K. Nielssen, F.R. de Boer, R. Boom, P.F. de Châtel, W. C. Mattens, and A. R. Miedema, *Calphad* **7**, 51 (1983).
20. A. P. Miodownik, *Calphad* **2**, 207 (1978).
21. K. Ishida, *Phil. Mag.* **32**, 663 (1975).
22. S. Crampin, K. Hampel, D. D. Vvedensky, and J. M. MacLaren, *J. Mater. Res.* **5**, 2107 (1990).
23. J.–H. Xu, W. Lin, and J. Freeman, *Phys. Rev.* **B43**, 2018 (1991).
24. M. Igarshi, M. Khantha, and V. Vitek, *Phil. Mag. B* **63**, 603 (1991).
25. P. B. Legrand, *Phil. Mag. B* **49**, 171 (1984).

26. F. Aldinger, in *Beryllium Science & Technology*, edited by D. Webster and G. J. London, Plenum Pub. Co., 1979. p. 7.
27. P. C. J. Gallagher, *Metall. Trans.* **1**, 2429 (1970).
28. T. F. Page and B. Ralph, *Phil. Mag.* **26**, 601 (1972).
29. L. E. Murr, *Scrip. Metal.*, **6**, 203 (1972).
30. A. P. Miodownik, *Calphad* **1**, 301 (1977).
31. A. T. Paxton, M. Methfessel, and H. M. Polatoglon, *Phys. Rev.* **B41**, 8127 (1990).
32. R. J. Weiss, *Phil. Mag. B* **40**, 425 (1979).
33. L. Kaufman, Man. Lab Tech. Rep. II 1959 Abstract; *Bull. Am. Phys. Soc.* **4**, 181 (1959).
34. D. Singh and D. A. Papaconstantopoulos, *Phys. Rev.* **B42**, (1990).
35. G. W. Fernando, R. E. Watson, M. Weinert, Y. I. Wang, and J. W. Davenport, *Phys. Rev.* **B41**, 11813 (1990).
36. V. L. Moruzzi, *Phys. Rev.* **B37**, 790 (1988).
37. G. Grimvall, *Thermophysical Properties of Materials*, North-Holland, Amsterdam, (1986). p. 112.
38. G. Grimvall, M. Thiessen, and A. F. Guillermet, *Phys. Rev.* **B36**, 7816 (1987).
39. R. E. Watson and M. Weinert, *Phys. Rev.* **B30**, 1641 (1984).
40. N. J. Saunders and A. P. Miodownik, *Appl. Phys. A* **36**, 189 (1985).
41. H. L. Skriver, *Phys. Rev.* **B31**, 1909 (1985).

Order-Disorder Kinetics Studied by the Path Probability Method

Tetsuo Mohri

Department of Metallurgical Engineering
Hokkaido University
Sapporo 060
Japan

Abstract

By employing the Path Probability Method, the time evolution of the cluster probabilities for distinctive atomic (spin) configurations on an octahedron cluster during an isothermal aging process following a quenching operation is investigated. A certain cluster shows a peak in the course of the relaxation, although the free energy of the system decreases monotonically towards the equilibrium value. The thermodynamic origin of this non-monotonic behavior is discussed based on the kinetic path in a thermodynamic configuration space and in terms of an interplay between cooling rate and the spin flip probability.

I. Introduction

It is generally recognized that one of the strategies for developing advanced materials is to achieve a metastable state brought about by various non-equilibrium processes which endow the material with new functional properties that are not expected in an ordinary equilibrium state. The stability of the metastable phase, which relates to the functional lifetime of the material, is clearly a central concern in such strategy.

Contrary to the rapid development of equilibrium thermodynamic theories, which now include first-principles calculations, less progress has been achieved for kinetic theories. Perhaps one of the primitive approaches often employed in the metallurgical community to predict the appearance of a metastable phase is the "Step Rule" proposed by Ostwald.[1] According to this rule, the order of appearance of metastable phases starting from an initial phase towards a final equilibrium phase is dictated by

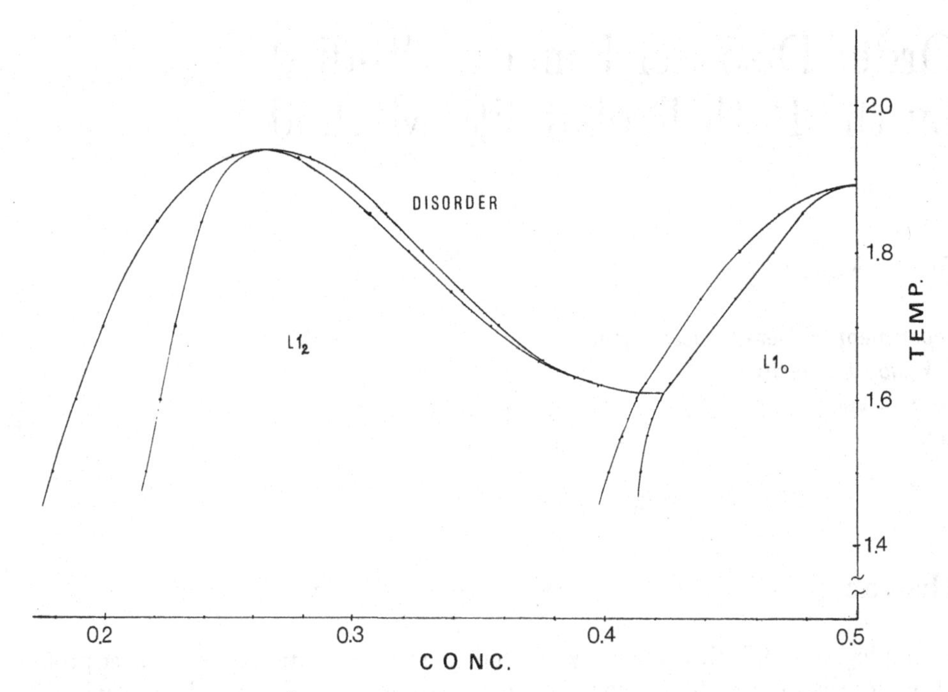

Figure 1. Phase diagram[8] for $L1_0$-$L1_2$-disorder calculated using the Tetrahedron approximation of the Cluster Variation Method. The temperature axis is normalized by the nearest neighbor effective pair interaction energy ν.

the relative magnitude of free energy of each phase. Although this rule is straightforward to apply since it requires no knowledge of complicated non-equilibrium thermodynamics, it does not provide information on the life time of each metastable phase which, as mentioned, is a central factor in material design. What is needed is a theory that explicitly incorporates the time as an independent variable. Motivated by such engineering needs, the author has been seeking theoretical models which are both reliable and tractable.

The Path Probability Method[2] (hereafter PPM) is the natural extension of the Cluster Variation Method[3] (hereafter CVM) to the time domain. As such, the PPM possesses several of the advantages inherent in the CVM. Over the last several years, the author's group has been employing the PPM to the study of metastable phase equilibria for order-disorder systems. Here we present preliminary results obtained for an *fcc*-based system.

The organization of this report is as follows. In the next section, the main theoretical aspects of the PPM are described through a few examples. Some preliminary applications of the PPM to metastable phenomena are introduced in the third section and followed by a summary in the final section. Throughout this brief report, the emphasis is placed on demonstrating general features of the PPM rather than on mathematical rigor. The reader interested in the formal aspects of the PPM should consult the original papers cited in the references.

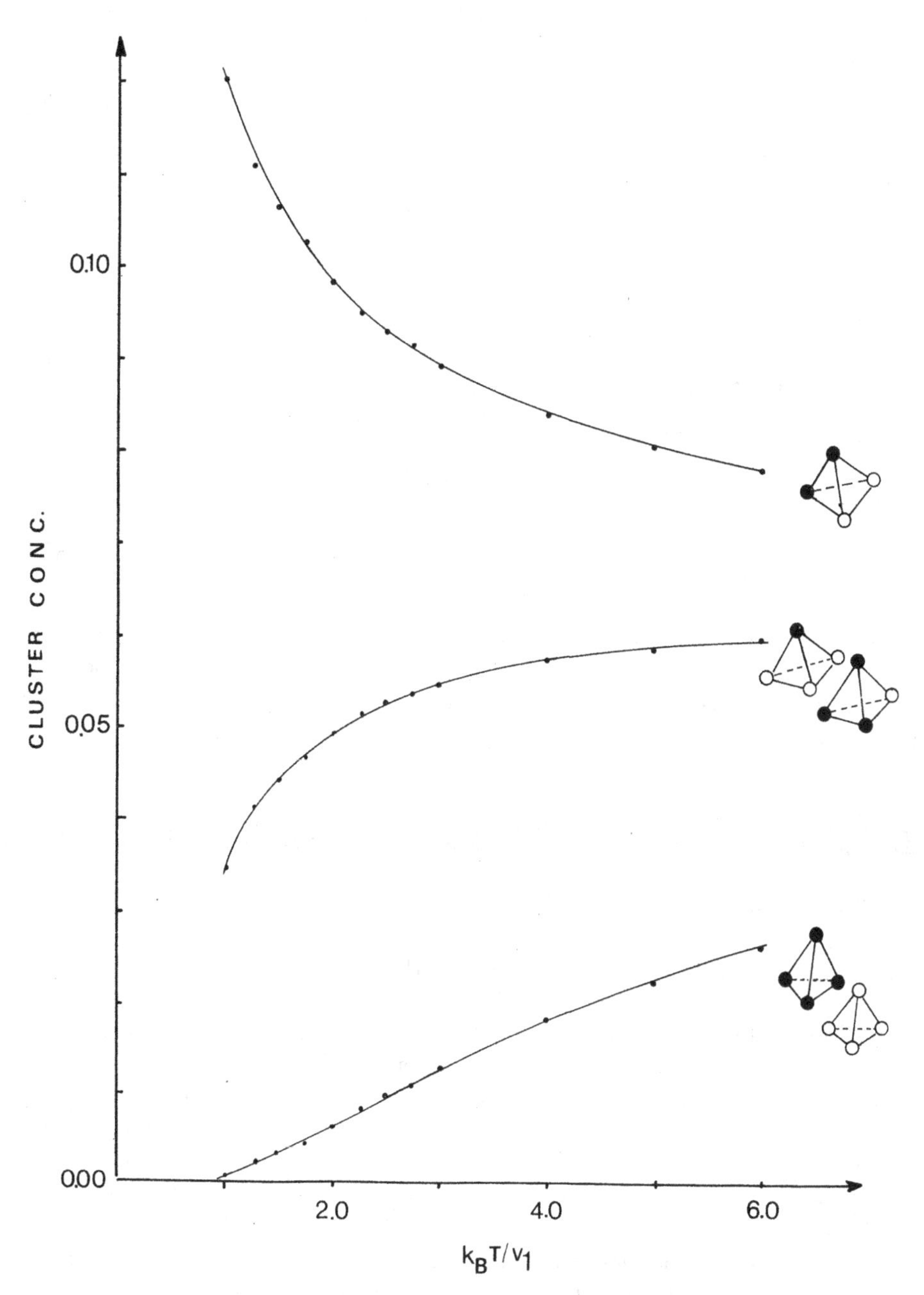

Figure 2. Temperature dependence of the cluster probability for the five distinct configurations on a tetrahedron cluster calculated using the Tetrahedron approximation of the CVM[8]. The open and solid circles indicate A and B atoms, respectively, for an alloy system and up and down spins, respectively, for a spin system.

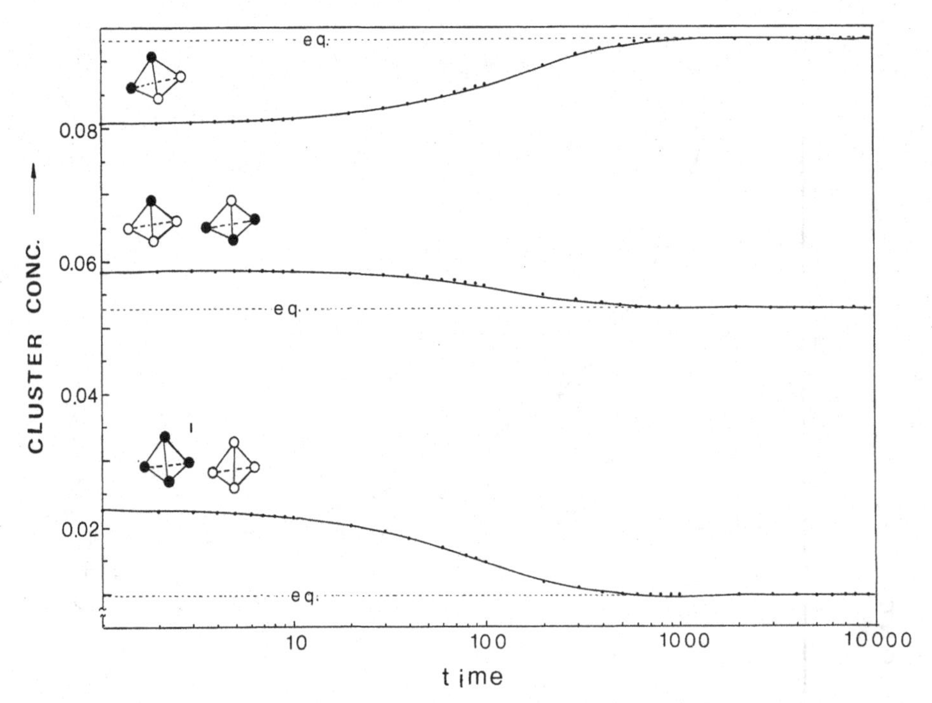

Figure 3. Time evolution of the cluster probabilities for the five distinct configurations on a tetrahedron cluster during an isothermal aging at $k_B T/v_1 = 2.5$ (see Ref. 8). Note that the time axis is normalized by the spin flip probability θ. Each dotted line indicates equilibrium value of the cluster probability independently calculated by the CVM for $k_B T/v_1 = 2.5$.

II. The Path Probability Method for Order-Disorder Kinetics

II.2 Equilibrium state

In order to focus on the replacive nature of the order-disorder phase transition, we assume a rigid lattice on which fixed interaction energies are assigned. Hence, the thermodynamic state of the system is described only by temperature, chemical potential and a set of configuration variables. In this report, the Tetrahedron approximation[4] or Tetrahedron-Octahedron (TO) approximation[5–7] of the CVM are employed to investigate the equilibrium properties. Then, for each temperature and chemical potential, the equilibrium state is obtained by minimizing the thermodynamic potential with respect to a set of independent configuration variables contained in the largest cluster in each approximation.

Shown in Fig. 1 is the phase diagram of $L1_0$, $L1_2$-disorder system calculated in the Tetrahedron approximation of the CVM.[8] The temperature axis is normalized by the nearest neighbor pair interaction energy and the concentration axis is only up to 50 at.% since the phase boundaries are symmetric. Shown in Fig. 2 is the temperature dependence[8] of the cluster probability for the five kinds of atomic configurations on a tetrahedron cluster at 50 at.%. One can see that only the probability for the tetrahedron cluster with equal number of both types of atoms increases with decreasing temperature. Note that this particular configuration is the one contained in the underlying $L1_0$ ordered phase.

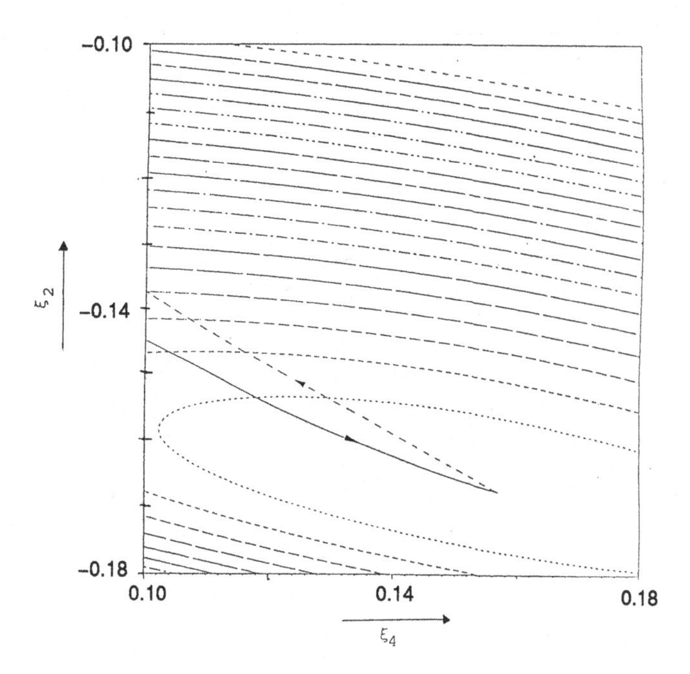

Figure 4. Kinetic path followed by the system during an isothermal aging process at $k_BT/v_1 = 2.5$ plotted in the thermodynamic configuration space spanned by the pair and tetrahedron correlation functions[8]. A dotted line indicates the path for an up-quenching treatment from $k_BT/v_1 = 2.5$ to 5.0.

In equilibrium thermodynamic theories, up and down spins are often assigned to A and B atoms, respectively. Therefore, an alloy system is regarded as equivalent to a classical Ising model. For a kinetic study, however, a distinction should be noted, since the phase transition in an alloy system involves the actual atomic migration while the flipping of spins on fixed lattice points is more appropriate for the study of magnetic phase transitions. Furthermore, the atomic concentration of an alloy system is a conservative quantity, while the ratio of both spins is not necessarily conserved. In this report, as a prototype investigation of the alloy system, the spin system is studied. In order to preserve the essential feature of the conservation of atomic concentration in the alloy, the present study is limited to the spin ratio of 50% which is conserved without additional constraints.

II.2 The Path Probability Method

When a system is quenched from high temperature to low temperature and is subjected to an isothermal aging treatment, the state of order relaxes towards equilibrium with time. Such relaxation kinetics is one of the main subjects studied by the PPM. In particular, we focus on the relaxation of cluster probabilities towards their equilibrium values shown, for example, in Fig. 2.

As the counterpart of the cluster probabilities of the CVM, the independent variables in the PPM are the set of path variables describing the transition probability from one cluster configuration to another during an infinitesimal time interval. Like-

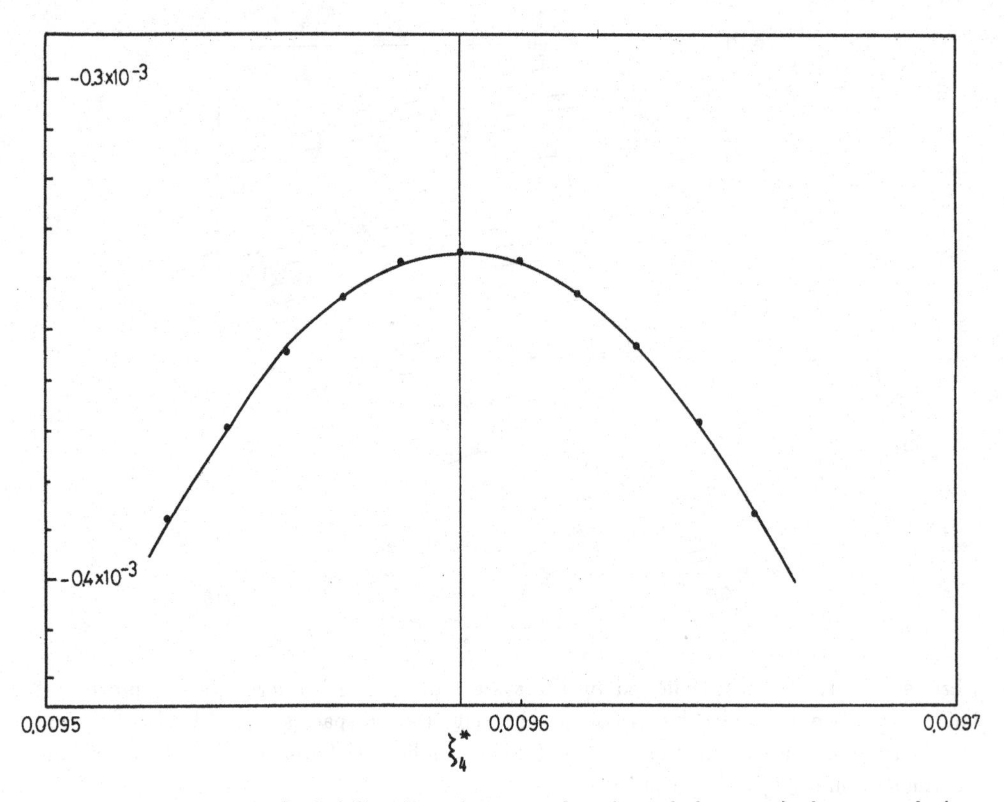

Figure 5. The Path Probability Function as a function of the tetrahedron correlation function at the isothermal aging stage t=1000 for $k_BT/v_1 = 2.5$.

wise, the counterpart of the free energy functional of the CVM is the Path Probability Function, P, which is given by the product of three terms:

$$P = P_1.P_2.P_3, \tag{1}$$

where P_1, P_2 and P_3 are, respectively, given by:

$$P_1 = (\theta.\Delta t)^{NX_{1,2}}(\theta.\Delta t)^{NX_{2,1}}(1-\theta.\Delta t)^{NX_{1,1}}(1-\theta.\Delta t)^{NX_{2,2}}, \tag{2}$$

$$P_2 = \exp\left(\frac{-\Delta E^*}{2k_BT}\right), \tag{3}$$

and

$$P_3 = \frac{\prod_{ijkl}(Y_{ij,kl}!)^6 N!}{(\prod W_{ijkl,mnop}!)^2 \prod_{ij}(X_{i,j}!)^5}. \tag{4}$$

The first term P_1 indicates a statistical average of non-correlated spin flip events over the entire lattice, θ is the spin flip probability per unit time which corresponds to the atomic exchange probability for an alloy system, N is the total number of lattice

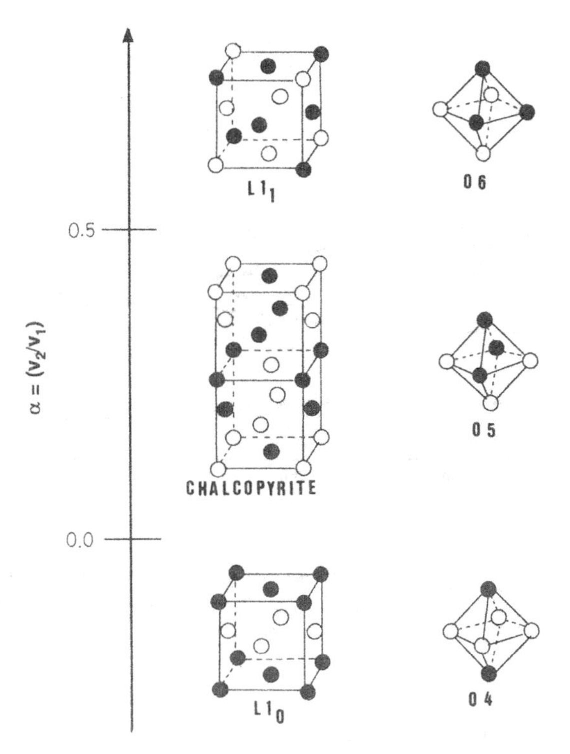

Figure 6. The stability of the $L1_0$, Charcopyrite, and $L1_1$ phases and corresponding atomic (spin) configuration on an octahedron cluster as a function of $\alpha(=v_2/v_1)$ value.[10,11]

points and $X_{i,j}$ are the path variables for a single site configuration from i-type spin (up or down) to j-type. The second term P_2 is the conventional thermal activation probability of gaining thermal energy from a heat bath. ΔE is the change of internal energy during an infinitesimal time Δt and k_B is the Boltzman constant. The main feature of the PPM is described by the last term P_3 in which $X_{i,j}$, $Y_{ij,kl}$ and $W_{ijkl,mnop}$ are path variables for the point, pair and tetrahedron cluster, and $ijk\ldots$ denotes a spin configuration on each cluster. We note the similarity with the configurational entropy of the CVM. In the PPM, this term describes the freedom of the path from one configuration to another.

The PPM prescribes that the most probable path for the time evolution of a system is given by the maximum of the Path Probability Function with respect to the path variables.

$$\frac{\partial P}{\partial \Xi_{\Phi,\Psi}} = 0, \tag{5}$$

where $\Xi_{\Phi,\Psi}$ stands for an independent set of path variables from configuration Φ to Ψ which are both specified by $ijk\ldots$ in Eq. (4). The condition above is the counterpart of the minimization condition of the free energy functional in the CVM.

The cluster probabilties $\chi_\Phi(t)$ at time t and $\chi_\Psi(t+\Delta t)$ at $t+\Delta t$ are related by the path variables $\Xi_{\Phi,\Psi}(t;t+\Delta t)$:

$$\chi_\Phi(t+\Delta t) = \chi_\Psi(t) + \Xi_{\Phi,\Psi}(t;t+\Delta t). \tag{6}$$

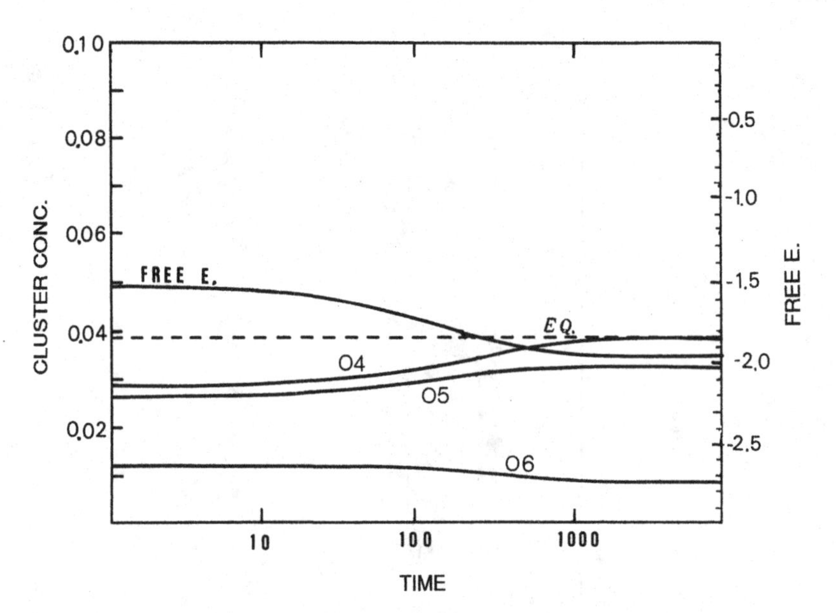

Figure 7. Time evolution of the cluster probabilities for the cluster O5 during an isothermal aging at $k_BT/v_1 = 2.5$ for $\alpha = -0.2$. (dotted line), 0.35 (broken line) and 0.65 (solid line). The pre-quenching temperature is $k_BT/v_1 = 5.0$.

The optimum path variables are calculated at each time using Eq. (5), and the time evolution of the cluster probabilties follows from Eq. (6). Shown in Fig. 3 [Ref. (8)] is the time dependence of the tetrahedron cluster probabilities for the five kinds of configurations when the system is quenched from $k_BT/v_1 = 5.0$ and is subjected to an isothermal aging at $k_BT/v_1 = 2.5$. Note that the system is in a disordered state at both pre-quenching temperature (5.0) and isothermal aging temperature (2.5), as can be seen from the phase diagram in Fig. 1. One can see that the probability for the cluster with equal number of opposite spin, which is characteristic of the underlying $L1_0$ ordered phase, increases with time while all the other probabilities decay. Most importantly, in the limit of infinite time, each cluster probability gradually converges to the equilibrium values obtained independently by the CVM for $k_BT/v_1 = 2.5$, which is one of the most attractive features of the PPM.

In both the CVM and PPM, the state of order of a given system is confined to a thermodynamic configuration space spanned by a set of independent configuration variables. Among various configuration variables, a set of correlation functions[5] is most conveniently employed to describe a system. In the tetrahedron approximation, four correlation functions, namely the point ξ_1, pair ξ_2, triangle ξ_3, and tetrahedron ξ_4 correlation functions are enough to describe the state of order of a disordered phase. It is noted that the cluster probabilities are uniquely described by the correlation functions as

$$\chi_{i_1 i_2 \cdots i_n} = \frac{1}{2^n}\left(1 + \sum_m V_{i_1 i_2 \ldots i_n}(m)\xi_m\right), \tag{7}$$

where the factor $V_{i_1 i_2 \ldots i_m}(n;m)$ is, in general,[5] a sum of m-order product involving the indices $i_1 i_2 \cdots$ ($i_m = +1$ and -1 for A and B atoms, respectively). It is noted

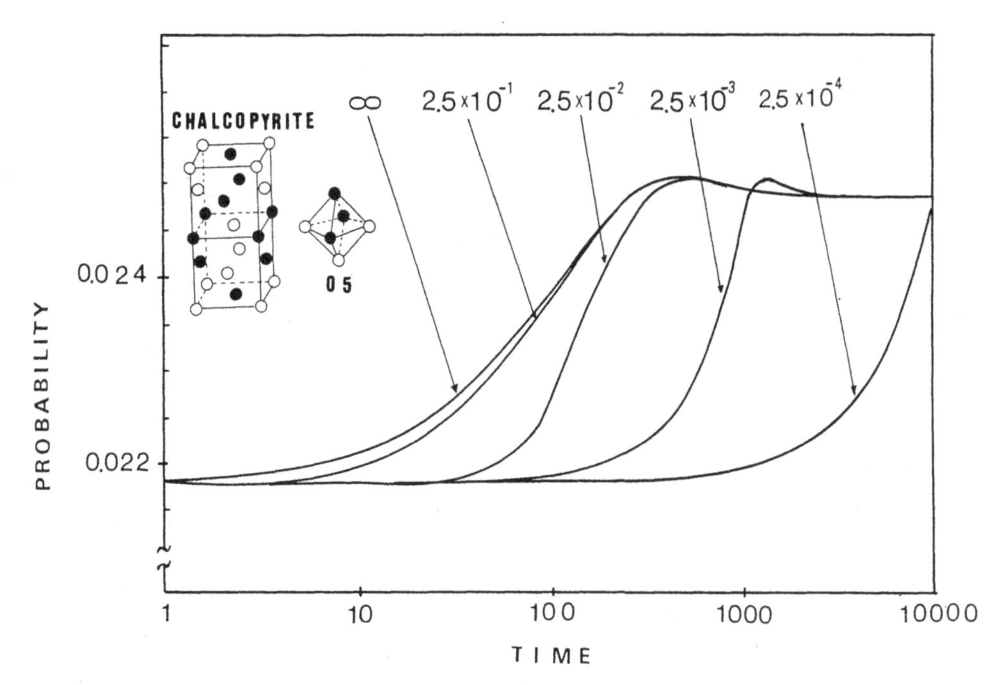

Figure 8. Time evolution of the cluster O5 for five kinds of cooling rate designated in the figure for a fixed spin flip probability $\theta = 0.001$ at $\alpha = 0.65$.

that both the point correlation and the triangle correlation functions vanish at 50% for a disordered phase. Then the thermodynamic configuration space is reduced to a two dimensional space spanned by the pair and tetrahedron correlation functions. Fig. 4 shows the most probable path[8] followed by the system during the isothermal aging process at $k_B T/v_1 = 2.5$. One sees that the path does not necessarily follow the steepest descent direction of the free energy contour, making evident the kinetic effect.

It is known that fluctuations play a central role in a phase transition. The calculation of the fluctuations of the correlation functions along the most probable path can be carried out in the following manner. For the calculation of the most probable path, starting from the most probable state at time t, which is specified by a set of cluster probabilities (or correlation functions), the Path Probability Function is maximized. For the fluctuation, starting with the most probable state at time t and an assigned state at time $t + \Delta t$, the Path Probability Function to the assigned state is maximized. The deviation of the assigned state from the most probable state is defined as the fluctuation. Hence an additional term P_4 is introduced in the Path Probability Function in order to describe constraints.

$$P_4 = \exp\left[\sum \lambda_{ijkl}\left\{\chi_{ijkl}(t+\Delta t) - \chi_{ijkl}(t) + C_{ijkl}\right\}\right], \tag{8}$$

where λ_{ijkl} is a Lagrange multiplier and C_{ijkl} designates assigned fluctuation and the sum is taken over independent configurations. Shown in Fig. 5 is the Path Probability Function at time $t = 1000$ for the isothermal aging temperature $k_B T/v_1 = 2.5$. The horizontal axis is the deviation (fluctuation) of the tetrahedron correlation function

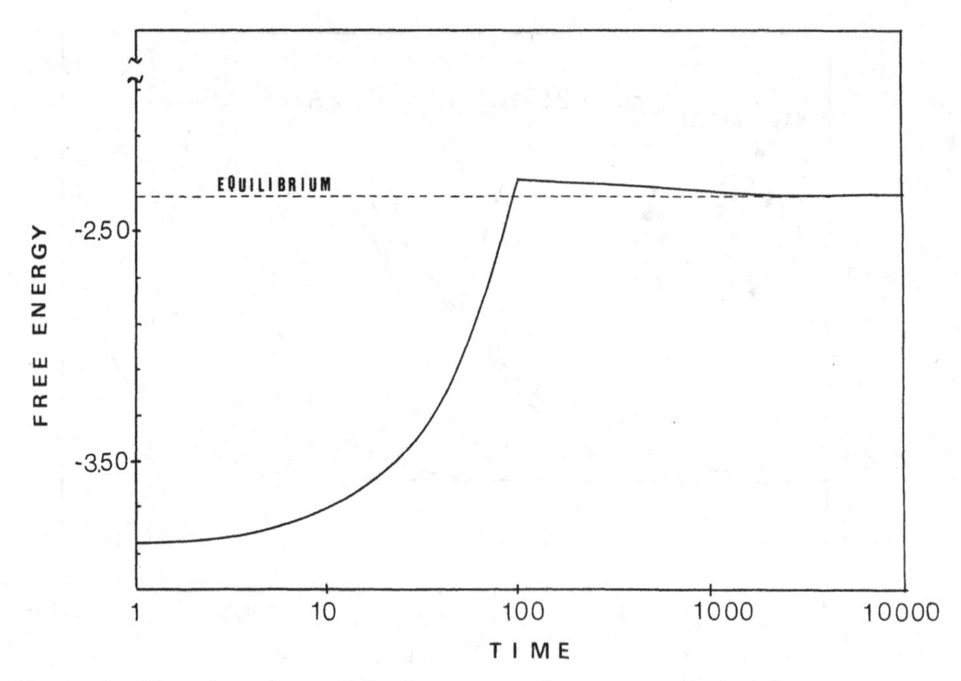

Figure 9. Time dependence of the free energy of a system quenched from temperature $k_BT/v_1 = 5.0$ to 2.5 with a cooling rate 2.5 for a fixed $\theta = 0.0001$. The equilibrium free energy is drawn by a dotted line.

from its most probable value designated by ξ_4^*. One may easily see that the shape of the fluctuation spectrum is approximately Gaussian. It is also confirmed that the half width becomes narrower by decreasing the aging temperature and the spectrum becomes sharper.

III. Application of the PPM to order relaxation phenomena

Based on the ground state analysis of an *fcc*-based system with first (v_1) and second (v_2) nearest neighbor pair interactions,[9] there are three ordered phases that may be stabilized at 50% depending upon the value of the ratio $\alpha = v_1/v_2$. These three ordered phases are $L1_0$ for $\alpha > 0.5$; Charcopyrite for $0.5 > \alpha > 0.0$ and $L1_0$ for $\alpha < 0.0$, each one characterized by a particular atomic configuration on the octahedron cluster O6, O5 and O4, respectively, shown in Fig. 6 (Refs. 10 and 11).

In this section, we present the calculated results for the time evolution of the octahedron cluster probabilities, which is expected to provide the kinetics of phase stability of the corresponding ordered phase up-quenched to the disordered phase region. In order to deal with the octahedron cluster, the PPM in a higher order approximation, the tetrahedron-octahedron[10,11] approximation, is employed. It should be noted that the particular atomic configuration on the octahedron cluster does not necessarily have a one to one correspondence to the ordered phase. In order to obtain more reliable information of an ordered phase, one needs to calculate the long-range order parameter characterizing each ordered phase. The difficulty of the PPM, how-

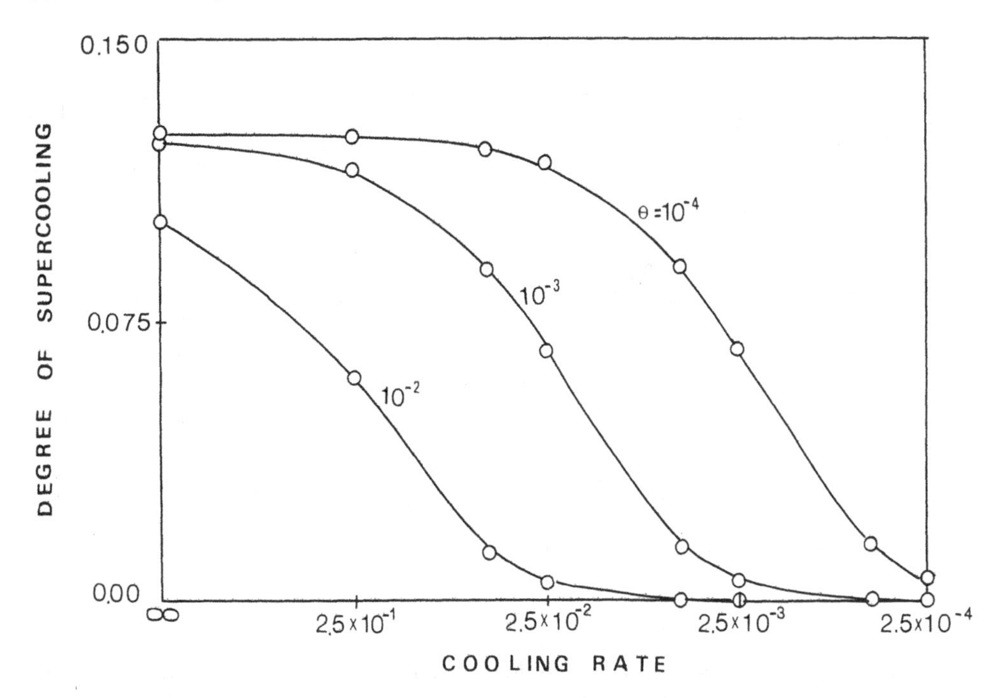

Figure 10. The degree of supercooling as a function of cooling rate for three kinds of spin flip probability designated in the figure.

ever, is that the number of the path variables becomes enormous with the increase of the cluster size and it is an intractable task even to formulate the basic equations for an ordered phase. Therefore, the present results are regarded as providing only preliminary information.

Fig. 7 shows the time evolution of the cluster probabilities for the clusters[10,11] during an isothermal aging at $k_BT/v_1 = 2.5$ following the quenching operation from $k_BT/v_1 = 5.0$ for $\alpha = -0.2$. We confirmed that the cluster probability in the limit of infinite time converges to the one independently calculated in the CVM. It is interesting to note that the stability of the ordered phase at the ground state (Fig. 6) for $\alpha = -0.2$ is reflected in the stability of the corresponding cluster in the high temperature disordered region.

A peculiar feature observed in the present study is that the probability for cluster O5 reaches a maximum and then decays towards its equilibrium value when α is 0.65, although the free energy of the system decreases monotonically. In order to clarify the nature of this non-monotonic behavior of the cluster probabilities, further calculations are performed by explicitly taking the cooling rate into account, while an infinite cooling rate is assumed for the calculations presented in the Fig. 7. Note that the cooling rate is defined as $(5.0 - 2.5)/\Delta t$, where Δt is the time required to reach the isothermal aging temperature $k_BT/v_1 = 2.5$ from the pre-quenching temperature $k_BT/v_1 = 5.0$. Shown in Fig. 8 is the time evolution of the cluster probability for the cluster O5 as a function of five different cooling rates designated in the figure.[12] One can see that the peak becomes less pronounced with decreasing cooling rate. The time dependence of the free energy of the system is also calculated[12] and is plotted in Fig. 9

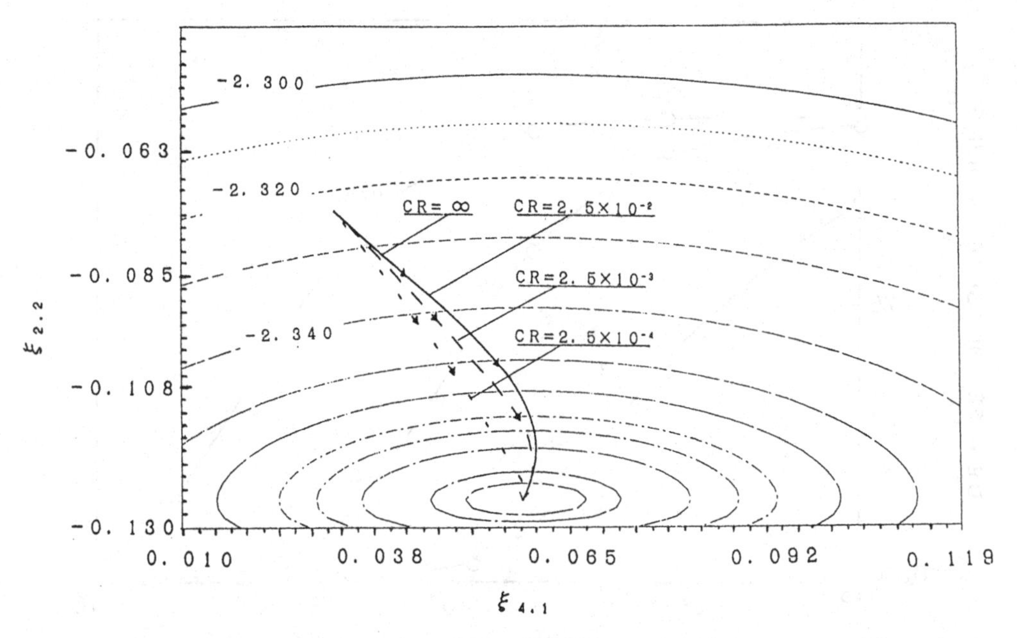

Figure 11. The kinetic path of the system for four kinds of cooling rate traced in a thermodynamic configuration space spanned by the regular tetrahedron correlation function $\xi_{4,1}$ (horizontal axis) and second nearest neighbor pair correlation function $\xi_{2,2}$ (vertical axis).

for a spin flip probability and a cooling rate 2.5×10^{-2}. One observes that the free energy overshoots the equilibrium value and then relaxes gradually to the equilibrium value. The peak value of the free energy is attained when the temperature has reached 2.5. Using the CVM, one can obtain the temperature at which the equilibrium free energy equals the peak value, and by subtracting it from 2.5, a notion of supercooling may be defined. The supercooling defined in this manner is plotted in Fig. 10 for various cooling rates and spin flip probabilities.[12] It is seen that the supercooling increases with a decrease in the spin flip probability and an increase in the cooling rate, indicating that the degree of the non-equilibrium character is controlled by an interplay between the cooling rate and spin flip probability. When a system does not have the intrinsic ability to accomodate the external change, the non-equilibrium state is realized.

Among the ten independent configuration variables in the TO approximaton, the regular tetrahedron and the second-nearest neighbor pair correlation functions are adopted to describe a two dimensional section of the thermodynamic configuration space, and the most probable path followed by the system during an isothermal aging at $k_B T/v_1 = 2.5$ for $\alpha = 0.65$ is plotted in Fig. 11 for four values of the cooling rate.[12] The spin flip probability is fixed at $\theta = 0.001$. One can observe that the curvature of the path becomes more pronounced with an increase in the cooling rate, although the free energy always decreases monotonically. The projection of the path onto the horizontal axis implies that the change of the tetrahedron correlation is not monotonic for the higher cooling rate, which is reflected in the non-monotonic behavior observed for the cluster probability O5.

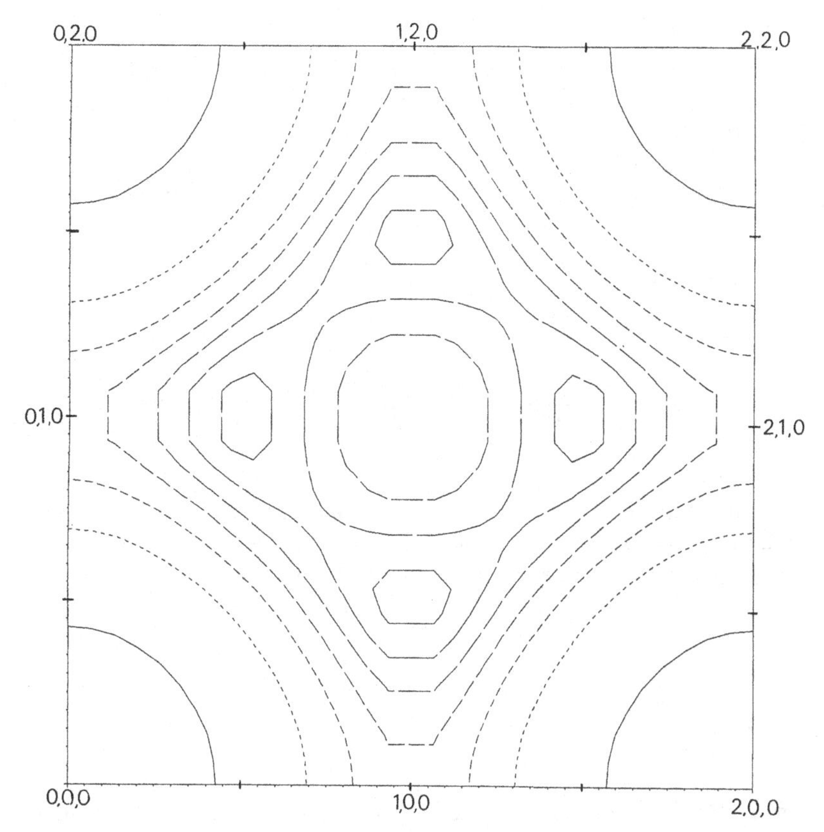

Figure 12. Short-range order diffuse intensity distribution at $t = 100$ in an (100) section of the k-space. The aging temperature is $k_B T/v_1 = 2.5$, spin flip probability is $\theta = 0.001$ and $\alpha = 0.25$.

The experimental investigations of the order-disorder kinetics have been centered around diffraction studies. It is, however, not possible to measure the cluster probabilities directly from diffraction experiments except for pair probabilities. The most efficient way to examine the present theoretical results may be to calculate the short range order diffuse intensity spectrum and to compare it with the one obtained by experiment. Although a detailed comparison with experimental results remains a subject of future work, we present here some preliminary results for the theoretical calculations of the time evolution of the short-range order diffuse intensity.

The Fourier transform of the pair correlation function provides the short-range order diffuse intensity distribution $I_{\mathrm{SRO}}(k)$ in k-space. The reader is referred to Ref. 13 for the procedure to calculate SRO diffuse intensity using the CVM free energy functional, and to Ref. 14 for applications to prototype *fcc*-based systems. In Fig. 12, the diffuse intensity distribution on the (100) reciprocal lattice plane is shown for $t = 100$, $\alpha = 0.25$ and $\theta = 0.001$. One can see that the highest intensity appears at (1,1/2,0) position which is known as the special point[15] for this α value. The time evolution of the inverse intensity $k_B T/I_{\mathrm{SRO}}$ is also calculated at (1,1/2,0) position and is shown in Fig. 13. We urge experimental confirmation of these results.

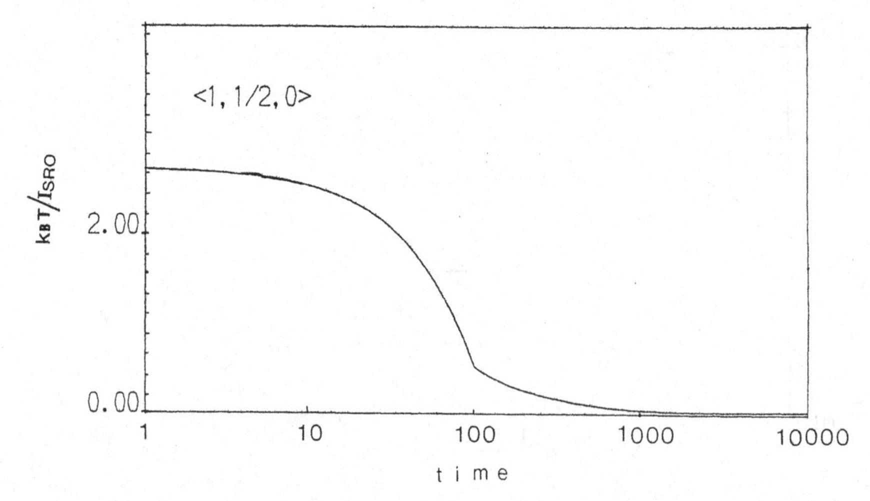

Figure 13. Time dependence of the inverse short-range order diffuse intensity at (1,1/2,0) position in k-space. The system is quenched from $k_BT/v_1 = 5.0$ to 1.96 which is just above the instability temperature with a cooling rate of (5.0–1.96)/100.

IV. Summary

By employing the Path Probability Method, the time evolution of the cluster probabilities for distinctive atomic (spin) configurations on an octahedron cluster during an isothermal aging process is investigated. The main findings are as follows:

(1) Each cluster probability in the limit of infinite time is exactly the one calculated independently by the Cluster Variation Method.

(2) A certain cluster shows non-monotonic time evolution behavior although the free energy of the system decreases monotonically towards its equilibrium value.

(3) The degree of the nonequilibrium state of the system is dominated by the interplay between the spin flip probability and the cooling rate.

(4) Short-range order diffuse intensity distribution in k-space is calculated as a function of time.

The extention of the present study to an alloy system requires more variables since actual atomic migration process adds more freedom to the path. The formulation has been undertaken and the results will be presented in a separate publication.

Acknowledgement

This work was partly supported by a Grant for International Research Project from the NEDO, Japan, and the Production Engineering Research Laboratory, Hitachi, Ltd., Yokohama, Japan.

References

1. W. Ostwald, *Z. Phys. Chem.* **22**, 289 (1987).
2. R. Kikuchi, *Prog. Th. Phys.*, Suppl. No.35, 1 (1966).
3. R. Kikuchi, *Phys. Rev.* **81**, 988 (1951).
4. R. Kikuchi, *J. Chem. Phys.* **60**, 1071 (1974).
5. J. M. Sanchez and D. de Fontaine, *Phys. Rev.* **B17**, 2926 (1978).
6. J. M. Sanchez and D. de Fontaine, *Phys. Rev.* **B21**, 216 (1980).
7. T. Mohri, J. M. Sanchez and D. de Fontaine, *Acta Metall.* **33**, 1171 (1985).
8. T. Mohri, *Acta Metall.* **38**, 2455 (1990).
9. M. J. Richard and J. W. Cahn, *Acta Metall.* **19**, 1263 (1971).
10. R. Kikuchi, T. Mohri and B. Fultz, *Proc. Materials Research Society*, Boston 1990, in press.
11. T. Mohri and Y. Sugawara, *Proc. Intern. Conf. Computer Appl. to Materials Science and Engineering* CAMSE 1990, Ed. by M. Doyama, ELSEVIER, Amsterdam, in press.
12. T. Mohri, *Proc. Sixth JIM Intern. Symp.* (JIMIS-6) Intermetallic Compounds, The Japan Institute of Metals, 209 (1991). In press.
13. J. M. Sanchez, *Physica* **111A**, 200 (1982).
14. T. Mohri, J. M. Sanchez and D. de Fontaine, *Acta Metall.* **33**, 1463 (1985).
15. D. de Fontaine, *Acta Metall.* **23**, 553 (1975).

Modeling of Invar Properties from Electronic Structure Calculations

Elio G. Moroni[1] and Thomas Jarlborg[2]

[1] *IPE, Université de Lausanne*
CH-1015 Lausanne
and IRRMA, PHB-Ecublens
CH-1015 Lausanne, Switzerland

[2] *DPMC, Université de Genève*
CH-1211 Genève, Switzerland

Abstract

It is shown how super cell (SC) total energy calculations can be instructive for the understanding of invar anomalies. The occurence of volume and magnetic instabilities in some invar system is explained from band theory results obtained at zero temperature, for magnetic and non-magnetic ground states. The total energy separation of different magnetic configurations near invar compositions is almost zero, but the difference in lattice spacing of each configurations plays an important role and is strongly related to the compositions and chemical constituents of the alloy. Moreover, it is found that anti-ferromagnetic ordering in some *fcc* iron based metallic systems prevents the formation of stable anti-invar materials. Anti-invar anomalies are predicted in Zr-V and Ti-V system.

I. Introduction

Several technological important alloys such as invar (from 'invariant' thermal expansion) systems have been the object of a large amount of investigations. The majority of invar systems, are multiphase system, presenting different lattice structures and different magnetic and non-magnetic groundstates. They show often anomalies as a function of temperature, composition, pressure and magnetic field. Experiments on invars have provided valuable information of the mechanism behind invar anomalies.[1,2] It is generally agreed that the invar properties are due to the closeness of and interplay between magnetic and non-magnetic (NM) ground states. The Weiss model[3] assumed thermal excitations between two electronic configurations of iron atoms (the anti-ferromagnetic (AF) and ferromagnetic (FM)) that have almost the same energy but different volumes. Band theory[4] confirmed that the total energies of NM and FM configurations are almost degenerate near the invar composition in the FeNi system. The most extraordinary invar feature is the vanishing thermal expansion coefficient α in a wide temperature range. It is believed that the magnetic contribution to α is negative and compensates for the normally positive α that is due to lattice vibrations. Also the bulk modulus show anomalous behaviour such as a hardening with increased temperature. The vivid variation of elastic properties with pressure, temperature and magnetic field, is characteristic for several invar systems.

Computer experiments have proved to be useful since they can model situations which are difficult or impossible to realize in a real world experiment. For instance, in the computer we can study metastable states, with varying magnetisation, at large or even negative pressure. We believe that first principle total energy and band-structure calculations are crucial for a better understanding of the magnetic and volume instabilities occuring in those system with respect to external pressure or mixing of different chemical constituents. The developments of the density-functional theory (DFT) within the local spin density approximation (LSDA) have permitted to analyze the bulk properties of several 3d transition metals and of some ordered alloys. The total energy and ground state properties at zero temperature of Fe-Ni invar system have been studied mainly using supercell (SC) methods where the computational limitations have limited the compositions to stoichiometric values and to few magnetic configurations.[4–6] Those calculations have shown that the difference in total energy of the NM and FM states is almost zero near invar compositions, while the lattice and elastic properties of the two configurations are different. Furthermore, combining those total energy results at $T = 0$ with models for the thermal excitations between the different states allows for a qualitative description of several invar effects.[5,6] The purpose of this work is to determine the electronic structure and the total energies of several invar materials, in order to test the generality of the proposed mechanism and models. The stability of different phases are studied and we also search for materials that may show anti-invar properties.

II. Method of calculation

The calculations have been carried out using the self-consistent Linear Muffin-Tin Orbital (LMTO) band method[7] within the local spin density approximation (LSDA). The basis set includes s-, p- and d-states. In order to allow for maximum convergence we include l=3 (f states) in the so called three-centre terms and tails. Core states are recalculated at each cycle of the self-consistent iterations and they include all relativistic effects. For the valence states our calculations include mass-velocity and Darwin corrections neglecting the much smaller spin-orbit coupling. In binary compounds we have chosen equal Wigner-Seitz radii for the two types of atoms. To achieve good convergence of the total energy calculations we use a small artificial thermal broadening of the bands, varying from 3 to 5 mRyd. In addition we use many k-points in the final iterations (corresponding to about 500 points in the Irreducible Brillouine Zone for a structure of high symmetry containing one atom). The tetrahedron integration scheme was used to obtain DOS functions.

The ground state properties have been obtained from a least square fit of the calculated total energies. In invar systems the separation of total energy between FM and

Table I.

Calculated ground-state properties for several ordered alloys.
ΔE is the total energy difference between NM and FM state.

Composition	Structure	a_{NM} (a.u.)	a_{FM} (a.u.)	B_{NM} (Mbar)	B_{FM} (Mbar)	ΔE (mRyd)
Fe_4	*fcc*	6.49	(6.75)	2.2		14.1
Fe_3Ni	$L1_2$	6.48	6.56	3.2	2.2	0.15
Fe_2Ni_2	$L1_0$	6.50	6.61	2.9	2.5	9.4
$FeNi_3$	$L1_2$	6.52	6.58	2.1	2.6	12.5
Ni_4	*fcc*	6.53	6.54	3.0	2.3	2.4
Fe_3Pt	$L1_2$	6.75	6.83	2.8	1.4	0.5
FeCo	$L1_0$	6.45	6.46	3.4	1.6	0.0
$FeCr_3$	*bcc*	5.32	5.35	2.6	2.3	0.3
FeCr	*bcc*	5.29	5.31	2.9	2.3	1.5
TiV	CsCl	5.81	5.82	1.5	1.6	−1.3
ZrV	CsCl	6.20	6.20	1.3	1.2	−0.1
ZrV_2	C15	13.83	13.9	0.6	0.6	−0.5

NM state is less than 1 mRyd per cell (as for Fe_3Ni) and we needed to calculate the total energies at different lattice volumes for each magnetic composition and at each fixed composition. The calculated total energies are then fitted to the Murnaghan's equation of state,

$$E(V) = E(V_0) + \frac{B_0 V}{B_0'(B_0' - 1)}[B_0'(1 - \frac{V_0}{V}) + (\frac{V_0}{V})^{B_0'} - 1], \tag{1}$$

where $E(V_0)$ is the minimum of total energy , V_0 the equilibrium volume, B_0 the bulk modulus and B_0' it is pressure derivative. We found that sensitive variations of B_0' do not much influence the stability of the predicted equilibrium values of V_0 and B_0. Therefore the calculated total energies can also be fitted to Birch equation of state where B_0' is assumed to be constant equal to 4. The Birch equation has proved to be good for different solids until $(V/V_0)_{min} \simeq 0.6$. Accurate least square fits are needed to determine the bulk modulus and its behaviour with pressure. The rms error of the least square fits are less than 0.8 mRyd. The variation of the bulk modulus with volume which follows from Eq. (1) is

$$B(V) = V\frac{d^2E}{dV^2} = B_0\,(\frac{V_0}{V})^{B_0'}\,. \tag{2}$$

The effect of the zero-point energy is to increase the total energy, to expand the lattice and to decrease the bulk modulus, and those effects are larger in the NM phases than the FM phase.

Our LD results for the different invar systems are consistent with other local density calculations[8,9] in predicting too small equilibrium lattice constants and too large bulk moduli compared to the experiment. (see Table I). Other details of our SC calculations are found in previous work.[7] Using cluster-expansion formalism the total energy of a particular configuration α, as a function of the volume, can be written as

$$\Delta E^{\alpha}(V) = \sum_{\gamma} \xi_{\gamma}^{\alpha} v_{\gamma}(V), \tag{3}$$

where $v_{\gamma}(V)$ are many-body cluster interactions and ξ_{γ}^{α} are multisite correlation functions. These correlation functions are defined as

$$\xi_{\gamma} = \frac{1}{N_{\gamma}} \sum_{p_i} \sigma_{p_1} \sigma_{p_2} \cdots \sigma_{p_{\gamma}}, \tag{4}$$

where σ_{p_i} takes the value +1 or −1 depending on the occupancy of site p, N_{γ} is the total number of γ-type clusters and the sum is over all γ-type clusters in the lattice. The expansion is useful if it converges rapidly *i.e.* if the maximum cluster that it considers is of low order. Following Connolly and Williams[9] it is possible to describe the local chemical environments occuring in an alloy by a cluster expansion containing only a few terms. For the *fcc* structure, the tetrahedron cluster expansion is considered. Provided that the α configurations are ordered structures, the correlation functions ξ_{γ}^{α} are determined explicitly[9] and Eq. (3) can be inverted in order to compute the cluster interaction energies $v_{\gamma}(V)$

$$v_{\gamma}(V) = \sum_{\alpha} (\xi_{\gamma}^{\alpha})^{-1} \Delta E^{\alpha}(V). \tag{5}$$

The volume dependence of the calculated total energies can be made more explicit by rewriting the Eq. (1) for each ordered structure α as

$$E(V) = E_0 + E_1 V + E_2 (\frac{1}{V})^{B_0' - 1}, \tag{6}$$

where

$$E_0 = E(V_0) - \frac{B_0 V_0}{(B_0' - 1)},$$

$$E_1 = \frac{B_0}{B_0'},$$

and

$$E_2 = \frac{B_0 (V_0)^{B_0'}}{B_0'(B_0' - 1)}.$$

The cluster interaction energies vary as a function of volume in the same way as the total energies.

III. Fe-Ni ordered compounds

III.1 Ground-state properties and heat of formation

The low temperature phase diagram of $Fe_{50}Ni_{50}$ has been studied by analysing the structure of meteorites (they have been cooled slowly) or of synthesized, metastable alloys where the defects are introduced by irradiation.[10] No ordering has been detected in the nickel rich region across the composition $Fe_3Ni_{1-x}Pt_x$ while in binary Fe-Pt alloys, ordering was identified unambiguosly.[11] In this region the ground-state is believed to be a two-phase mixture. In the composition range 69–77% of Ni the phase diagram is characterized by a order-disorder transition between the ordered $L1_2$ and disordered

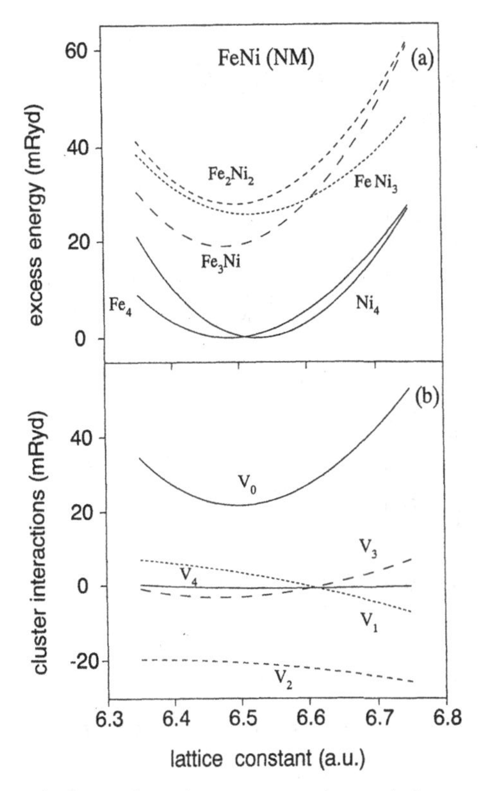

Figure 1. (a) Lattice variations of total excess energies and the extracted (b) cluster interactions energies of NM configurations for ordered Fe_nNi_{1-n} phases in the *fcc* structure.

A1 phases. The respective two-phase boundaries have been determined experimentally by Mössbauer spectroscopy.[12] The difficulties to obtain experimental data concerning ground-state properties for Fe-Ni binary alloy and the existence of very few *ab initio* calculations of ordering energies, motivated us to undertake some calculations of total energies of the NM and FM phases, and to study the magnetic contributions at $T = 0$ in these systems.

Since order-disorder effects are not important for invar properties we consider the Fe_nNi_{4-n} (n=0,1,2,3,4) *fcc* ordered compouds as a representative model of the Fe-Ni system. The total energy calculations verified that the FM and NM state are almost degenerate near invar composition.[4-6] For *fcc* structure we find such compositions when the number of electrons per atom (e/a) is near 8.5 (*i.e.* Fe_3Ni, Fe_3Pt and FeCo). We find that pure Ni and Fe_3Ni have small energy separations, but in Ni the difference of equilibrium volume of the NM and FM phase is much more less than in Fe_3Ni (see Table 1). This indicates that near the invar compositions, a small pressure is able to induce a transition from the FM to the NM state, which is in agreement with the observation of a collapse of the magnetic moment at around 60 Kbar.[13] Increasing the nickel compositions, in FeNi and $FeNi_3$ the separation between the E(V) curves of the FM and NM phase increases, and these systems show small magnetic Grüneisen parameters. The calculated ground-state properties are summarized in Table 1.

It is clear that the calculated differences in total energies are very sensitive parameters in models of phase stability. All approximations that are used, such as the

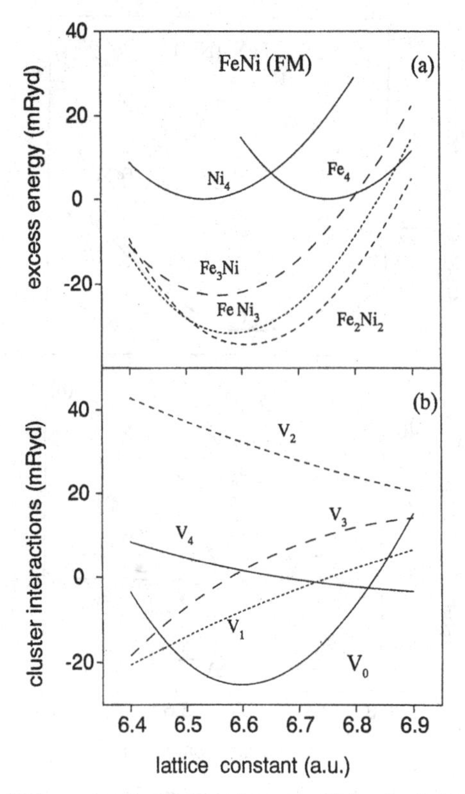

Figure 2. As in fig 1 but for FM calculations

spherical muffin-tin (MT) and local density (LD) approximations, should if possible be eliminated. From comparison of our results with LD calculations in pure Fe using Full potential methods[14] we believe that non-MT corrections are of minor importance in these close packed materials. Corrections to LDA have been considered by use of gradient corrections to the exchange-correlation potential for pure elements Fe and Ni[15] and for Fe_3Ni.[16] But it is found that the separation between the FM and NM states is increased both in total energy and volume. This inhibits the usual invar mechanism. Therefore we conclude that the effects of gradient corrections are exaggerated for the invar alloys and that LDA still is the best tested method.

The volume dependence of the calculated total energies per unit cell (total excess energies), defined as

$$\Delta E^n(V) = E[Fe_nNi_{1-n}, V] - nE_{Fe}[V_{Fe}] - (4-n)E_{Ni}[V_{Ni}] \quad (7)$$

for the NM and FM phases of the five stochiometric ordered compounds, are reported in Figs. 1a and 2a. The heat of formation of the ordered compounds is given by

$$\Delta H^{(n)} = \Delta E^n(V_{\text{eq}}). \quad (8)$$

The calculated $\Delta H^{(n)}$ are positive in the NM excited phase showing a tendency to segregation, while for the the lower FM ground-state phase, the calculated heat of formation is negative for the three different ordered alloys Fe_3Ni, Fe_2Ni_2 and $FeNi_3$. This is in good agreement with experimental evidence of complete mixing.

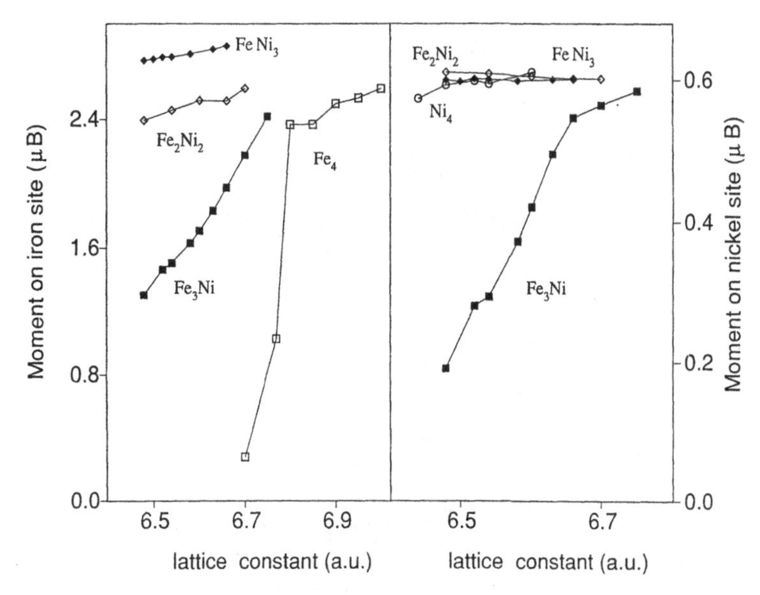

Figure 3. Calculated local magnetizations for the FM Fe_nNi_{1-n} configurations on Ni and Fe atoms as a function of the lattice constant.

Figures 1b and 2b report the calculated cluster interactions energies [see Eq. (5)] as a function of lattice constant for the NM and FM phases. The fundamental assumption that the allowed configurations can be constructed from units of nearest neighbour configurations is simple, but the rapid decrease of the calculated interaction energies with cluster complexity (v_3 and v_4 are much smaller and less important than the other v_n) indicates that the expansion is convergent. The tendency for segregation in NM phases shows up as a negative pair interaction potential (v_2) while in FM phases it is positive, indicating ability of compound formation. Assuming that pair- and higher order correlations functions can be expressed as products of point correlations functions ($\xi_n^{\text{dis}} = (1-2x)^n$), the heat of formation of random Fe_xNi_{1-x} alloy are calculated for the two states. The heat of mixing for the random Fe-Ni alloy for the FM state is found to be negative with a minimum around 44 % of nickel. Moreover in the FM state, the ordered phases are more stable compared to the random alloy. The calculated ordering energies (per cell) for the Fe_3Ni, Fe_2Ni_2 and $FeNi_3$ are respectively -2.5, -9.2 and -17.0 mRyd.

II.2 Electronic and magnetic properties

The calculated local magnetic moment on each site of Fe or Ni type for the different Fe_nNi_{4-n} ($n = 0, \cdots, 4$) are shown as a function of lattice constant in Fig. 3. In pure *fcc* Fe_4 different stable and metastable magnetic configurations are found[17] as a function of volume. We have not investigated the region where this metamagnetic behaviour of *fcc* iron occurs and we report the calculated magnetic moment of Fe for lattice constants larger than 6.75 a.u. (Fig. 3). The calculated values of magnetic moment near iron rich compositions produce large value of the magnetic Grüneisen parameter, indicating magnetic instabilities as a function of composition and pressure. For these ordered

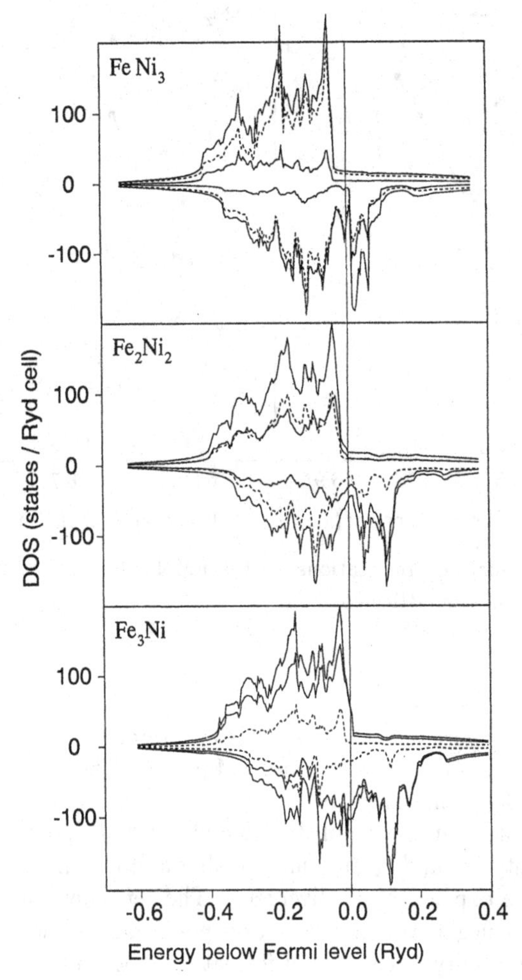

Figure 4. Calculated majority and minority total and component densities of states of FM *fcc* Fe_3Ni, Fe_2Ni_2 and $FeNi_3$.

phases we find the highest Fe moments not in the iron rich region but with 25% of Ni (see Fig. 3). Our electronic structure results show that this system can be considered as strong magnets with mainly minority states at the Fermi level (see Fig. 4). There is no volume dependence of the magnetization on Ni site, at 50% ,75% and 100% of Ni (the value of the magnetic moment is stable at 0.6 μ_B) and only in the invar region there is a collapse of magnetic moment at low volume induced by the collapse of the moment on the iron site. These features show that the magnetism in Ni plays no direct role, but that the variations of magnetization with volume are strongly related to the magnetism on iron site.

Figure 4 shows the calculated spin-polarized DOS of the ordered Fe-Ni compounds and it is seen that the majority bands hybridize more in comparison to the minority bands which are quite different for the two atoms. In Fe_3Ni at the calculated equilibrium (a_{eq}= 6.56 a.u.) as well as at expanded lattice parameter (a=6.75 a.u. as in Fig. 4) the majority band is not filled, in contrast to $FeNi_3$ and Ni where the spin-up d-band

Table II

Magnetic properties in *fcc* Fe_3Pt. Calculated total spin, valence d and interstitial magnetic moment on Fe and Pt site, hyperfine field on Fe -all versus lattice constant.

lattice constant (a.u.)	m(3d)	Fe M_{tot} μ_B	M_{int}	m(5d)	Pt M_{tot} μ_B	M_{int}	Fe HF (KG)
7.0	2.44	2.42	-0.014	0.45	0.37	-0.025	-339
6.95	2.35	2.33	-0.013	0.41	0.33	-0.025	-328
6.85	1.98	1.97	-0.01	0.25	0.18	-0.02	-280
6.8	1.57	1.56	-0.00	0.15	0.10	-0.02	-225
6.7	0.19	0.19	-0.00	0.01	0.00	-0.00	-33

lies entirely below the Fermi level. These SC results differ from coherent-potential-approximation (CPA) results,[18] where the majority spin DOS are completely filled near the invar region. This suggests stronger volume dependence in ordered than in disordered compounds. Moreover, near the invar region the majority DOS at the Fermi level is larger in ordered calculations. From the decomposed spin polarized DOS in Fe-Ni the minority-spin electrons are splitted quite substantially with the Ni d-bands at low energies and the d-bands of iron at high energies.

Several Mössbauer experiments (partly done at high pressure) indicate that the FM ground state in Fe-Ni invar systems shows a few percent of locally inhomogenous regions where 30% of the spins are aligned oppositely to the FM matrix. In our DFT calculations we study only such configurations that are allowed within the unit cell, *i.e.* much fewer magnetic configurations are considered. However, the theoretical prediction of the pressure induced magnetic transition is found to be in good agreement with experiments and prove quite unambiguosly the validity of total energy band calculations.

IV. FCC Fe-Based Invar alloys

IV.1 Fe-Pt system

Experimental measurements on magnetization of Fe-Pt system have shown that at 75% of Fe the *fcc* ordered phase is stable even at 4.2 K[19] and is strongly ferromagnetic. These alloys display invar characteristics (as anomalies in the thermal expansion coefficient) below the Curie temperatures in the ordered as well in the disordered case, but differently to the 'weak' invar system such as $Fe_{65}Ni_{35}$, the magnetic moment is stable against pressure and composition variations. Like the Fe-Ni system, the Fe-Pt alloys undergo a martensitic transformation at low temperature which is correlated with a anomalous softening of the C_{11}-C_{12} shear elastic constant. The experimental pressure derivatives of the elastic stiffness parameters are large negative quantities and at room temperature these materials become softer.

The total energy (relative to the minimum energy of the FM state) of the Fe_3Pt ordered compounds are shown in Figure 5. As for *fcc* Fe_3Ni, in Fe_3Pt the FM phase is the most stable, with equilibrium lattices larger than the NM minima, but the rigid band model does not apply since in Fe_3Pt the ΔE and the $\Delta V/V$ separations are larger than in Fe_3Ni. This fact puts Fe_3Pt in the limit for giving a net invar behaviour in the

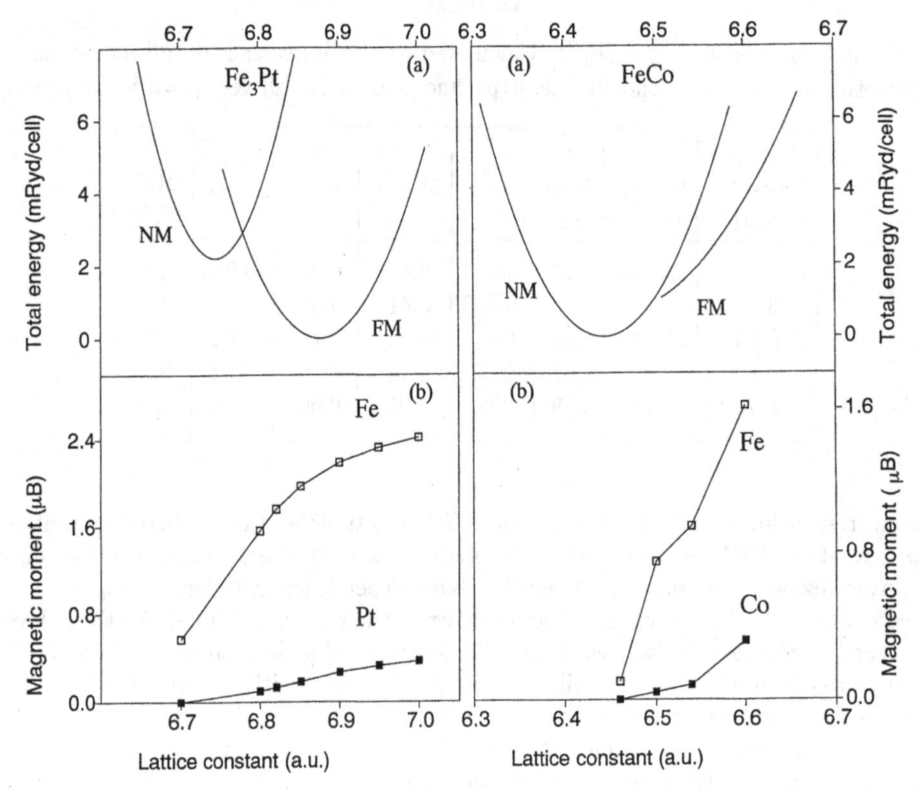

Figure 5. (a) Total energies versus lattice constant for ordered NM and FM calculations in *fcc* Fe_3Pt and FeCo. (b) Site-decomposed magnetic moments versus lattice constant for the two ordered invar alloys of (a).

thermal expansion. The calculated ground-state properties are in good agreement with experiments.

In Fe_3Pt the d-bands of Fe and Pt are splitted, and this is to a large extent due to the relativistic downshift of Pt s-bands. At the Fermi level, the states in both spin have essentialy Fe character. The moment on Fe is more stable in this system than in the isoelectronic Fe_3Ni because the equilibrium lattice parameter is larger. The magnetic properties of Fe_3Pt at different compressed volumes are summarized in Table II and these values should be compared with the experimental values of 2.7 μ_B for Fe and 0.5 for Pt[20] (the experimental lattice constant in Fe_3Pt is a=7.09 a.u.).

IV.2 Fe-Co system

The electronic structure and its dependence on the local environment have been studied for several ordered Fe-Co structures (CsCl, Fe_3Al and NaTl)[21]. Investigations of the magnetic phase diagram for *fcc* CoFe, have shown that only in a very little region (0-25 % of Fe) the *fcc* phase is stable. In this region the Curie-temperature decreases on addition of Fe in the same way as in Fe-Ni alloys. Studies of these systems are motivated by the interest in the related ternary Fe-Ni-Co alloys, and their super invar properties (around 0-18 % of Co). Differently to Mn rich or Cr based invar alloys, AF or spin-glass magnetic ordering are not present in Co-based systems. In *fcc* FeCo (having the same e/a ratio as Fe_3Pt and Fe_3Ni) the ΔE and the $\Delta V/V$ values are smaller than in Fe_3Ni, and the fluctuations can not be described well by the two state model that we have developed for Fe_3Ni[5] (see Fig. 5). The magnetic properties show interesting features as a function of volume. Passing from $a = 6.46$ a.u. (the equilibrium lattice constant) to $a = 6.62$ a.u. the moment on the Fe site varies from zero to μ_{Fe}=1.62 μ_B, while on the Co site the moment varies from zero to μ_{Co}=0.32 μ_B. The system shows low bulk modulus for FM state . In *fcc* Fe_3Co the calculated ΔE is much too large (5 mRyd/atom) to make a NM-FM transition at a reasonable T and no invar features are possible.

V. BCC Fe-Cr system

In spite of the difference in the crystalline structure, the experimental magnetic phase diagram of *bcc* Fe-Cr system shows analogies with that of invar Fe-Ni *fcc* alloys.[1,22] The negative magnetic contributions to the thermal expansion is observed from 75% to 95% of chromium content over a temperature range from 4.2K to above T_N.[23] Pure Fe and Cr have different magnetic interactions, but similar *bcc* lattice structures and Fe-Cr alloys retain the *bcc* structure. The chemical arrangements in the alloy depend strongly on heat treatment and ordered phases are not observed. The ferromagnetism in invar alloys often constitutes an obstacle for practical applications and therefore Fe-Cr alloys are interesting materials because they can be considered low-magnetic invar alloys near the rich Cr region, and the susceptibility decreases gradually with increasing the temperature. In this invar region the correlations between magnetism and the elastic and cohesive properties are less known and a theoretical study of the ground-state properties with varying the composition and the lattice parameter is missing.

Although the existence of the ordered alloys is still hypothetical, we have studied the ground-state properties for the ordered NM and FM phases of Fe_xCr_{1-x} with $x = 0.5$ and $x = 0.75$. The system with $x = 0.5$ is represented with a structure with two atoms per unit cell (CsCl structure) and $x = 0.75$ by a structure with four atoms per unit cell, at positions (0,0,0), (0.5,0.5,0.5), (0.5,0.5,−0.5) and (1,0,0). The latter structure has three inequivalent sites and the lattice is of high symmetry and has a Brillouin zone of a *fcc* lattice. The total energies per unit cell calculated for the NM and FM phases are shown in Fig. 6 together with the local moments. The equilibrium lattice constants are given in Table I. We find almost degenerate total energies (0.3 mRy/atom) in the *bcc* system $FeCr_3$ when e/a is 6.5, while a larger separation (1.5 mRy/atom) is found for FeCr. The former separation allows for the two state model, that we developed for Fe_3Ni, to get the $\alpha(T)$ and $B(T)$ variations. In ordered $FeCr_3$ we find a tendency to form an antiferromagnetic state (AFM). This concerns the Cr site which is most distant from the Fe site, while the Cr adjacent to Fe is polarized FM. Moreover, the Fe atomic moment varies strongly with the lattice parameter, differently to Cr moments that are found to be stable. In FM FeCr, at the calculated equilibrium minimum a =5.3 a.u., the local moments are respectively μ_{Fe}=1.22 μ_B and μ_{Cr}=0.5 μ_B, and increased volumes give increased moment on the Fe site, while the smaller Cr moment decreases. The latter fact may be a signature of a close AFM state.

Effects of disorder on the DOS have been studied using a new self-consistent[24] method related to LMTO for calculations of the local electronic structure (LES) of one site (Fe or Cr) distributed in an averaged scattering field of the other sites. The effect

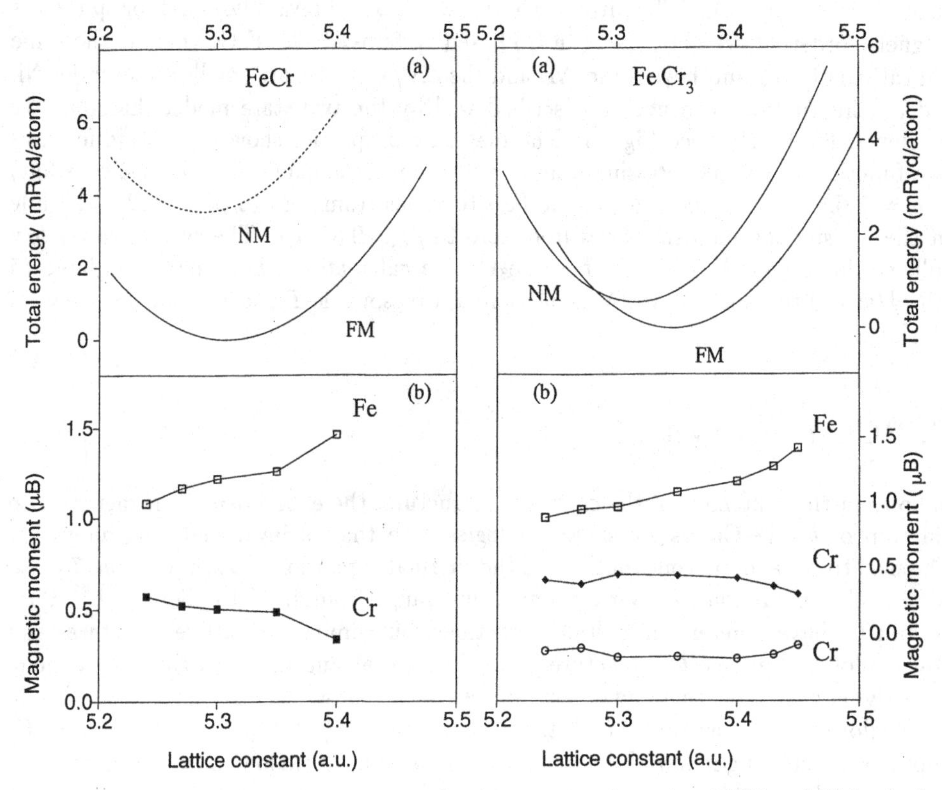

Figure 6. (a) Total energies curves versus lattice constant for ordered NM and FM calculations in *bcc* $FeCr_3$ and FeCr. (b) Site-decomposed magnetic moments versus lattice constant for the two ordered alloys of (a).

of disorder seems not to be strong for the DOS for this system. In NM $FeCr_3$, the large Cr DOS at E_F that is found in SC results is shifted to higher energy in LES, while the Fe DOS at E_F remains large and is important for the magnetic stability. Total energies are sensitive to calculate and the method has to be tested further before total energies can be compared.

VI. Anti-invar systems

VI.1 Fe-Co-Mn

On the basis of the calculated results and a qualitative comprehension of a two-state model, it is possible to visualize an 'anti-invar' effect in a system in which the NM state has a lower energy than the FM one and smaller equilibrium volume. At low temperature only the NM state would be occupied and at higher temperature, the presence of the FM state with larger volume, will increase the thermal expansion of the mixed system. Moruzzi[8] has suggested anti-invar behaviour in the concentration region between 8.0 (*fcc* Fe) and 8.6 (invar) e/a. By additions of Mn, Cr or V to invar Fe-Ni materials it is possible to avoid the martensitic transformation and to reduce the e/a ratio. But the magnetic ordering has a complex behaviour and the related elastic effects are complicated to analyze. The spin polarized calculations on Fe_2CoMn and $FeCo_2Mn$ were started from a weak FM configuration. During self-consistency an AFM moment was developed on the Mn sites, and the converged moments at a lattice constant $a = 6.48$ a.u. are the following: in Fe_2CoMn $\mu_{Fe} = 0.57\ \mu_B$, $\mu_{Co} = 0.01\ \mu_B$ and $\mu_{Mn} = -1.27\ \mu_B$; while in $FeCo_2Mn$ $\mu_{Fe} = 0.81\ \mu_B$, $\mu_{Co} = 0.09\ \mu_B$ and $\mu_{Mn} = -1.27\ \mu_B$. Thus, both compounds show a very small magnetization on the cobalt site while the Fe site is large (majority) and strong negative spin magnetization is present on Mn. This is in agreement with experimental findings.[25] These results underline again the fact that rigid bands do not apply for these materials.

From band filling according to the e/a ratio one cannot predict an AFM state. It also perturbs simple predictions of the possible physical properties in doped systems, for instance, the expected anti-invar effect. From extrapolations of the total energy results in 'invar' compositions it seemed plausible that 'anti-invar' properties should occur in Fe_2CoMn and $FeCo_2Mn$, with a NM ground state. But our calculated total energies for the two super-cells mentioned earlier show that AFM ordering is prefered on Mn, with lower energy than the NM state. The calculations give 8 mRyd in Fe_2CoMn for difference in total energy between NM and AFM states, near the calculated equilibrium volume. Similar results are found in $FeCo_2Mn$. Therefore, in this case there is no longer a condition left for the anti-invar behaviour.

VI.2 Zr-V systems

Another candidate for anti-invar behaviour of the described type could be found in the Zr-V system. The reason is that a peak is found in the DOS in the hypotetical ZrV (and TiV) compositions of CsCl structure, that makes the system very Stoner enhanced. Also ZrV_2 of C15-structure has this property,[26] and it is found that the FM state becomes stable at large volumes. In Fig. 7 we show the calculated total energies for these systems. As seen, the NM states are the ground states with FM configurations prefered only at extended volumes. However, it is possible to obtain weak FM solutions near the ground state. These FM states are metastable since they appear to be separated from the NM solutions and they could be populated via thermal excitations. In order to test this possibility we have performed FM electronic structure calculations for the real C15 system ZrV_2 with application of magnetic fields. We found metastable FM states separated by a small, 'total energy barrier' from the NM ground state at the equilibrium volume. The evaluation in terms of a two state model is more difficult, since the metastable state is not very sharp but is extended over a wider magnetization range. The difference in the bulk properties of the two states is not so evident as in Fe based systems. Qualitatively one would expect a tendency towards anti-invar behaviour such as rather strong thermal expansion at low temperatures.

VII. Conclusion

In conclusion, a qualitative understanding of invar properties in several systems is obtained from the SC calculations. The FM and NM states are almost degenerate near the invar compositions, independently of the structure (*fcc* or *bcc*) or of the atomic ingredients. Often the separation between NM and FM states is such that a quantitative description can be done from a two-state model and predictions of anti-invar proper-

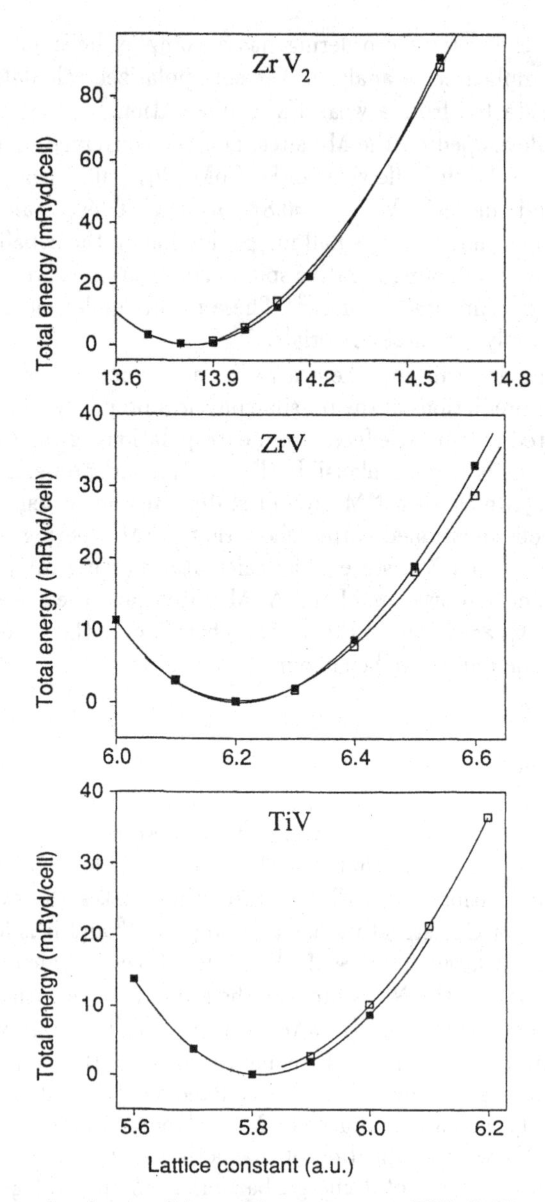

Figure 7. Total energies for possible anti-invar systems TiV, ZrV and ZrV_2 in the NM (fillbox) and FM (box) phase. The two former are of hypotetical CsCl structure while the latter exists in C15 structure.

ties are possible in some enhanced paramagnetic systems. The role of disorder and non-stochiometry is important in real systems and makes CPA-like or LES calculations necessary. An efficient method that is based on LMTO (LES) shows rather small effects of disorder on the DOS. However, further development is required for comparisons of total energies.

Concerning computations of *ab-initio* x-T phase diagrams, some approximations are done to obtain the entropies and free energies according to the Cluster Variation

Method. Assuming small clusters, these calculations predict good mixing properties over the entire FeNi concentration range. This is for FM ground states but for magnetic systems the cluster size could increase due to the complicated magnetic configurations that may form.[27,28] The electronic structure for such metastable configurations can be studied via the fixed spin moment technique (FSM)[17] or by addition of fields. Such developments are possible future refinements of methods for computing magnetic phase diagrams.

Acknowledgements

We are grateful to Prof. S. Steinemann for fruitful discussions. This work was supported in part by the Swiss National Science Foundation under Grants No. 20–5446.87.

References

1. E. F. Wasserman, *Phys. Scr.* **T25**, 209 (1989).
2. S. G. Steinemann, *J. Magn. Magn. Mat.* **7**, 84 (1978).
3. R. J. Weiss, *Proc. Phys. Soc.*, **82**, 281 (1963).
4. A. R. Williams, V. L. Moruzzi, C. D. Gelatt, J. Kubler and K. Schwarz, *J. Appl. Phys.* **53**, 2019 (1980).
5. E. G. Moroni and T. Jarlborg, *Phys. Rev.* **B41**, 9600 (1990).
6. P. Mohn, K. Schwarz, and D. Wagner, *Phys. Rev.* **B**, (1991).
7. G. Arbman and T. Jarlborg , *J. Phys.* **F7**, 1635 (1977).
8. V. L. Moruzzi, *Phys. Rev.* **B41**, 6939 (1990).
9. J. W. D. Connolly and A. R. Williams, *Phys. Rev.* **B27**, 5169 (1983).
10. A. Chamberod, J. Laughier and J. M. Penisson, *J. Magn. Magn. Mat.* **10**, 139 (1979).
11. A. P. Miodownik, *J. Magn. Magn. Mat.* **10**, 126 (1979).
12. J. K. Van Deen and F. Van Der Woude, *Acta Metall.* **29**, 1255 (1981).
13. M. M. Abd-Elmeguid and H. Micklitz, *Physica B* **161**, 17 (1989).
14. C. S. Wang, B. M. Klein, and H. Krakauer, *Phys. Rev. Lett.* **54**, 1852 (1985).
15. B. Barbiellini, E. G. Moroni, and T. Jarlborg, *J. Phys. Cond. Matt.* **2**, 7597 (1990).
16. B. Barbiellini, E. G. Moroni and T. Jarlborg, *Helv. Phys. Acta* **64**, 164 (1991).
17. V. L. Moruzzi, P. M. Marcus and J. Kübler, *Phys. Rev.* **B39**, 6957 (1989).
18. D. D. Johnson, F. J. Pinski and G. M. Stocks, *J. Appl. Phys.* **57**, 1 (1985).
19. K. Sumiyama, M. Shiga, and Y. Nakamura, *J. Magn. Magn. Mater.* **31–34**, 111 (1983).
20. O. Caporaletti and G. M. Graham, *J. Magn. Magn. Mater.* **22**, 25 (1980).
21. K. Schwarz, P. Mohn, P. Blaha, and J. Kübler, *J. Phys.* **F14**, 2659 (1984).
22. S. K. Burke and B. D. Rainford, *J. Phys.* **F8**, L239 (1978).
23. V. E. Rode, S. A. Finkelberg and A. I. Lyalin, *J. Magn. Magn. Mater.* **31–34**, 293 (1983).
24. E. G.Moroni and T. Jarlborg, in *Proceedings of the ICM'91 Conference*, to appear in *J. Magn. Magn. Mater.*, (1991).
25. K. Adachi *et al.*, *IEEE Trans. Magnetics* **5**, 693 (1972).

26. T. Jarlborg and A. J. Freeman, *Phys. Rev.* **B22**, 2332 (1980).
27. S. S. Peng and H. J. F. Jansen, *Phys. Rev.* **B43**, 3518 (1991).
28. A. Kootte, C. Haas, and R. A. de Groot, *J. Phys.: Condens. Matter.* **3**, 1133 (1991).

Angularly Dependent Many-Body Potentials Within Tight Binding Hückel Theory

D. G. Pettifor and M. Aoki*

Department of Mathematics
Imperial College of Science, Technology and Medicine
London SW7 2BZ
England

Abstract

Angularly-dependent, many-body potentials for the bond order of saturated and unsaturated bonds are derived within Tight Binding Hückel theory. These potentials should prove invaluable for the atomistic simulation of semi-conductors and transition metals over the wide range of co-ordination observed from clusters through to the bulk.

I. Introduction

The need for simple, yet reliable many-body potentials for modelling the behaviour of materials is widespread (see, for example, Vitek and Srolovitz[1]). The effective cluster interactions of the Generalized Perturbation Method (see, for example, Bieber and Gautier[2]) or the Connolly-Williams scheme (see, for example, Carlsson[3]) have been very successful in predicting phase stability with respect to a fixed underlying lattice such as *fcc* or *bcc*. However, apart from the pioneering work of Moriarty[4] in developing explicit many-body potentials by doing perturbation theory about a given atomic site, no potentials exist for semi-conductors and non-*fcc* transition metals which are reliable over a wide range of co-ordination. In this paper an angularly-dependent many-body potential for the bond order is derived within Tight Binding (TB) Hückel theory by embedding the bond rather than the atom within its local atomic environment.[5–7]

* Present Address: Department of Physics, Gifu University, Gifu 501-11, Japan.

Structural and Phase Stability of Alloys
Edited by J.L. Morán-López *et al.*, Plenum Press, New York, 1992

II. The TB Hückel Model

The semi-empirical Tight Binding Hückel model is the simplest scheme for describing the energetics of semi-conductors and transition metals within a quantum mechanical framework. The total binding energy is written in the form

$$U = U_{\text{rep}} + U_{\text{bond}} + U_{\text{prom}}, \tag{1}$$

where U_{rep} is a semi-empirical pairwise repulsive contribution, namely

$$U_{\text{rep}} = \frac{1}{2} \sum_{i,j}{}' \phi(R_{ij}), \tag{2}$$

and U_{bond} is the covalent bond energy which results from evaluating the local density of states $n_{i\alpha}(E)$ associated with orbital α on site i within the two-centre, orthogonal TB approximation. That is,

$$U_{\text{bond}} = \sum_{i\alpha} \int^{E_F} (E - E_{i\alpha}) n_{i\alpha}(E)\, dE, \tag{3}$$

where $E_{i\alpha}$ is the effective atomic energy level of orbital α at site i and E_F is the Fermi energy. The third contribution in Eq. (1) is the promotion energy which for the case of sp orbitals takes the form

$$U_{\text{prom}} = (E_p - E_s) \sum_i \Delta N_p^i = E_{sp} \sum_i \Delta N_p^i \tag{4}$$

where ΔN_p^i gives the change in p occupancy on going from the free atom state to atom i in a given bonding situation. In practice, Eq. (1) gives the binding energy with respect to some *reference* free atom state which usually differs from the *true* atomic ground state due to, for example, the neglect of spin-polarization or the shift in atomic energy levels arising from the renormalization of the wave functions in the bonding situation (see, for example, Sankey and Niklewski[8]).

The form of Eq. (1) may be derived from first principles (see Sutton *et al.*, Ref. 9 and references therein). The pairwise nature of the repulsive term follows directly from the Harris-Foulkes approximation to density functional theory (Harris[10], Foulkes and Haydock[11]), whereas the Hückel-type two-centre orthogonal form of the matrix elements may be justified in principle within Anderson's[12] chemical pseudopotential theory (see Skinner and Pettifor, Ref. 13, for a detailed application to hydrogen). Eq. (1) is evaluated under the constraint of local charge neutrality which is achieved by adjusting the local atomic energy levels with respect to one another. The importance of using the *bond* energy rather than the *band* energy within semi-empirical schemes is discussed fully in §3 of Ref. 6.

The repulsive pair potential $\phi(R)$ and the two-centre hopping integrals are obtained by *fitting* to known bandstructure and Local Density Functional (LDF) binding energy curves for different structure types. For sp-bonded systems the bond integrals are assumed for simplicity to display the same functional dependence on interatomic distance $h(R)$, namely

$$\left.\begin{array}{l} ss\sigma(R) \\ pp\sigma(R) \\ pp\pi(R) \\ sp\sigma(R) \end{array}\right\} = \left\{\begin{array}{c} -1 \\ p_\sigma \\ -p_\pi \\ \sqrt{p_\sigma} \end{array}\right\} h(R) \tag{5}$$

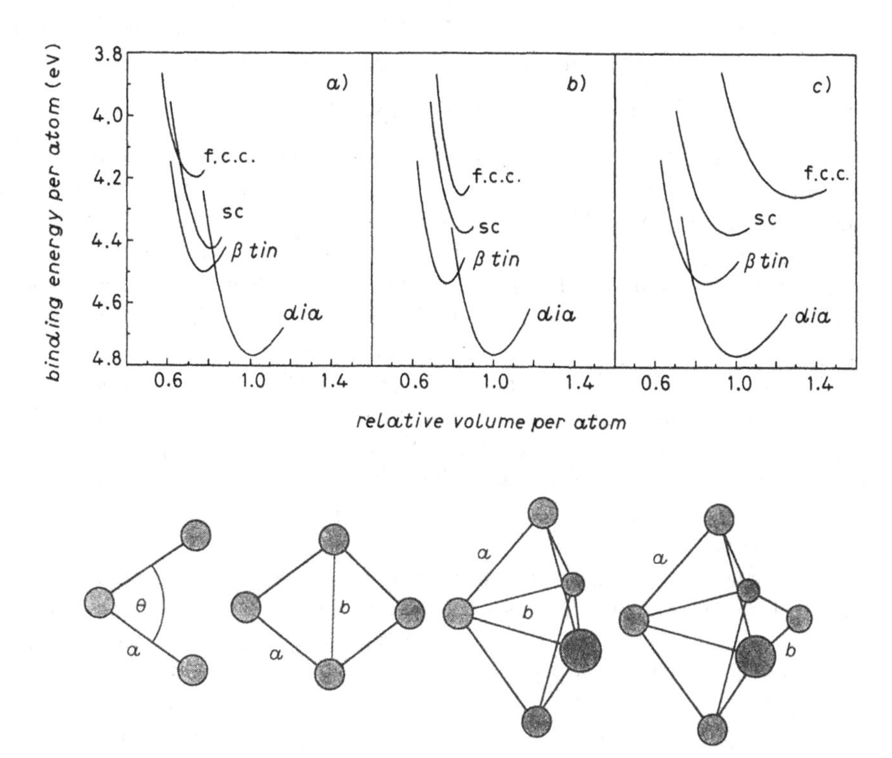

Figure 1. The upper panel compares the LDF binding energy curves of silicon (a) with the rescaled TB curves (b) and the unrescaled TB curves (c). The lower panel shows the lowest energy configurations of Si_3, Si_4, Si_5, and Si_6 clusters which are predicted by the TB parameters.[18]

where p_σ and p_π are positive constant and $sp\sigma$ is taken as the geometric mean of $|ss\sigma|$ and $pp\sigma$. This constraint of identical distance dependences $h(R)$ could be relaxed if necessary to take explicit account of the differences in behaviour of the $ss\sigma$, $pp\sigma$, and $pp\pi$ bonds (see Fig. 4 of Allen *et al.*[14]). This distance dependence is important for obtaining reasonable *bandstructure* fits over the large range of co-ordination from diamond through to *fcc*. For example, Papaconstantopoulos[15] found $p_\sigma \sim 1.3$ for *sp* bonded diamond[15] lattices, whereas $p_\sigma \sim 3$ for *fcc* lattices. The *binding energy* curves will be fitted neglecting the distance dependence of p_σ and p_π as thin will lead to simpler functional forms for the many-body potentials presented later [compare, for example, Eqs. (30)–(34) for $G^\sigma_{ij}(R_{ik}, \theta_{jk})$ with Eqs. (74)–(78) of Ref. 6]. Moreover, Eq. (5) has given a good account of the structural *trends* within the *sp* bonded elements.[16]

The upper panel in Fig. 1 shows that a reasonable fit to the LDF binding energy curves of silicon[17] may be obtained within TB Hückel theory by using a logical rescaling method.[18] They assumed *short-range* transferable TB parameters of the form

$$\phi(R) = A(R_\circ/R)^n f^n(R), \tag{6}$$

and

$$h(R) = B(R_\circ/R)^m f^m(R), \tag{7}$$

where $A = 3.46$ eV, $B = 1.82$ eV, $R_0 = 2.35$ Å (the equilibrium bond length in silicon with the diamond structure), and $f(R)$ is a smooth cut-off function, namely

$$f(R) = \exp[-(R/R_c)^p]/\exp[-(R_o/R_c)^p], \tag{8}$$

where $R_c = 3.67$ Å and $p = 6.48$. The power law exponents n and m were taken as $n = 4.54$ and $m = 2$, the latter being suggested by Harrison's[19] canonical form. We note that for simplicity $\phi(R)$ has been chosen proportional to $[h(r)]^{n/m}$. Harrison's values[20] of p_σ and p_π were chosen, namely $p_\sigma = 1.68$ and $p_\pi = 0.48$. [Note that Goodwin *et al.*[18] took $sp\sigma = 1.08h(R)$ rather than $1.30h(R)$ given by the geometric mean in Eq. (5). The latter constraint is retained here because it simplifies the σ bond order potential dramatically; see Eq. (65) of Ref. 6.] Finally, the *sp* atomic energy level separation was taken as 8.295 eV, which is 18 % larger than the free atom value of 7.03 eV, in order to move the close packed binding energy curve closer to tetrahedrally co-ordinated diamond (see Fig. 6.3 of Skinner[21]). This increase in E_{sp} is just sufficient to close the gap in four-fold co-ordinated silicon.[22] If a gap is required in addition to reproducing the *fcc* binding energy curves, then the distance dependence of p_σ and p_π must be included at the outset.

The lower panel in Fig. 1 shows the lowest energy configurations of Si_3, Si_4, Si_5 and Si_6 clusters which are predicted by these TB parameters.[18] The geometries are in good agreement with the LDF predictions of Raghavachari and Logovinsky.[23] For example, Si_3 is predicted by TB to have a bond length of 2.25 Å and a bond angle of 77.9° compared to the LDF values of 2.17 Å and 77.8°, respectively. Recently, Wang *et al.*[24] have shown that the above TB parameterization yields an excellent radial distribution function for liquid silicon and a good description of point defects.

Thus, it appears as though it is possible to find a set or short-range, transferable TB Hückel parameters that can describe the energetics of the silicon bond over a wide range of co-ordination. In this paper we show that these TB calculations can be simplified still further by deriving explicit angularly-dependent many-body potentials for the bond order.

III. The Bond Order

The bond energy in Eq. (3) is given directly in terms of the local density of states associated with the individual *atoms*. It may be rewritten in terms of the contributions from the individual *bonds* (see, for example the Ref. 9 and references therein) as

$$U_{\text{bond}} = \frac{1}{2} \sum_{i,j}{}' U_{\text{bond}}^{ij}, \tag{9}$$

where

$$U_{\text{bond}}^{ij} = 2 \sum_{\alpha,\beta} H_{i\alpha,j\beta} \Theta_{j\beta,i\alpha}, \tag{10}$$

and where the pre-factor 2 accounts for spin-degeneracy. This has a particularly transparent form. H is the Slater-Koster[25] *bond integral* matrix linking the orbitals on sites i and j together. Θ is the corresponding *bond order* matrix whose elements give the difference between the number of electrons in the bonding $\frac{1}{\sqrt{2}}|i\alpha + j\beta\rangle$ and anti-bonding $\frac{1}{\sqrt{2}}|i\alpha - j\beta\rangle$ states. For the particular choice $sp\sigma = (|ss\sigma|pp\sigma)^{1/2}$ the *sp*-bond energy reduces to the form (*cf.* Eqs. (65)-(67) of Ref. 6)

$$U_{bond}^{ij} = -2(1 + p_\sigma)h(R_{ij})\Theta_{j\sigma,i\sigma} - 4p_\pi h(R_{ij})\Theta_{j\pi,i\pi}, \tag{11}$$

where the hybrid σ orbitals $|i\sigma\rangle$ and $|j\sigma\rangle$ are defined by

$$|i\sigma\rangle = (|is\rangle + \sqrt{p_\sigma}|iz\rangle)/\sqrt{1+p_\sigma}, \tag{12}$$

and

$$|j\sigma\rangle = (|js\rangle - \sqrt{p_\sigma}|jz\rangle)/\sqrt{1+p_\sigma}, \tag{13}$$

choosing the z-axis along $\mathcal{R}_{\sim ij}$ and the π bond order by

$$\Theta_{j\pi,i\pi} = \frac{1}{2}(\Theta_{jx,ix} + \Theta_{jy,iy}). \tag{14}$$

Although Eq. (11) gives the bond energy between a given pair of atoms i and j it is *not* pairwise because the bond order itself depends on the local atomic environment. We display this dependence explicity by using the recursion method of Haydock *et al.*[26] to write the bond order as an integral over the difference of two continued fractions:

$$\Theta_{i\alpha,j\beta} = -\frac{1}{\pi}\Im m \int^{E_F} [G^{+}_{oo}(E) - G^{-}_{oo}(E)]\, dE, \tag{15}$$

where $\Im m$ is the imaginary part of the bonding and anti-bonding Green's functions which are given by

$$\begin{aligned} G^{\pm}_{oo}(E) &= \langle u^{\pm}_o|(E-H)^{-1}|u^{\pm}_o\rangle \\ &= \cfrac{1}{(E-a^{\pm}_o) - \cfrac{(b^{\pm}_1)^2}{(E-a^{\pm}_1) - \cdots}} \end{aligned} \tag{16}$$

where $|u^{\pm}_o\rangle = \frac{1}{\sqrt{2}}|i\alpha \pm j\beta\rangle$. The coefficients are determined by the Lanczos recursion algorithm, namely

$$b^{\pm}_{n+1}|u^{\pm}_{n+1}\rangle = H|u^{\pm}_n\rangle - a^{\pm}_n|u^{\pm}_n\rangle - b^{\pm}_n|u^{\pm}_{n-1}\rangle, \tag{17}$$

with the boundary condition that $|u^{\pm}_{-1}\rangle$ vanishes. The Hamiltonian H is, therefore, tridiagonal with respect to the recursion basis $|u^{\pm}_n\rangle$, having non-zero elements

$$\langle u^{\pm}_n|H|u^{\pm}_n\rangle = a^{\pm}_n, \tag{18}$$

and

$$\langle u^{\pm}_{n+1}|H|u^{\pm}_n\rangle = b^{\pm}_{n+1}. \tag{19}$$

The dependence of the recursion coefficients on the local atomic environment about the bond ij may be obtained by using the well-known relationship between the recursion coefficients $a^{\pm}_n$, $b^{\pm}_n$ and the moments $\mu^{\pm}_n = \langle u^{\pm}_o|H^n|u^{\pm}_o\rangle$, namely

$$\mu^{\pm}_o = 1, \tag{20}$$

$$\mu^{\pm}_1 = a^{\pm}_o, \tag{21}$$

$$\mu^{\pm}_2 = (a^{\pm}_o)^2 + (b^{\pm}_1)^2, \tag{22}$$

$$\mu^{\pm}_3 = (a^{\pm}_o)^3 + 2a^{\pm}_o(b^{\pm}_1)^2 + a^{\pm}_1(b^{\pm}_1)^2, \tag{23}$$

and

$$\begin{aligned} \mu^{\pm}_4 = {} & (a^{\pm}_o)^4 + 3(a^{\pm}_o)^2(b^{\pm}_1)^2 + 2a^{\pm}_o a^{\pm}_1(b^{\pm}_1)^2 \\ & +(a^{\pm}_1)^2(b^{\pm}_1)^2 + (b^{\pm}_1)^2(b^{\pm}_2)^2 + (b^{\pm}_1)^4. \end{aligned} \tag{24}$$

Taking $\mu_1 = (\mu^{+}_1 + \mu^{-}_1)/2$ as the energy zero, it follows that

$$a^{\pm}_o = \pm H_{i\alpha,j\beta}, \tag{25}$$

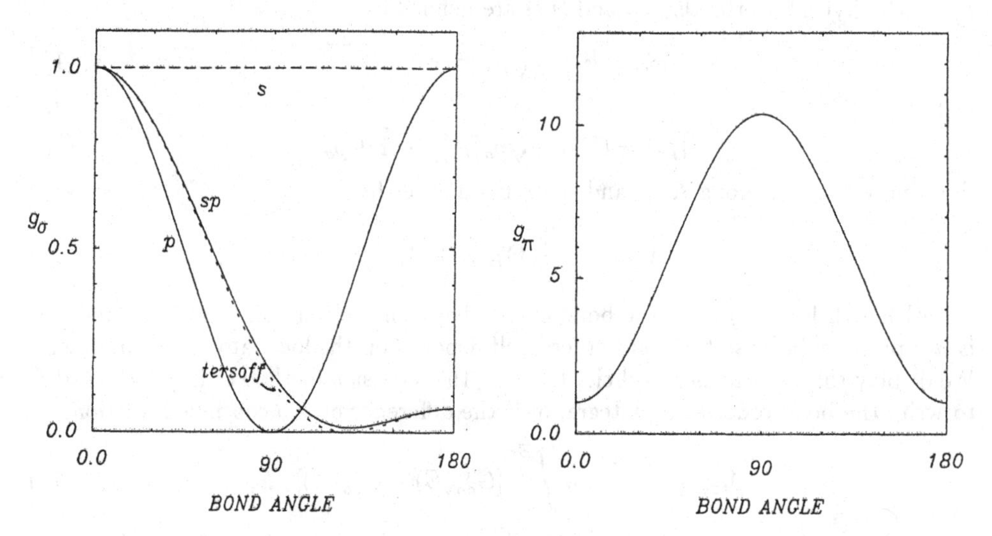

Figure 2. The angular dependence of the predicted embedding functions $g_\sigma(\theta)$ and $g_\pi(\theta)$ with the semi-empirical[29] for comparison.

and

$$(b_1^{\pm})^2 = b_1^2 \pm \sum_{k\neq i,j;\gamma} H_{i\alpha,k\gamma} H_{k\gamma,j\beta}, \tag{26}$$

where

$$b_1^2 = \frac{1}{2}\left[\sum_{k\neq j;\gamma} H^2_{i\alpha,k\gamma} + \sum_{k\neq i;\gamma} H^2_{j\beta,k\gamma}\right]. \tag{27}$$

Thus, the influence of the local atomic environment first enters the continued fraction through $b_1^{\pm}$. The explicit analytic form of b_1 has already been derived in Ref. 6 for $sp-$ and $sd-$ valent systems with zero E_{sp} and E_{sd} by using the Slater-Koster[25] bond integral matrix H. For the general case $E_{sp} \neq 0$ the $sp\,\sigma$ bond will be characterized by the zero of energy $\mu_1 = (E_s + p_\sigma E_p)/(1+p_\sigma)$ so that $E_s = -p_\sigma E_{sp}/(1+p_\sigma)$ and $E_p = E_{sp}/(1+p_\sigma)$. Writing $H_{ij}^\sigma = H_{i\sigma,j\sigma}$ it follows from Eq. (27) that

$$(b_1^\sigma/H_i^\sigma)^2 = p_\sigma E_{sp}^2/[(1+p_\sigma)^2 h(R_{ij})]^2 + (\hat{b}_1^\sigma/H_{ij}^\sigma)^2, \tag{28}$$

where the square of the normalized *embedding function* for the $ij\,\sigma$ bond is given by

$$(\hat{b}_1^\sigma/H_{ij}^\sigma)^2 = \sum_{k\neq i,j} \frac{1}{2}\left[G_{ij}^\sigma(R_{ik},\theta_{jik}) + G_{ji}^\sigma(R_{jk},\theta_{ijk})\right], \tag{29}$$

where θ_{jik} and θ_{ijk} are the appropiate bond angles and

$$G_{ij}^\sigma(R_{ik},\theta) = [h(R_{ik})/h(R_{ij})]^2 g_\sigma(\theta), \tag{30}$$

with

$$g_\sigma(\theta) = c + d\cos\theta + e\cos 2\theta, \tag{31}$$

$$c = 1 - d - e, \tag{32}$$

$$d = 2p_\sigma/(1+p_\sigma)^2, \tag{33}$$

and

$$e = p_\sigma^2/2(1+p_\sigma)^2 - p_\sigma p_\pi^2/2(1+p_\sigma)^3. \tag{34}$$

The $sp\pi$ bond will be characterized by a different zero of energy, namely $E_p = 0$. It has the associated embedding function $\hat{b}_1^\pi = b_1^\pi$ whose square is given by

$$(b_1^\pi/H_{ij}^\pi)^2 = \sum_{k\neq i,j} \frac{1}{2}\left[G_{ij}^\pi(R_{ik},\theta_{jik}) + G_{ji}^\pi(R_{jk},\theta_{ijk})\right], \tag{35}$$

where

$$G_{ij}^\pi(R_{ik},\theta) = [h(R_{ik})/h(R_{ij})]^2 g_\pi(\theta), \tag{36}$$

with

$$g_\pi(\theta) = C + E\cos 2\theta, \tag{37}$$

$$C = 1 - E, \tag{38}$$

and

$$E = \frac{1}{4} - p_\sigma(1+p_\sigma)/4p_\pi^2. \tag{39}$$

Fig. 2 shows the angular dependence of the embedding functions $g_\sigma(\theta)$ and $g_\pi(\theta)$ using the Goodwin *et al.*[18] values of p_σ and p_π. The pure $s\sigma$ bond (corresponding to $p_\sigma = 0$) displays no angular dependence as expected, the influence of a neighboring atom k on the ij bond strength being independent of the bond angle. On the hand, the pure $p\sigma$ bond (corresponding to vanishing $ss\sigma$ and $pp\pi$ falling to zero for a bond angle of 90° since there will be no coupling to this neighboring atom if p_π is zero. The sp hybrid has a minimum in $g_\sigma(\theta)$ around 130° and takes a value less tan 0.1 for $\theta \geq 100°$ as the hybrid orbitals, Eqs. (12) and (13), have very little weight in these directions, so that neighbors may be added in this range without affecting the strength of the original σ bond. thus, graphite and diamond with bond angles of 120° and 109° respectively will have nearly *saturated* σ bonds (for $E_{sp} = 0$). In contrast the angular dependence of the π bond with its lobes extending in planes perpendicular to the bond axis leads to *unsaturated* behavior as any neighbor will drastically reduce the strength of the original dimeric bond, as can be seen in Fig. 2 where $g_\pi(\theta)$ rises to a value an order of magnitude larger than that for angularly independent s orbitals.

The order recursion coefficients in the continued fraction may be determined from a knowledge of the moments $\mu_n^\pm$ through relations such as Eqs. (22)–(24). It is helpful to display the difference between the bonding and anti-bonding moments explicitly by writing

$$\mu_n^\pm = \mu_n \pm \zeta_{n+1}, \tag{40}$$

where it follows that since $|u_\circ^\pm\rangle = \frac{1}{\sqrt{2}}|i\alpha \pm j\beta\rangle$ we have that μ_n is the *average* nth moment with respect to the appropiate orbitals on sites i and j, namely

$$\mu_n = \frac{1}{2}[\langle i\alpha|H^n|i\alpha\rangle + \langle j\beta|H^n|j\beta\rangle] = \frac{1}{2}(\mu_n^{i\alpha} + \mu_n^{j\beta}), \tag{41}$$

and ζ_{n+1} is the *interference* term, namely

$$\zeta_{n+1} = < i\alpha|H^n|j\beta\rangle. \tag{42}$$

The expressions for μ_n and ζ_{n+1} rapidly become very complicated for $n > 2$ if we simply multiply together the Slater-Koster matrices H which are given in terms of the direction cosines of the bond directions with respect to a fixed set of axes. Rather we choose co-ordinates natural to the hopping path as illustrated in Fig. 3 for the four connected sites i, j, k and l which define the bond angles θ_j and θ_k and the dihedral

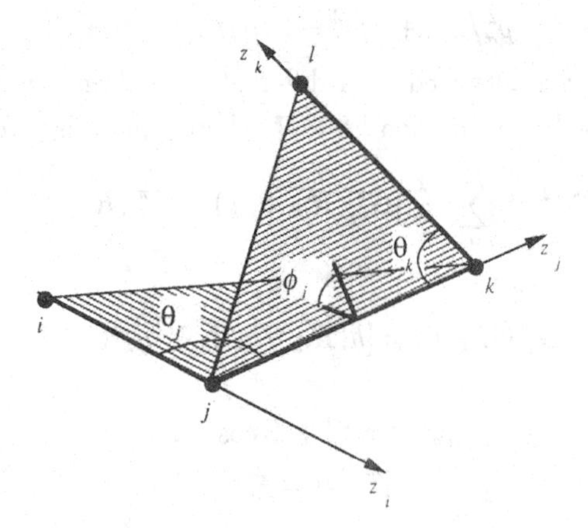

Figure 3. The four connected atomic sites i, j, k, and l which define the bond angles θ_j and θ_k and the dihedral angle ϕ_j.

angle ϕ_j. Let us choose Oz_i along $\mathcal{R}_{ij}$ and Oy_i perpendicular to the plane ijk in the direction $\mathcal{R}_{ij} \times \mathcal{R}_{jk}$. Similarly, let Oz_j be along $\mathcal{R}_{jk}$ and Oy_j perpendicular to the plane jkl in the direction $\mathcal{R}_{jk} \times \mathcal{R}_{kl}$ etc. Therefore, in moving from the $(x_i,\ y_i,\ z_i)$ co-ordinate system to the $(x_j,\ y_j,\ z_j)$ co-ordinate system we require the following rotation matrix for sp valent systems, namely

$$\Re_j = \begin{pmatrix} 1 & 0 & 0 & 0 \\ 0 & -\cos\theta_j & \sin\theta_j & 0 \\ 0 & -\sin\theta_j\cos\phi_j & -\cos\theta_j\cos\phi_j & \sin\phi_j \\ 0 & \sin\theta_j\sin\phi_j & \cos\theta_j\sin\phi_j & \cos\phi_j \end{pmatrix}. \tag{43}$$

The hopping from j to k will then be performed by the bond integral matrix $\hat{H}h(R_{kj})$ where

$$\hat{H} = \begin{pmatrix} -1 & -\sqrt{p_\sigma} & 0 & 0 \\ \sqrt{p_\sigma} & p_\sigma & 0 & 0 \\ 0 & 0 & -p_\pi & 0 \\ 0 & 0 & 0 & -p_\pi \end{pmatrix}. \tag{44}$$

This allows us to derive explicit expressions for the *ring* terms, for example, about a given bond, say $i = 0$, $j = 1$. The n-membered ring contribution, which involves hopping from $1 \to 2 \to 3 \to \cdots (n-1) \to n = 0$ where all the sites are distinct, can be written

$$\zeta_n^{\text{ring}} = \sum_{2,3,\cdots(n-1)} h(R_{n(n-1)}) \cdots h(R_{32})h(R_{21})\langle 0|\Re_n\hat{H}\cdots\Re_2\hat{H}\Re_1|1\rangle. \tag{45}$$

We find the following expressions for the three and four membered ring terms about a given sp bond:

$$\zeta_3^{\text{ring},\sigma} = \sum_2 g_\sigma^{(3)}(\theta_1,\theta_2,\theta_3)h(R_{12})h(R_{23}) \tag{46}$$

where

$$g_\sigma^{(3)}(\theta_1,\theta_2,\theta_3) = \{(1+p_\sigma\cos\theta_2)(1+p_\sigma\cos\theta_3) + p_\sigma p_\pi[\sin\theta_1\sin\theta_2(1+p_\sigma\cos\theta_3)+\sin\theta_2\sin\theta_3(1+p_\sigma\cos\theta_1)] - p_\sigma p_\pi^2\sin\theta_1\sin\theta_3\cos\theta_2\}/(1+p_\sigma),\tag{47}$$

$$\zeta_3^{\mathrm{ring},\pi} = \sum_2 g_\pi^{(3)}(\theta_1,\theta_2,\theta_3)h(R_{12})h(R_{23})\tag{48}$$

where

$$g_\pi^{(3)}(\theta_1,\theta_2,\theta_3) = \{\ p_\sigma\sin\theta_1\sin\theta_3(1+p_\sigma\cos\theta_2) - p_\sigma p_\pi\sin\theta_2\sin(\theta_1+\theta_3) + p_\pi^2(1-\cos\theta_1\cos\theta_2\cos\theta_3)\}/2,\tag{49}$$

$$\zeta_4^{\mathrm{ring},\sigma} = \sum_{2,3} g_\sigma^{(4)}(\theta_1,\theta_2,\theta_3,\theta_4,\phi_1,\phi_2,\phi_3)h(R_{12})h(R_{23})h(R_{24})\tag{50}$$

where

$$(1+p_\sigma)g_\sigma^{(4)}(\theta_1,\theta_2,\theta_3,\theta_4,\phi_1,\phi_2,\phi_3)$$

$$= \left\{\begin{array}{l} -(1+p_\sigma\cos\theta_1)(1+p_\sigma\cos\theta_2)(1+p_\sigma\cos\theta_3)(1+p_\sigma\cos\theta_4) \\ -p_\sigma p_\pi\left[\begin{array}{l}(1+p_\sigma\cos\theta_1)(1+p_\sigma\cos\theta_4)\sin\theta_3\sin\theta_2\cos\phi_2 \\ +(1+p_\sigma\cos\theta_1)(1+p_\sigma\cos\theta_2)\sin\theta_4\sin\theta_3\cos\phi_3 \\ +(1+p_\sigma\cos\theta_4)(1+p_\sigma\cos\theta_3)\sin\theta_1\sin\theta_2\cos\phi_1\end{array}\right] \\ +p_\sigma p_\pi^2\left[\begin{array}{l}(1+p_\sigma\cos\theta_4)\sin\theta_3\sin\theta_1(\cos\theta_2\cos\phi_2\cos\phi_1+\sin\phi_2\sin\phi_1) \\ +(1+p_\sigma\cos\theta_1)\sin\theta_4\sin\theta_2(\cos\theta_3\cos\phi_3\cos\phi_2+\sin\phi_3\sin\phi_2)\end{array}\right] \\ -p_\sigma^2 p_\pi^2\sin\theta_4\sin\theta_3\sin\theta_2\sin\theta_1\cos\phi_3\cos\phi_1 \\ -p_\sigma p_\pi^3\sin\theta_4\sin\theta_1\left[\begin{array}{l}\cos\theta_3\cos\phi_3(\cos\theta_2\cos\phi_2\cos\phi_1+\sin\phi_2\sin\phi_1) \\ +\sin\phi_3(\cos\theta_2\sin\phi_2\cos\phi_1-\cos\phi_2\sin\phi_1)\end{array}\right]\end{array}\right.\tag{51}$$

and

$$\zeta_4^{\mathrm{ring},\pi} = \sum_{2,3} g_\pi^{(4)}(\theta_1,\theta_2,\theta_3,\theta_4,\phi_1,\phi_2,\phi_3,\phi_4)h(R_{12})h(R_{23})h(R_{34})\tag{52}$$

where

$$
\begin{aligned}
&2g_\pi^{(4)}(\theta_1,\theta_2,\theta_3,\theta_4,\phi_1,\phi_2,\phi_3,\phi_4)\\
&= \left\{
\begin{array}{l}
- p_\sigma(1+p_\sigma\cos\theta_2)(1+p_\sigma\cos\theta_3)\sin\theta_4\sin\theta_1\cos\phi_4\\
- p_\sigma p_\pi \left[\begin{array}{l}
(1+p_\sigma\cos\theta_3)\sin\theta_4\sin\theta_2(\cos\theta_1\cos\phi_4\cos\phi_1+\sin\phi_4\sin\phi_1)\\
+(1+p_\sigma\cos\theta_2)\sin\theta_1\sin\theta_3(\cos\theta_4\cos\phi_4\cos\phi_3+\sin\phi_4\sin\phi_3)\\
- p_\sigma\sin\theta_4\sin\theta_3\sin\theta_2\sin\theta_1\cos\phi_4\cos\phi_2
\end{array}\right]\\
- p_\sigma p_\pi^2 \left[\begin{array}{l}
\sin\theta_4\sin\theta_3\cos\theta_1\cos\phi_4\,(\cos\theta_2\cos\phi_2\cos\phi_1+\sin\phi_2\sin\phi_1)\\
+\sin\theta_1\sin\theta_2\cos\theta_4\cos\phi_4\,(\cos\theta_3\cos\phi_2\cos\phi_3+\sin\phi_2\sin\phi_3)\\
+\sin\theta_2\sin\theta_3\cos\theta_4\cos\phi_3\,(\cos\theta_1\cos\phi_1\cos\phi_4+\sin\phi_4\sin\phi_1)\\
+\sin\theta_1\sin\theta_2\sin\phi_4\,(\cos\theta_3\cos\phi_2\sin\phi_3-\sin\phi_2\cos\phi_3)\\
+\sin\theta_4\sin\theta_3\sin\phi_4\,(\cos\theta_2\cos\phi_2\sin\phi_1-\sin\phi_2\cos\phi_1)\\
+\sin\theta_3\sin\theta_2\sin\phi_3\,(\cos\theta_1\cos\phi_1\sin\phi_4-\sin\phi_1\cos\phi_4)
\end{array}\right]\\
- p_\pi^2 \left[\begin{array}{l}
\cos\theta_4\cos\theta_1\cos\phi_4\left[\begin{array}{l}
\cos\theta_3\cos\theta_2\cos\phi_3\cos\phi_2\cos\phi_1-\cos\phi_2\sin\phi_1\sin\phi_3\\
+\cos\theta_3\cos\phi_3\sin\phi_2\sin\phi_1+\cos\theta_2\cos\phi_1\sin\phi_2\sin\phi_3
\end{array}\right]\\
+\cos\theta_1\cos\theta_3\sin\phi_4\sin\phi_3(\cos\theta_2\cos\phi_2\cos\phi_1+\sin\phi_2\sin\phi_1)\\
+\cos\theta_4\cos\theta_2\sin\phi_4\sin\phi_1(\cos\theta_3\cos\phi_2\cos\phi_3+\sin\phi_2\sin\phi_3)\\
-\cos\theta_1\sin\phi_4\cos\phi_3(\cos\theta_2\sin\phi_2\cos\phi_1-\cos\phi_2\sin\phi_1)\\
-\cos\theta_4\sin\phi_4\cos\phi_1(\cos\theta_3\sin\phi_2\cos\phi_3-\cos\phi_2\sin\phi_3)\\
-\cos\theta_3\cos\theta_2\cos\phi_4\sin\phi_3\cos\phi_2\sin\phi_1+\cos\phi_4\cos\phi_3\cos\phi_2\cos\phi_1\\
+\cos\phi_4\sin\phi_2(\cos\theta_3\sin\phi_3\cos\phi_1+\cos\theta_2\sin\phi_1\cos\phi_3)
\end{array}\right]
\end{array}\right.
\end{aligned}
\tag{53}
$$

Corresponding expressions for the *sd* ring terms have been derived and will be published elsewhere.[27]

IV. Many-Body Potential for Saturated Bonds

The σ bond in *sp* valent systems with the graphite or diamond structure will be nearly saturated for $E_{sp} = 0$ because of the very weak coupling of any given σ bond to the local atomic environment through such bond angles (*cf.* Fig. 2). The four eigenvalues of the *isolated sp* dimer with σ character may be obtained analytically[28] and are given by

$$
\begin{aligned}
\epsilon_1^\pm &= \frac{1}{2}(E_s+E_p)-\frac{1}{2}(1+p_\sigma)h \pm \frac{1}{2}[E_{sp}^2+2E_{sp}(1-p_\sigma)h+(1+p_\sigma)^2h^2]^{1/2}\\
\epsilon_2^\pm &= \frac{1}{2}(E_s+E_p)+\frac{1}{2}(1+p_\sigma)h \pm \frac{1}{2}[E_{sp}^2-2E_{sp}(1-p_\sigma)h+(1+p_\sigma)^2h^2]^{1/2}.
\end{aligned}
\tag{54}
$$

We note that if $|ss\sigma| = pp\sigma$ *i.e.* $p_\sigma = 1$, then these eigenvalues are symmetric with respect to the average energy $(E_s + E_p)/2$.

In order to recover this eigenspectrum in the limit of zero coupling to the environment, the continued fraction in Eq. (16) must be taken to four levels. We shall neglect the asymmetry introduced by $p_\sigma \neq 1$ and write

$$G_{oo}^{\pm}(E) = \cfrac{1}{E - a_o^{\pm} - \cfrac{b_1^2}{E - a_1^{\pm} - \cfrac{b_2^2}{E + a_1^{\pm} - \cfrac{b_1^2}{E + a_o^{\pm}}}}}, \tag{55}$$

where the continued fraction has been terminated by $a_2^{\pm} = -a_1^{\pm}$, $b_3^{\pm} = b_1$, and $a_3^{\pm} = -a_o^{\pm}$ in order to guarantee symmetric poles with respect to the reference energy $(E_s + p_\sigma E_p)/(1 + p_\sigma)$. The bonding and anti-bonding Green's functions have identical poles $\pm E_1$, $\pm E_2$ provided $a_o^{\pm} = \pm a_o$, $a_1^{\pm} = \pm a_1$ when they are given by

$$E_1 = [\alpha - (\alpha^2 - \beta^2)^{1/2}]^{1/2}, \tag{56}$$

and

$$E_2 = [\alpha + (\alpha^2 - \beta^2)^{1/2}]^{1/2}, \tag{57}$$

with

$$\alpha = \frac{1}{2}(2b_1^2 + b_2^2 + a_o^2 + a_1^2), \tag{58}$$

and

$$\beta = (b_1^4 - 2a_o a_1 b_1^2 + a_o^2 b_2^2 + a_o^2 a_1^2)^{1/2}. \tag{59}$$

For the isolated dimer these poles $\pm E_1$, $\pm E_2$ of the continued fraction are equivalent to $\epsilon_1^{\pm}$, $\epsilon_2^{\pm}$ for the symmetric case $p_\sigma = 1$, since it follows from Eqs. (21)-(24) that $a_o^{\pm} = \mp 2h$, $b_1^{\pm} = \frac{1}{2}E_{sp}$, $a_1^{\pm} = 0$, and $b_2^{\pm} = 0$. Substituting into Eqs. (56)–(59) we have that the squares of E_1 and E_2 are given by

$$E_{1,2}^2 = (E_{sp}/2)^2 + 2h^2 \mp 2h[(E_{sp}/2)^2 + h^2]^{1/2}. \tag{60}$$

This can be seen to be identical to Eq. (54) [with respect to the reference energy $(E_s + E_p)/2)$] for the case $p_\sigma = 1$ by rewriting the right hand side of Eq. (60) as

$$E_{1,2}^2 = [h \mp [(E_{sp}/2)^2 + h^2]^{1/2}]^2. \tag{61}$$

The σ bond order may be obtained by substituting Eq. (55) into Eq. (15). For a half-full bond we find the bond order per spin associated with the hybrid σ orbitals, Eqs. (12) and (13), is given by

$$\Theta_\sigma(N_\sigma = 1) = (1 + \gamma)/[2(\alpha + \beta)/a_o^2]^{1/2}, \tag{62}$$

where

$$\gamma = [b_2^2 + a_1^2 - (a_1/a_o)b_1^2]/\beta, \tag{63}$$

with $a_o = H_{ij}^\sigma = -(1 + p_\sigma)h(R)$. In the absence of three membered rings $b_1 = b_1^\sigma$ through Eq. (28) and neglecting the skewing due to $p_\sigma \neq 1$ it follows from Eq. (23) that

$$a_1 = \zeta_4^{\text{ring},\sigma}/b_1^2, \tag{64}$$

with $\zeta_4^{\text{ring},\sigma}$ given by Eqs. (50) and (51). Finally, if in addition there are no four or five membered rings then it follows from Eq. (24) that b_2 is given in terms of all three-atom self-retracing paths of length two which start at and end on the given bond atoms i or j, namely

$$b_2^2 = \frac{1}{2}(\hat{\mu}_4^i + \hat{\mu}_4^j)/b_1^2, \tag{65}$$

where

$$\hat{\mu}_4^j = \sum_{k,1 \neq i,j} \hat{g}_\sigma^{(3)}(\theta_j, \theta_k, \phi_j) \left[h(R_{jk})h(R_{kl})\right]^2, \tag{66}$$

with

$$(1+p_\sigma)\hat{g}_\sigma^{(3)}(\theta_j, \theta_k, \phi_j) = \begin{cases} (1+p_\sigma)\left[(1+p_\sigma\cos\theta_j)(1+p_\sigma\cos\theta_k) + p_\sigma p_\pi \sin\theta_j \sin\theta_k \cos\phi_j\right]^2 \\ +p_\sigma p_\pi^2 \left[(1+p_\sigma\cos\theta_j)\sin\theta_k - p_\pi \sin\theta_j \cos\theta_k \cos\phi_j\right]^2 \\ +p_\sigma p_\pi^4 \left[\sin\theta_j \sin\phi_j\right]^2. \end{cases} \tag{67}$$

The bond angles θ_j and θ_k and dihedral angle ϕ_j are defined in Fig. 3.

This expression for the saturated σ bond order simplifies for the graphite and diamond lattices if we assume that $b_2 = 0$ because the coupling with the lattice is weak. Then, since a_1 also vanishes for these two lattices, we have that $\gamma = 0$ and the bond order may be written

$$\Theta_\sigma(N_\sigma = 1) = \left\{1 + 4\hat{E}_{sp}^2 + 2\sum_{k \neq i,j}\left[G_{ij}^\sigma(R_{ik}, \theta_{jik}) + G_{ji}^\sigma(R_{jk}, \theta_{ijk})\right]\right\}^{-1/2}, \tag{68}$$

with

$$\hat{E}_{sp}^2 = [p_\sigma/(1+p_\sigma)^4][E_{sp}/h(R_\circ)]^2, \tag{69}$$

so that the bond order is fully saturated with $\Theta = 1$ for $E_{sp} = 0$ when there is zero coupling to the local environment.

The bond order potentials are used under the contraint that the *sp* energy level separation E_{sp} is ajusted to maintain constancy in the number of s and p electrons so that there is no change in the promotion energy Eq. (4). This is valid to first order [*cf.* Eqs. (60)–(63) of Ref. 6]. For the isolated dimer this would be achieved by regarding $[E_{sp}/h(R)]$ as constant. Thus, $\hat{E}_{sp}^2$ is a constant with $h(R_\circ)$ evaluated at the equilibrium bond length $R_\circ$ of the ground state structure. It takes the value 0.68 for the Goodwin *et al.*[18] parameterization of silicon, so that the isolated dimer has a σ bond order of 0.52. This reflects the decrease in the hybrid bond energy for increasing *sp* atomic energy level mismatch.

Expression (68) for the saturated σ bond order is the central result of this section. It has a form which is not too different from the semi-empirical Tersoff[29] potential, so that it is not surprising that this angular dependence which was obtained by *fitting* to the bulk properties of silicon follows very closely that *predicted* within TB Hückel theory (see Fig. 2). However, it is important to note that Eq. (68) was derived from Eq. (62) under the assumption that there are no four membered ring terms ($a_1 = 0$) and that the coupling with the lattice is weak ($b_2 = 0$). These two assumptions break down for other lattices such as simple cubic and *fcc* where the presence of four membered ring terms destabilize these lattices with respect to graphite and diamond for half-full bands. The relevance of this to structural binding energy curves such as those in the upper panel of Fig. 1 will be presented elsewhere.[30]

V. Many-Body Potential for Unsaturated Bonds

The bonds in s valent, π bonded, or close-packed σ bonded systems will be unsaturated due to a sizeable overlap of the given bond with the surrounding atomic neighborhood. In this case of $\zeta_{n+1}/\mu_2^{n/2}$ being small, we may do first order perturbation theory about the bond (see Refs. 5–7), to derive a many-body expansion for the bond order, which has been shown to converge rapidly to the known TB σ, π, and Δ bond orders of *bcc, fcc*, and *hcp* transition metals.[27]

This expression can be simplified within the Ring Approximation [see Eqs. (2.36)-(2.40) of Ref. 7] to give an explicit cluster expansion for the bond order, namely

$$\Theta(N) = 2\sum_{n=2}^{\infty} \hat{\chi}_n(N)\zeta_n^{\text{ring}}/b^{n-1}. \tag{70}$$

For $E_{sp} = 0$ or $E_{sd} = 0$ b is the appropiate embedding function $\hat{b}_1$ such as is given by Eqs. (29) or (35) but with the interbond hopping ij also included [see, for example, Eq. (13) of Ref. 5]. $\hat{\chi}_n(N)$ is the reduced susceptibility characterizing the n-membered ring contribution, namely

$$\hat{\chi}_n(N) = \frac{1}{\pi}\left[\frac{\sin(n-1)\phi_F}{n-1} - \frac{\sin(n+1)\phi_F}{n+1}\right], \tag{71}$$

where ϕ_F is fixed by the number of valence electrons per spin per bond N through

$$N = (2\phi_F/\pi)\left[1 - (\sin 2\phi_F)/2\phi_F\right]. \tag{72}$$

This bond order potential has been shown to explain the relative stability of s valent clusters[5,31] and to allow for the negative Cauchy which are observed in some transition metals and intermetallic compounds.[7] It has also been used to explain the oscillatory behavior of the effective cluster interactions which are obtained within the Connolly-Williams scheme.[32] Further details may be found elsewhere, in a recent publication.[7]

VI. Conclusion

These newly derived bond order potentials for saturated and unsaturated bonds should prove invaluable for the atomistic simulation of semi-conductors and transition metals where both the angular character and the many-body nature of the potential are important. Their derivation from TB Hückel theory within a set of well-defined approximations leads to the hope that they will prove reliable over the wide range of co-ordination displayed from clusters through to the bulk.

Acknowledgements

We wish to thank the U. S. Department of Energy, Energy Conversion and Utilization Technologies (ECUT) Materials Program, for financial support under subcontract No. 19X–55992V through Martin Marietta Energy Systems Inc. We also thank Parvaneh Alinaghian and Mark Datko for providing Figs. 2 and 3, respectively. We thank Adrian Sutton for bringing Kohyama's recent paper to our attention.

References

1. V. Vitek and D. G. Srolovitz, *Atomistic Simulation of Materials Beyond Pair Potentials*, Plenum Pub. Co., New York, 1989.
2. A. Bieber and F. Gautier, *Acta Metall.* **34**, 2291 (1986).
3. A. E. Carlsson, *Phys. Rev.* **B40**, 912 (1989).
4. J. A. Moriarty, *Phys. Rev.* **B42**, 1609 (1990).
5. D. G. Pettifor, *Phys. Rev. Lett.* **63**, 2480 (1989).
6. D. G. Pettifor, *Springer Proc. Phys.* **48**, 64 (1990).
7. D. G. Pettifor and M. Aoki, *Phil. Trans. R. Soc. Lond.* **A334**, 439 (1991).
8. O. F. Sankey and D. J. Niklewski, *Phys. Rev.* **B40**, 3979 (1989).
9. A. P. Sutton, M. W. Finnis, D. G. Pettifor, and Y. Ohta, *J. Phys. C: Solid State Phys.* **21**, 35 (1988).
10. J. Harris, *Phys. Rev.* **B31**, 1770 (1985).
11. W. M. C. Foulkes and R. Haydock, *Phys. Rev.* **B39**, 12520 (1989).
12. P. W. Anderson, *Phys. Rev. Lett.* **21**, 13 (1968).
13. A. J. Skinner and D. G. Pettifor, *J. Phys.: Condens. Matter* **3**, 2029 (1991).
14. P. B. Allen, J. Q. Broughton, and A. K. McMahan, *Phys. Rev.* **B34**, 859 (1986).
15. S. A. Papaconstantopoulos, *Handbook of the Bandstructure of Elemental Solids*, Plenum Pub. Co., New York, 1986.
16. J. C. Cressoni and D. G. Pettifor, *J. Phys.: Condens. Matter* **3**, 495 (1991).
17. M. T. Yin and M. L. Cohen, *Phys. Rev.* **B26**, 5668 (1982).
18. L. Goodwin, A. J. Skinner, and D. G. Pettifor, *Europhys. Lett.* **9**, 70 (1989)
19. W. A. Harrison, *Electronic Structure and the Properties of Solids*, Freeman, San Francisco, 1980.
20. W. A. Harrison. *Phys. Rev.* **B27**, 3592 (1983).
21. A. J. Skinner, PhD thesis, University of London, 1989.
22. M. Kohyama, *J. Phys.: Condens. Matter* **3**, 2193 (1991).
23. K. Raghavachari and V. Logovinsky, *Phys. Rev. Lett.* **55**, 2853 (1985).
24. C. Z. Wang, C. T. Chan, and K. M. Ho 1991 (preprint).
25. J. C. Slater and G. F. Koster, *Phys. Rev.* **94**, 1498 (1954).
26. R. Haydock, V. Heine, and M. J. Kelly, *J. Phys. C: Solid State Phys.* **5**, 2845 (1972).
27. M. Aoki and D. G. Pettifor 1991 (to be published).
28. J. C. Cressoni, PhD thesis, University of London, 1989.
29. J. Tersoff, *Phys. Rev.* **B38**, 9902 (1988).
30. P. Alinaghian and D. G. Pettifor 1991 (to be published).
31. M. Shah and D. G. Pettifor 1991 (to be published).
32. J. W. D. Connolly and A. R. Williams, *Phys. Rev.* **B27**, 5169 (1983).

Structure and Thermodynamics of SiGe Alloys from Computational Alchemy

Stefano Baroni[1,2], Stefano de Gironcoli[2], and Paolo Giannozzi[2]

[1] *SISSA – Scuola Internazionale Superiore di Studi Avanzati*
via Beirut 4
I-34014 Trieste
Italy

[2] *IRRMA – Institut Romand de Recherche Numérique en Physique des Matériaux*
PHB – Ecublens
CH-1015 Lausanne
Switzerland

Abstract

We present a new method for studying theoretically the structural properties of semiconductor alloys. The alloy is considered as a perturbation with respect to a periodic virtual crystal, and the relevant energies calculated by density-functional perturbation theory. We show that —up to second order in the perturbation— the energy of the alloy is equivalent to that of a lattice gas with only two-body interactions. The interaction constants of the lattice gas are particular linear response functions of the virtual crystal, which can be determined from first principles. Once the interaction constants have been calculated, the finite-temperature properties of the alloy can be studied rather inexpensively by MonteCarlo simulations on the lattice gas. As an application, we consider the case of Si_xGe_{1-x}. A comparison with traditional self-consistent calculations for some simple ordered structures demonstrates that the accuracy of the perturbative approach is in this case of the same order as that of state-of-the-art density-functional calculations. Ignoring lattice relaxation, the range of the interactions is very short. Atomic relaxation renormalizes the interactions and makes them rather long range, propagating mainly along the bond chains. Monte Carlo simulations show that Si_xGe_{1-x} is a model random alloy with a miscibility gap below ≈ 170 K. The bond length distribution displays three well defined peaks whose positions

Structural and Phase Stability of Alloys
Edited by J.L. Morán-López *et al.*, Plenum Press, New York, 1992

depend on composition, but not on temperature. The resulting lattice parameter follows very closely Vegard's law.

I. Introduction

The last decade has witnessed a spectacular growth of our ability to calculate the physical properties of *real materials* (such as metals, semiconductors, and insulators) without resorting to semiempirical models, but starting directly from a quantum description of the microscopic constituents (electrons and ions) and their interactions. Density-Functional Theory [1] (DFT) within the Local-Density Approximation [2] (LDA) has proved to be a valuable theoretical tool for predicting the electronic ground-state properties* of a large class of materials. [3,4] Technically, the implementation of a DFT calculation amounts to the solution of the Schrödinger equation for a system of independent electrons whose potential depends on its own eigenfunctions. Despite the enormous conceptual and practical simplification achieved by reducing the many-body problem to one of noninteracting particles, the solution of the one-electron Schrödinger equation for an extended system still remains a formidable task. The introduction of linearized all-electron techniques [5,3] (such as LAPW and LMTO) and of accurate norm-conserving pseudopotentials [6,4] has allowed to solve some of the difficulties related to a proper treatment of core electrons. The availability of powerful vector and parallel machines, together with significant algorithmic advances obtained using iterative techniques for diagonalizing large matrices [7,8] and/or global-optimization strategies for minimizing the energy functional in DFT, [9,8] have considerably widened the scope of first-principles calculations for real systems. In spite of all these achievements, the computational complexity of a DFT calculation for a system of N inequivalent atoms scales as N^3, and the maximum size of a system which can be studied with state-of-the art techniques is of the order of a few tens of atoms. Because of this, the study of disordered systems has long been considered beyond the scope of *ab-initio* investigations, due to the large number of inequivalent atoms necessary to simulate disorder. It is by now clear that the study of such complex systems requires not only powerful computer resources, but especially new theoretical ideas and computational techniques.

Semiconductor alloys are substitutionally disordered systems in which the atoms occupy the lattice sites of a diamond or zincblende lattice. In general, the pure constituents do not have the same lattice parameter, and the underlying diamond or zincblende lattice is distorted in order to accommodate for macroscopic and internal strains. At low temperature, an alloy can either separate into the pure constituents or form an ordered structure. As the temperature is raised, ordered or segregated phases transform into a homogeneous disordered structure in which lattice sites are occupied randomly by one or the other of the elements. The great variety of possible ordered structures and the additional degrees of freedom with respect to the pure crystalline solids suggest the possibility of tuning and controlling the physical properties of crystals by alloying. For this reason, semiconductor alloys have a great technological interest in view of their application in the design of microelectronic devices. Besides their technological interest, semiconductor alloys also deserve attention from a more

* By *electronic ground-state properties* we mean all those properties which can be obtained from the ground-state wavefunction of the electrons in the field of *fixed* nuclei. In the Born-Oppenheimer approximation, these include structural and vibrational properties, as well as static response functions (such as dielectric, piezoelectric properties etc.).

fundamental point of view. These systems are characterized by a positive formation enthalpy:[10,11] so they tend to segregate into pure components at low temperature, and they have long been considered as model random alloys at room temperature.[12] However, it has been recently found that most of them exhibit long-range order when grown homogeneously by gas-phase epitaxy in certain temperature ranges.[13–17] Though it has been proposed that kinetic and/or surface effects may play an important role in the ordering,[17–19] the mechanisms responsible for it are not yet fully clarified.

The microscopic description of alloys has until now relied on semiempirical models,[20] in which the alloy energy is assumed to depend on a few many-body short-range interactions whose value is usually fitted to some experimentally known properties or to DFT calculations performed for some ordered structures. This approach has been given a firm theoretical basis by Connoly and Williams,[21] and recently applied to semiconductor alloys by Zunger and coworkers.[11] Once the energy has been so parameterized, the thermodynamics of the alloys is often studied by the Cluster Variation Method of Kikuchi.[22]

The substitutional character of the disorder as well as the chemical similarity between the constituent atoms make semiconductor alloys ideal candidates for being studied by perturbative methods. This allows to greatly simplify the treatment of disorder, and makes these systems accessible to computational study, through recently established techniques.[23–25] In this paper we show that whenever an alloy is described as a perturbation with respect to an ordered reference system, its energetics can be studied rather inexpensively using appropriate response functions of the reference system. Conceptually, our procedure is similar to that used to study simple metals (both ordered and disordered) via pseudopotential perturbation theory[26,27] starting from the homogeneous electron gas as the reference system. The valence charge distribution in semiconductors is however rather inhomogeneous, and jellium is a poor reference system to start from. Therefore, we describe pure semiconductors without any substantial approximation but the LDA, and we use perturbation theory only to describe the small differences between the alloy and some suitably chosen average periodic crystal (*virtual crystal*). The response of the virtual crystal to the *alchemical* perturbation consisting in the replacement of a virtual ion with a real one provides all the information necessary to calculate the relevant energy differences up to second order in the perturbation. The energy of any given microscopic realization of the alloy is described–*exactly*, to second order in the perturbation–from an expression similar to that of a lattice gas with only two-body interactions.[28] Once the equivalence between the alloy and a lattice gas has been established, the latter can be used to calculate very inexpensively the energy of any microscopic realization of the alloy, and hence to calculate its thermodynamical properties by standard MonteCarlo (MC) simulations.

As a first application, we consider the case of $\mathrm{Si}_x\mathrm{Ge}_{1-x}$, which is the simplest example of lattice-mismatched semiconductor alloy, and whose atomic constituent are chemically similar enough to allow their difference to be accurately treated by perturbation theory. MC simulations of the finite-temperature properties of bulk $\mathrm{Si}_x\mathrm{Ge}_{1-x}$ confirm that this system is a model random alloy above $T_c \approx 170$ K, thus indicating that the origin of the observed ordered phases in samples grown by molecular-phase epitaxy[14–16] should be searched in surface and/or kinetic effects.[17]

This paper is organized as follows: In Sec. II we introduce the computational alchemy approach to disorder; in Sec. III we show how the relevant response functions can be calculated from density-functional perturbation theory; in Sec. IV we present our results for $\mathrm{Si}_x\mathrm{Ge}_{1-x}$ alloys, including some details of our MonteCarlo simulations; Sec. V finally contains our conclusions.

II. Thermodynamics of Substitutional Alloys from Computational Alchemy

The thermodynamic properties of any system are given by standard Boltzmann averages:

$$\langle \mathcal{A} \rangle = \frac{\sum_C \mathcal{A}(C) \exp(-\frac{E(C)}{k_B T})}{\sum_C \exp(-\frac{E(C)}{k_B T})}, \tag{1}$$

where C denotes a microscopic configuration of the system, $E(C)$ its energy, $\mathcal{A}(C)$ the corresponding value of the relevant observable, T is the temperature, and k_B is the Boltzmann constant. Within DFT-LDA, the energy of a given microscopic configuration of the alloy is determined by the self-consistent solution of the Kohn-Sham equations: [2]

$$H_{\text{SCF}} \psi_n(\mathbf{r}) \equiv \left(-\frac{\hbar^2}{2m}\Delta + V_{\text{bare}}(\mathbf{r}) + V_{\text{ee}}(\mathbf{r}) \right) \psi_n(\mathbf{r}) = \epsilon_n \psi_n(\mathbf{r}), \tag{2a}$$

$$V_{\text{ee}}(\mathbf{r}) = e^2 \int \frac{n(\mathbf{r}')}{|\mathbf{r} - \mathbf{r}'|} d\mathbf{r}' + \mu_{\text{xc}}(n(\mathbf{r})), \tag{2b}$$

$$n(\mathbf{r}) = 2 \sum_n |\psi_n(\mathbf{r})|^2 \theta(\epsilon_{\text{F}} - \epsilon_n), \tag{2c}$$

where V_{bare} is the sum of the (pseudo) potentials of the bare ions acting on the electrons, $\mu_{\text{xc}}(n)$ is the exchange-correlation contribution to the chemical potential of a homogeneous electron gas at density n, θ is the unit step function, and ϵ_{F} is the Fermi energy of the system, which is determined by the charge-neutrality condition. The computational labor and computer memory necessary to solve Eqs. (2) scale as the third and second power of the number of inequivalent atoms, respectively. The maximum size of systems which can be studied by current computers and algorithms through Eqs. (2) is of the order some tens, and only a few configurations can be sampled. In order to study larger systems and to sample a statistically significant number of configurations, one has to resort to some methods which avoid the explicit solution of Eqs. (2) for each configuration.

A microscopic configuration of a substitutional A/B alloy is fully specified by a set of Ising-like variables, $C \equiv \{\sigma_{\mathbf{R}}\}$, whose value is $\sigma_{\mathbf{R}} = \pm 1$ if the lattice site $\mathbf{R}$ is occupied by an atom of species A or B respectively. The most general expression for the configurational energy of a system of N such spins is:

$$\begin{aligned} E(C) \equiv E[\{\sigma_{\mathbf{R}}\}] = E_0 + K \sum_{\mathbf{R}} \sigma_{\mathbf{R}} + \overbrace{\frac{1}{2} \sum_{\mathbf{R}\mathbf{R}'} J(\mathbf{R} - \mathbf{R}') \sigma_{\mathbf{R}} \sigma_{\mathbf{R}'}}^{\text{two-body}} + \\ \underbrace{\frac{1}{6} \sum_{\mathbf{R}\mathbf{R}'\mathbf{R}''} L(\mathbf{R} - \mathbf{R}', \mathbf{R}' - \mathbf{R}'', \mathbf{R}'' - \mathbf{R}) \sigma_{\mathbf{R}} \sigma_{\mathbf{R}'} \sigma_{\mathbf{R}''}}_{\text{three-body}} + \quad \cdots \quad + \text{N-body terms.} \end{aligned} \tag{3}$$

The number of independent many-spin interactions constants appearing in Eq. (3) is equal to the number of independent configurations, *i.e.* 2^N. However, if some assumptions are made concerning the maximum order of the interactions (e.g. only interactions up to a certain number of spins are retained) and on their range, Eq. (3) can be used to *parameterize* the energetics of the alloy, and the relevant interactions can be fitted to some known experimental data, or obtained by comparison with DFT calculations made for a few ordered configurations of the alloy.[21] Once the interaction constants have been so obtained, Eqs. (1) and (3) can be used for determining the thermodynamical properties of the alloy through some approximate statistical-mechanical scheme (such as e.g. the cluster variation method[22]) or by direct Monte Carlo simulation. This approach has been recently pursued by Zunger and collaborators,[11] and applied to a variety of semiconductor alloys. In this paper we follow a different but related approach, where one does not make any assumptions on the order and range of spin-spin interactions, but only on the *strength* of the alchemical perturbation which "transforms" the ions of the atomic species A into B ions. Though our approach could be formulated in principle in an all-electron framework, the difference between the potentials of chemically similar ions can be most easily treated by perturbation theory using pseudopotentials to eliminate core electrons, and we will stick to a pseudopotential scheme throughout the present paper.

II.1. Mapping the alloy onto a lattice gas

Let us ignore for the moment the fact that the AA, BB, and AB bonds may have different lengths, and let us assume therefore that all the ions lie at the knots of a periodic lattice, $\{\mathbf{R}\}$. Under this hypothesis, the bare alloy potential reads:

$$
\begin{aligned}
V_{\mathrm{bare}}(\mathbf{r}) &= \sum_{\mathbf{R}} \left(\frac{1+\sigma_{\mathbf{R}}}{2} v_A(\mathbf{r}-\mathbf{R}) + \frac{1-\sigma_{\mathbf{R}}}{2} v_B(\mathbf{r}-\mathbf{R}) \right) \\
&= \underbrace{\sum_{\mathbf{R}} \frac{1}{2}\Big(v_A(\mathbf{r}-\mathbf{R}) + v_B(\mathbf{r}-\mathbf{R}) \Big)}_{V_0(\mathbf{r})} + \underbrace{\sum_{\mathbf{R}} \sigma_{\mathbf{R}} \frac{1}{2}\Big(v_A(\mathbf{r}-\mathbf{R}) - v_B(\mathbf{r}-\mathbf{R}) \Big)}_{\Delta V(\mathbf{r}) \equiv \sum_{\mathbf{R}} \sigma_{\mathbf{R}} \Delta v(\mathbf{r}-\mathbf{R})}, \qquad (4)
\end{aligned}
$$

where $v_{A,B}$ are the ionic pseudopotentials of the two atomic species. Eq. (4) naturally suggests the partition of the alloy potential into an unperturbed, periodic contribution, $V_0(\mathbf{r})$ (the *virtual crystal* potential), and a perturbation, $\Delta V(\mathbf{r})$, which is the sum of localized contributions of opposite signs, according to the chemical species of the atom sitting at $\mathbf{R}$. In order to keep the notation simple, in Eq. (4) and in subsequent related equations we consider only Bravais lattices and suppose that the electron-ion potential is local. The extension to lattices with a basis and to non local potentials is straightforward.

The energy difference between the alloy and the reference virtual crystal can be calculated from the derivative with respect to the strength of the perturbation through the Hellmann-Feynman theorem.[29] Suppose that the *bare* external potential acting on the electrons, $V_{\boldsymbol{\lambda}}(\mathbf{r})$, depends continuously on some parameters $\boldsymbol{\lambda} \equiv \{\lambda_i\}$. The Hellmann-Feynman theorem states that the "force" associated with the variation of the external parameters $\boldsymbol{\lambda}$ is given by the ground-state expectation value of the derivative of $V_{\boldsymbol{\lambda}}$:

$$
\frac{\partial E_{\boldsymbol{\lambda}}}{\partial \lambda_i} = \int n_{\boldsymbol{\lambda}}(\mathbf{r}) \frac{\partial V_{\boldsymbol{\lambda}}(\mathbf{r})}{\partial \lambda_i}\, d\mathbf{r}, \qquad (5)
$$

where $E_{\boldsymbol{\lambda}}$ is the electron ground-state energy relative to given values of the $\boldsymbol{\lambda}$ parameters, and $n_{\boldsymbol{\lambda}}(\mathbf{r})$ is the corresponding electron density distribution. Total energy variations are obtained from Eq. (5) by integration. In order to have energy variations correct up to second order in $\boldsymbol{\lambda}$, it is necessary that the right hand side of Eq. (5) is correct to linear order:

$$\frac{\partial E_{\boldsymbol{\lambda}}}{\partial \lambda_i} = \int \left(n_0(\mathbf{r}) \frac{\partial V_{\boldsymbol{\lambda}}(\mathbf{r})}{\partial \lambda_i} + \sum_j \lambda_j \frac{\partial n_{\boldsymbol{\lambda}}(\mathbf{r})}{\partial \lambda_j} \frac{\partial V_{\boldsymbol{\lambda}}(\mathbf{r})}{\partial \lambda_i} + n_0(\mathbf{r}) \sum_j \lambda_j \frac{\partial^2 V_{\boldsymbol{\lambda}}(\mathbf{r})}{\partial \lambda_i \partial \lambda_j} \right) d\mathbf{r} + \mathcal{O}(\boldsymbol{\lambda}^2), \quad (6)$$

all the derivatives being calculated at $\boldsymbol{\lambda} = 0$. Integration of Eq. (6) gives:

$$E_{\boldsymbol{\lambda}} = E_0 + \sum_i \lambda_i \int n_0(\mathbf{r}) \frac{\partial V_{\boldsymbol{\lambda}}(\mathbf{r})}{\partial \lambda_i} d\mathbf{r} + \frac{1}{2} \sum_{ij} \lambda_i \lambda_j \underbrace{\int \left(\frac{\partial n_{\boldsymbol{\lambda}}(\mathbf{r})}{\partial \lambda_j} \frac{\partial V_{\boldsymbol{\lambda}}(\mathbf{r})}{\partial \lambda_i} + n_0(\mathbf{r}) \frac{\partial^2 V_{\boldsymbol{\lambda}}(\mathbf{r})}{\partial \lambda_i \partial \lambda_j} \right) d\mathbf{r}}_{\frac{\partial^2 E_{\boldsymbol{\lambda}}}{\partial \lambda_i \partial \lambda_j}} + \mathcal{O}(\boldsymbol{\lambda}^3). \quad (7)$$

According to Eq. (7), the linear response $\partial n / \partial \lambda$ yields energy variations up to second order in $\boldsymbol{\lambda}$ (actually, up to *third* order, as shown in Ref. 30). Suppose now that the $\boldsymbol{\lambda}$ parameters represent ion substitutions, $\sigma_{\mathbf{R}}$. Inserting Eq. (4) into Eq. (7) we obtain:

$$E_{\text{alloy}}[\{\sigma\}] = E_0 + K \sum_{\mathbf{R}} \sigma_{\mathbf{R}} + \frac{1}{2} \sum_{\mathbf{R}\mathbf{R}'} J(\mathbf{R} - \mathbf{R}') \sigma_{\mathbf{R}} \sigma_{\mathbf{R}'} + \mathcal{O}(\Delta v^3), \quad (8)$$

where:

$$K = \int n_0(\mathbf{r}) \Delta v(\mathbf{r}) d\mathbf{r}, \quad (9a)$$

$$J(\mathbf{R}) = \int \Delta v(\mathbf{r} - \mathbf{R}) \Delta n(\mathbf{r}) d\mathbf{r}, \quad (9b)$$

and $\Delta n(\mathbf{r})$ is the linear density response to $\Delta v(\mathbf{r})$. Eq. (8) shows that the alloy problem can be mapped onto a lattice gas with only two-body interactions (not necessarily short range), the mapping being *exact* up to second order in the potential differences between the constituent atoms. The interaction constants of the lattice gas are particular response functions of the virtual crystal (Eq. (9b)).

II.2 Including lattice relaxation

So far, we have neglected all bond-length fluctuations which arise from chemical disorder. The problem of finding the bond-length distribution in a substitutional disordered system has been solved in the linear regime long time ago.[31–33] Here we use a similar approach, and extend it to determine the contribution of atomic relaxation to the energy of the alloy. Let us indicate by $\{\mathbf{u}_{\mathbf{R}}\}$ the displacements of the atoms from their ideal lattice positions. The virtual crystal is in equilibrium with respect to atomic displacement; furthermore, the equilibrium values of the $\mathbf{u}$'s in the alloy are to leading order linear in Δv. As a consequence, the most general expression of the alloy energy as a function of the σ's and the $\mathbf{u}$'s reads:

$$
\begin{aligned}
E[\{\sigma_{\mathbf{R}}\},\{\mathbf{u}_{\mathbf{R}}\}] = E_0 + K\sum_{\mathbf{R}}\sigma_{\mathbf{R}} + \frac{1}{2}\sum_{\mathbf{R}\mathbf{R}'} J(\mathbf{R}-\mathbf{R}')\sigma_{\mathbf{R}}\sigma_{\mathbf{R}'} \\
+\frac{1}{2}\sum_{\mathbf{R}\mathbf{R}'} \mathbf{u}_{\mathbf{R}}\cdot\mathbf{\Phi}(\mathbf{R}-\mathbf{R}')\cdot\mathbf{u}_{\mathbf{R}'} - \sum_{\mathbf{R}\mathbf{R}'}\mathbf{u}_{\mathbf{R}}\cdot\mathbf{F}(\mathbf{R}-\mathbf{R}')\,\sigma_{\mathbf{R}'} + \mathcal{O}(\Delta v^3),
\end{aligned}
\tag{10}
$$

where the J's are the interaction constants defined by Eq. (9); $\mathbf{\Phi}$ is the matrix of the virtual crystal interatomic force constants[24]; and $\mathbf{F}(\mathbf{R})$ is the force acting on the $\mathbf{R}$-th ion of the virtual crystal when the virtual ion at the origin is replaced by a real one (the $\mathbf{F}$'s are sometimes referred to as Kanzaki's forces[31]). Of course, the $\mathbf{u}$'s are not independent variables. Their actual values are those that–for a given microscopic configuration $\{\sigma_{\mathbf{R}}\}$–minimize the energy given by Eq. (10):

$$
-\frac{\partial E}{\partial \mathbf{u}_{\mathbf{R}}} = -\sum_{\mathbf{R}'}\mathbf{\Phi}(\mathbf{R}-\mathbf{R}')\cdot\mathbf{u}_{\mathbf{R}'} + \sum_{\mathbf{R}'}\mathbf{F}(\mathbf{R}-\mathbf{R}')\sigma_{\mathbf{R}'} = 0. \tag{11}
$$

Eq. (11) simply states that at equilibrium the total force (Kanzaki's plus harmonic) acting on individual atoms vanish. Solving Eq. (11) with respect to the $\mathbf{u}$'s gives:

$$
\mathbf{u}_{\mathbf{R}} = \sum_{\mathbf{R}'\mathbf{R}''}\mathbf{\Phi}^{-1}(\mathbf{R}-\mathbf{R}')\cdot\mathbf{F}(\mathbf{R}'-\mathbf{R}'')\sigma_{\mathbf{R}''}. \tag{12}
$$

Inserting Eq. (12) into Eq. (10), one arrives at an expression for the energy similar to Eq. (8)—where lattice relaxation was neglected—in which relaxation does not appear explicitly, but the spin-spin interactions are renormalized by it. The final expression reads:

$$
E[\{\sigma_{\mathbf{R}}\}] = E_0 + K\sum_{\mathbf{R}}\sigma_{\mathbf{R}} + \frac{1}{2}\sum_{\mathbf{R}\mathbf{R}'}\widehat{J}(\mathbf{R}-\mathbf{R}')\sigma_{\mathbf{R}}\sigma_{\mathbf{R}'}, \tag{13}
$$

where

$$
\widehat{J}(\mathbf{R}-\mathbf{R}') = J(\mathbf{R}-\mathbf{R}') - \sum_{\mathbf{R}''\mathbf{R}'''}\mathbf{F}(\mathbf{R}''-\mathbf{R})\cdot\mathbf{\Phi}^{-1}(\mathbf{R}''-\mathbf{R}''')\cdot\mathbf{F}(\mathbf{R}'''-\mathbf{R}'). \tag{14}
$$

The quantities entering Eq. (14) are second derivatives of the energy with respect to ion substitutions $(\partial/\partial\sigma_{\mathbf{R}})$ and displacements $(\partial/\partial\mathbf{u}_{\mathbf{R}})$. According to Eq. (7), they can be calculated from the linear response of the virtual crystal to each of the two perturbations separately. Let $\Delta v^D_\alpha(\mathbf{r})$ and $\Delta v^S(\mathbf{r})$ be the bare perturbations due to the displacement in the α-th direction and to the substitution of a single atom in the virtual crystal respectively, and $\Delta n^D_\alpha(\mathbf{r})$ and $\Delta n^S(\mathbf{r})$ the corresponding density responses. According to Eq. (7), the relevant interaction constants are given by:

$$
\begin{aligned}
J(\mathbf{R}-\mathbf{R}') &= \int \Delta v^S(\mathbf{r}-\mathbf{R})\Delta n^S(\mathbf{r}-\mathbf{R}')\,d\mathbf{r} \\
\Phi_{\alpha\beta}(\mathbf{R}-\mathbf{R}') &= \int \Delta v^D_\alpha(\mathbf{r}-\mathbf{R})\Delta n^D_\beta(\mathbf{r}-\mathbf{R}')\,d\mathbf{r} \qquad (\mathbf{R}\neq\mathbf{R}') \\
F_\alpha(\mathbf{R}-\mathbf{R}') &= -\int \Delta v^D_\alpha(\mathbf{r}-\mathbf{R})\Delta n^S(\mathbf{r}-\mathbf{R}')\,d\mathbf{r} \\
&= -\int \Delta v^S(\mathbf{r}-\mathbf{R})\Delta n^D_\alpha(\mathbf{r}-\mathbf{R}')\,d\mathbf{r}.
\end{aligned}
\tag{15}
$$

According to Eq. (7), the expression of $\Phi_{\alpha\beta}$ for $\mathbf{R} = \mathbf{R}'$ is more complicated than reported in Eq. (15); the missing term for $\mathbf{R} = \mathbf{R}'$ is easily recovered by explicitly imposing translational invariance. Using translational invariance, Eqs. (12-14) are most easily calculated in reciprocal space, where convolutions are replaced by simple products:

$$\widetilde{\mathbf{u}}_{\mathbf{q}} = \widetilde{\mathbf{\Phi}}^{-1}(\mathbf{q}) \cdot \widetilde{\mathbf{F}}(\mathbf{q})\widetilde{\sigma}_{\mathbf{q}}, \tag{16}$$

$$\widehat{J}(\mathbf{q}) = \widetilde{J}(\mathbf{q}) - \widetilde{\mathbf{F}}(\mathbf{q})^{+} \cdot \widetilde{\mathbf{\Phi}}^{-1}(\mathbf{q}) \cdot \widetilde{\mathbf{F}}(\mathbf{q}), \tag{17}$$

$$E[\{\sigma_{\mathbf{R}}\}] = E_0 + KN\sigma_{\mathbf{q}=0} + \frac{1}{2}\sum_{\mathbf{q}} \widehat{J}(\mathbf{q})\widetilde{\sigma}_{\mathbf{q}}\widetilde{\sigma}^{*}_{\mathbf{q}}, \tag{18}$$

where the tilde indicates Fourier transform.

II.3. Macroscopic effects of lattice mismatch

The same mechanism which determines fluctuations of the interatomic distances on the *microscopic* scale, also determines the dependence of the alloy equilibrium volume upon composition (*macroscopic* relaxation) and the two kinds of relaxation could be treated on the same footing by our method. In fact, the $\mathbf{q} \to 0$ limits of $\widetilde{\mathbf{\Phi}}(\mathbf{q})$ and of $\widetilde{\mathbf{F}}(\mathbf{q})$ are related to the elastic constants and to the macroscopic stress induced by an alchemical perturbation, respectively. Therefore, Eq. (18) can be used directly for dealing with macroscopic relaxation, using parameters computed at a given reference volume. This procedure is perfectly consistent, and correct when anharmonic effects are negligible. However, it is more practical and accurate to deal with microscopic and macroscopic relaxation effects separately, according to the procedure given below which also accounts for some anharmonic effects.

In order to treat the effects of macroscopic relaxation, we define the alloy formation energy, $\Delta E_{\mathrm{alloy}}[\{\sigma_{\mathbf{R}}\}, \Omega]$, as the difference between the energy of the alloy in a given configuration $\{\sigma_{\mathbf{R}}\}$ and at a given molar volume Ω, and the appropriate average of those of the pure components at their respective equilibrium volumes:

$$\Delta E_{\mathrm{alloy}}[\{\sigma_{\mathbf{R}}\}, \Omega] = E_{\mathrm{alloy}}[\{\sigma_{\mathbf{R}}\}, \Omega] - xE_A[\Omega_A] - (1-x)E_B[\Omega_B], \tag{19}$$

where $E_{A,B}$ and $\Omega_{A,B}$ are the equilibrium energies and volumes of the pure materials, and $x = \frac{\langle\sigma\rangle+1}{2}$ is the molar composition of the A_xB_{1-x} alloy. It is now convenient to split Eq. (19) into an elastic contribution, $\Delta E_{\mathrm{elast}}$ -which does not depend on the microscopic configuration— plus a configurational contribution, $\Delta E_{\mathrm{config}}$:

$$\Delta E_{\mathrm{alloy}}[\{\sigma_{\mathbf{R}}\}, \Omega] = \Delta E_{\mathrm{elast}}[x, \Omega] + \Delta E_{\mathrm{config}}[\{\sigma_{\mathbf{R}}\}, \Omega], \tag{20a}$$

where

$$\Delta E_{\mathrm{elast}}[x, \Omega] = x\big(E_A(\Omega) - E_A(\Omega_A)\big) + (1-x)\big(E_B(\Omega) - E_B(\Omega_B)\big) \tag{20b}$$

is the energy paid to bring the two pure materials from their equilibrium volumes to the volume of the alloy, and

$$\Delta E_{\mathrm{config}}[\{\sigma_{\mathbf{R}}\}, \Omega] = E_{\mathrm{alloy}}[\{\sigma_{\mathbf{R}}\}, \Omega] - xE_A(\Omega) - (1-x)E_B(\Omega) \tag{20c}$$

is the formation energy of the alloy at a *fixed volume*. The elastic energy $\Delta E_{\mathrm{elast}}$ can be easily calculated from the equations of state for the pure components, whereas the configurational energy ΔE_{config} is given by an expression similar to Eq. (18), where the interaction constants appropriate to the actual volume Ω are used, and the term $\mathbf{q} = 0$ is not renormalized since the macroscopic relaxation is already accounted for exactly in ΔE_{elast}.

III. Density-Functional Perturbation Theory

Eqs. (15) show that the relevant interaction constants necessary to map the alloy onto a lattice gas can be calculated from the linear density response of the virtual crystal to appropriate *isolated* perturbations. A direct *real-space* calculation of such responses could be done by considering a system of many perturbing centers arranged on a periodic superlattice whose spacing is larger that the spatial range of the responses. Such a calculation is impractical because it would require a numerical effort proportional to the third power of the volume of the superlattice unit cell. A more convenient approach is to work in reciprocal space, taking advantage of the translational invariance of the unperturbed system (*i.e.* the virtual crystal). Let $\Delta v^{\mathbf{q}}(\mathbf{r})$ be the component of the bare localized perturbation, with wavevector $\mathbf{q}$ in the Brillouin Zone (BZ) of the virtual crystal:

$$\Delta v^{\mathbf{q}}(\mathbf{r}) = e^{i\mathbf{q}\cdot\mathbf{r}} \sum_{\mathbf{G}} \Delta\widetilde{v}(\mathbf{q}+\mathbf{G}) e^{i\mathbf{G}\cdot\mathbf{r}}, \tag{21}$$

where $\Delta\widetilde{v}(\mathbf{q}+\mathbf{G})$ is the Fourier transform of $\Delta v(\mathbf{r})$, and the $\mathbf{G}$'s are reciprocal-lattice vectors of the virtual crystal. Because of symmetry, the corresponding electron-density response and variation in the self-consistent potential have the same periodicity:

$$\Delta n^{\mathbf{q}}(\mathbf{r}) = e^{i\mathbf{q}\cdot\mathbf{r}} \sum_{\mathbf{G}} \Delta\widetilde{n}(\mathbf{q}+\mathbf{G}) e^{i\mathbf{G}\cdot\mathbf{r}}, \tag{22a}$$

$$\begin{aligned} \Delta V^{\mathbf{q}}_{\text{SCF}}(\mathbf{r}) &= \Delta v^{\mathbf{q}}(\mathbf{r}) + \Delta V^{\mathbf{q}}_{\text{ee}}(\mathbf{r}) \\ &= e^{i\mathbf{q}\cdot\mathbf{r}} \sum_{\mathbf{G}} \Delta\widetilde{v}_{\text{SCF}}(\mathbf{q}+\mathbf{G}) e^{i\mathbf{G}\cdot\mathbf{r}}, \end{aligned} \tag{22b}$$

$$\begin{aligned} \Delta V^{\mathbf{q}}_{\text{ee}}(\mathbf{r}) &= e^2 \int \frac{\Delta n^{\mathbf{q}}(\mathbf{r}')}{|\mathbf{r}-\mathbf{r}'|} d\mathbf{r}' + \Delta n^{\mathbf{q}}(\mathbf{r}) \left[\frac{d\mu_{\text{xc}}}{dn}\right]_{n=n_0(\mathbf{r})} \\ &= e^{i\mathbf{q}\cdot\mathbf{r}} \sum_{\mathbf{G}} \Delta\widetilde{v}_{\text{ee}}(\mathbf{q}+\mathbf{G}) e^{i\mathbf{G}\cdot\mathbf{r}}. \end{aligned} \tag{22c}$$

Eqs. (22b,c) yield the variation of the selfconsistent potential, once the electron-density response is known. The latter is given as an explicit functional of the former, by an RPA-like[34] expression[23]:

$$\Delta\widetilde{n}(\mathbf{q}+\mathbf{G}) = \frac{4}{N\Omega} \sum_{\mathbf{k}} \sum_{c,v} \frac{\langle \psi_{v,\mathbf{k}} | e^{-i(\mathbf{q}+\mathbf{G})\mathbf{r}} | \psi_{c,\mathbf{k}+\mathbf{q}} \rangle \langle \psi_{c,\mathbf{k}+\mathbf{q}} | \Delta V^{\mathbf{q}}_{\text{SCF}} | \psi_{v,\mathbf{k}} \rangle}{\epsilon_{v,\mathbf{k}} - \epsilon_{c,\mathbf{k}+\mathbf{q}}}, \tag{23}$$

where Ω is the volume of the unit cell, v and c indicate valence and conduction bands respectively, and the sum over $\mathbf{k}$ covers the first BZ. Eqs. (22b,c) and (23) form a system which can be solved iteratively.

For computational convenience, it is desirable to avoid the sum over conduction bands of Eq. (23). This can be achieved by rewriting Eq. (23) in the following way:

$$\Delta\tilde{n}(\mathbf{q}+\mathbf{G}) = \frac{4}{N\Omega}\sum_{\mathbf{k}}\sum_{v}\langle\psi_{v,\mathbf{k}}|e^{-i(\mathbf{q}+\mathbf{G})\mathbf{r}}P_c G(\epsilon_{v,\mathbf{k}})P_c\Delta V_{\mathrm{SCF}}^{\mathbf{q}}|\psi_{v,\mathbf{k}}\rangle, \quad (24)$$

where P_c is the projector over the conduction-state manifold, and $G(\epsilon) = 1/(\epsilon - H_{\mathrm{SCF}})$ is the one-electron Green's function of the unperturbed system. To evaluate Eq. (24), we further rewrite it as:

$$\Delta\tilde{n}(\mathbf{q}+\mathbf{G}) = \frac{4}{N\Omega}\sum_{\mathbf{k}}\sum_{v}\langle\psi_{v,\mathbf{k}}|e^{-i(\mathbf{q}+\mathbf{G})\mathbf{r}}P_c|\Delta\psi_{v,\mathbf{k}+\mathbf{q}}\rangle, \quad (25)$$

where $\Delta\psi_{v,\mathbf{k}+\mathbf{q}}$ is a solution of the linear system:

$$\left[\epsilon_{v,\mathbf{k}} - H_{\mathrm{SCF}}\right]|\Delta\psi_{v,\mathbf{k}+\mathbf{q}}\rangle = P_c\Delta V_{\mathrm{SCF}}^{\mathbf{q}}|\psi_{v,\mathbf{k}}\rangle. \quad (26)$$

The linear system (26) has an infinite number of solutions because the determinant of $\left[\epsilon_{v,\mathbf{k}} - H_{\mathrm{SCF}}\right]$ vanishes, and the vector on the left hand site is orthogonal to the null space of $\left[\epsilon_{v,\mathbf{k}} - H_{\mathrm{SCF}}\right]$. In practice, $\Delta\psi_{v,\mathbf{k}+\mathbf{q}}$ is defined within a multiple of $\psi_{v,\mathbf{k}}$. As $\Delta\psi_{v,\mathbf{k}+\mathbf{q}}$ enters Eq. (26) only through its projection onto the conduction-state manifold, such an indeterminacy does not affect the final result. Depending on the size of the basis set, Eq. (26) can be solved either by factorization techniques, or by iterative methods. If factorization techniques are used, the best choice is to tridiagonalize first H_{SCF}, and solve then the linear system in the basis where it is tridiagonal. The advantage is that this allows to perform just one factorization for *all* linear systems corresponding to different values of $\epsilon_{v,\mathbf{k}}$. In both cases, the calculation of all the $\Delta\psi$ functions requires a numerical labor comparable to that needed for a single SCF iteration for the *unperturbed* system.

When calculating the response to an isolated perturbation, all $\mathbf{q}$ vectors in the BZ must be sampled. Sampling the $\mathbf{q}$'s on a regular mesh spaced by Δq is mathematically equivalent to consider a superlattice of localized perturbations whose real-space spacing is $2\pi/\Delta q$: as a consequence, the number of $\mathbf{q}$ points in the virtual-crystal BZ is proportional to the volume of the equivalent real-space supercell. The use of reciprocal-space techniques thus allows to obtain the response to a localized perturbation with a numerical effort (CPU time and memory requirement) which scales *linearly* with the volume of the necessary supercell. Real-space interactions are finally calculated by Fourier transforming their reciprocal-space counterparts.

IV. Results for Si_xGe_{1-x}

As an application of the preceeding ideas we have considered the Si_xGe_{1-x} alloy. Si_xGe_{1-x} has long been believed to be a model random alloy at room temperature.[35,36] Recently, it has been observed that Si_xGe_{1-x} displays long-range order along the (111) direction, when grown by Molecular Beam Epitaxy over a Si substrate, at temperatures of the order of 450 C. Ordering was first attributed to the stabilizing effects of strain,[37] but it has been later observed also in unstrained samples.[16] Although a *reversible* order/disorder phase transition has been reported by the authors of Ref. 14, other groups were unable to detect such a reversible transition.[16,17] It is worth noticing that — among the two structures compatible with the symmetry of the ordered phase (denoted by 'R1' and 'R2' in Table I)— the one which corresponds to the observed intensity pattern of the electron diffraction data has the higher energy ('R2').[16] This seems to

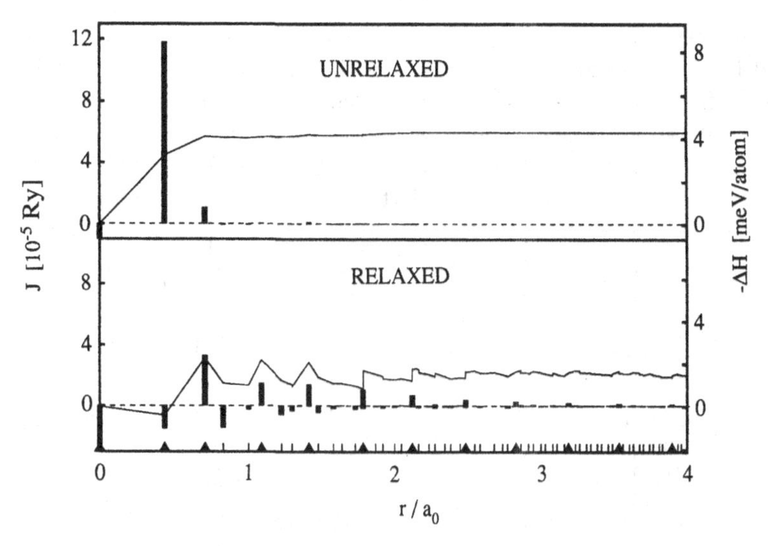

Figure 1. Atomic interaction constants (thick vertical bars) calculated neglecting ("unrelaxed") and including ("relaxed") lattice relaxation, as functions of the interatomic separation. Thin vertical bars on the horizontal azis indicate shells of neighbors. Shells lying on the same bond chain as the atom at the origin are indicated by solid triangle. The continuous line indicates the formation energy of the random alloy in the regular solution approximation, obtained by truncating the interactions at a given shell of neighbors.

rule out the bulk thermodynamical origin of the ordered phase, as suggested in Ref. 17. To complete the rather confusing picture of the status of the experimental knowledge, we mention a recent paper by von Känel and collaborators,[38] reporting that the ordered structure observed during growth ('R2') is irreversibly destroyed by heating, whereas the 'R1' structure appears and disappears reversibly by subsequent annealing cycles. In this paper we readdress the question of the possible bulk thermodynamical origin of the ordered phase of Si_xGe_{1-x} by performing accurate Monte Carlo simulations, along the lines sketched above.

The technical ingredients of our calculations are as follows. We have used norm-conserving pseudopotentials and plane-wave basis sets up to a kinetic-energy cutoff of 12 Ry. The sum over BZ has been performed with six Chadi-Cohen points in the irreducible wedge. Interaction constants have been calculated in reciprocal space on a grid of $\mathbf{q}$ points as explained above, and then Fourier-transformed to real space. Our reciprocal-space grid is such to allow the calculation of $\mathbf{J}$, $\mathbf{F}$, and $\mathbf{\Phi}$ in real space up to the 22nd complete shell of neighbors. Other technical details are the same as in Ref. 25

In Fig. 1 we display the calculated interaction constants as functions of the interatomic distance, calculated at the lattice parameter given by Vegard's law for $x = \frac{1}{2}$. Neglecting lattice relaxation (upper panel), the interactions are quite short range, and practically vanish beyond the 3rd shell of neighbors. Lattice relaxation makes the interactions propagate rather far along the bond chains (lower panel). The reason for such behavior is that interatomic force constants themselves tend to be rather long-ranged along the bond chains, as already recognized in Ref. 39. The solid lines indicate the value of the configurational energy of the random alloy, calculated using the regular-solution approximation and truncating the interaction constants at different shells of

Table I

Comparison between the configurational energies of several superlattice structures, calculated by density-functional perturbation theory (DFPT), and full self-consistent calculations (SCF), both neglecting ("Unrelaxed"), and including ("Relaxed") lattice relaxation [meV/atom]. The $n+m$ notation indicates n Si, and m Ge layers. For [111] SL's, the "R1" and "R2" labels refer to Si-Ge–Ge-Si and Si-Si–Ge-Ge stacking respectively. All the calculations are performed at the lattice constant given by Vegard's law for $x=\frac{1}{2}$.

Structure	Unrelaxed		Relaxed	
	DFPT	SCF	DFPT	SCF
ZB	−6.4	−6.5	−6.4	−6.5
$[001]_{2+2}$	−4.6	−4.6	−9.4	−9.7
$[001]_{3+1}$	−3.9	−3.9	−6.3	−6.4
$[001]_{1+3}$	−3.9	−3.9	−6.3	−6.5
$[001]_{3+3}$	−3.1	−3.2	−9.2	−9.5
$[111]_{2+2}$ R1	−5.9	−5.7	−9.9	−10.0
$[111]_{2+2}$ R2	−2.5	−2.5	−6.1	−6.3
$[111]_{3+3}$	−2.9	−3.0	−7.6	−7.9
$[110]_{2+2}$	−2.7	−2.7	−10.4	−10.7
$[110]_{3+1}$	−2.5	−2.5	−7.5	−7.8
$[110]_{1+3}$	−2.5	−2.5	−7.5	−7.9

neighbors. In the unrelaxed case the convergence is very rapid and a few interaction constants suffice. When relaxation is included the convergence is slower and peaks are observed in correspondence with shells containing atoms belonging to the same bond chain as the atom at the origin. This indicates that a larger number of interaction constants have to be considered.

As a check of the reliability of our perturbative approach, we compare in Table I the configurational contribution to the formation energies of several ordered structures of Si_xGe_{1-x} —both neglecting and including lattice relaxation— as obtained by the present approach and by accurate DFT self-consistent (SCF) calculations. Inspection of this table shows that the typical accuracy achievable by the present approach is of the order of 5 %, and often better than this, thus giving confidence in its predictive power.

Hence we have performed constant-pressure ($P = 0$), constant-chemical-potential MC simulations on a system of 1024 atoms (corresponding to an *fcc* supercell of linear dimension 8 times larger than the primitive diamond cell). In our simulations a MC step is given by an attempt to change the nature of a given atom (reversing its $\sigma_{\mathbf{R}}$) followed by an attempt to change the volume of the cell (of a random amount $\Delta\Omega$): to this end, interaction constants for an arbitrary volume are obtained by quadratic interpolation of the values calculated for three different volumes (near the equilibrium volumes of pure Si, pure Ge, and $Si_{\frac{1}{2}}Ge_{\frac{1}{2}}$, as given by Vegard's law). The two trial moves (alchemical and volumic) are made independently, the acceptance ratio of each being weighted with the proper Boltzmann factor.

In Fig. 2 we display the average concentration x, as a function of the chemical potential,[40] μ, obtained for three different temperatures (above, below, and at the critical temperature, T_c). For each considered temperature a sequence of chemical potentials has been sampled in both directions: once starting from a Ge-rich phase and increasing the chemical potential, once starting from a Si-rich phase and decreasing

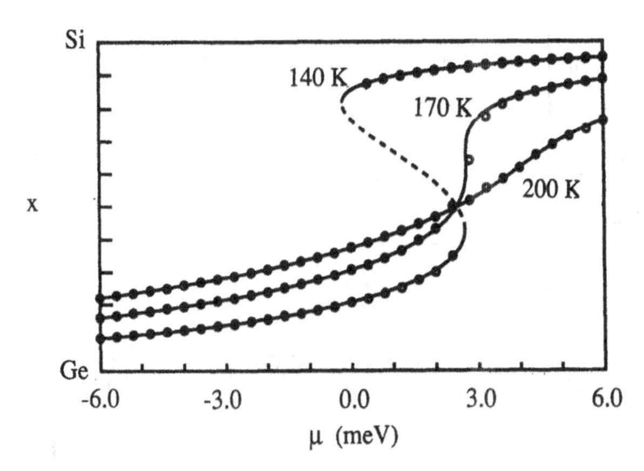

Figure 2. Average Si concentration of the alloy as a function of the chemical potential, for three different temperatures (above, below and at the critical temperature T_c). The statistical errors are smaller than the dimensions of the dots. The lines are obtained by a fitting procedure, as described in the text. The dashed portion of the line at 140 K corresponds to the instability region inside the spinodal line.

it. At each value of μ a thousand MC steps per site have been performed resulting in statistical errors on the average composition smaller than the dimension of the displayed circles. The two independent runs give identical results in the low- and high-concentration regions, but at sufficiently low temperature they show hysteresis- the signature of the presence of miscibility gap. In fact, above T_c, the concentration is a continuous function of the chemical potential, thus indicating complete miscibility for any concentration. As the temperature decreases below T_c a discontinuity appears and we have sudden jumps in the concentration when the chemical potential is increased (decreased) at the lower (upper) edge of the unstable region.

To extract the phase diagram from these data we need an expression for the free energy of the system. The free energies of the Ge- and Si-rich phases are obtained by integration: to this end, the dependence of the chemical potential upon concentration is fitted by the function

$$\mu(x,T) = kT\log\big(x/(1-x)\big) + \frac{\partial P(x,T)}{\partial x}, \tag{27}$$

where $P(x,T)$ is a fourth-order polynomial with respect to x vanishing at $x=1$ and $x=0$, so that the correct high- and low-concentration limits are enforced. The analytic curves thus obtained are reported as solid lines in Fig. 2 (dashed in the unstable region where it is unphysical). The free energy can then be expressed as:

$$f(x,T) = kT\big(x\log(x) + (1-x)\log(1-x)\big) + P(x,T). \tag{28}$$

The miscibility gap is found corresponding to the values of chemical potential where the two phases have the same free energy, whereas the maximum and minimum of $\mu(x,T)$ appearing below T_c correspond to the spinodal line.

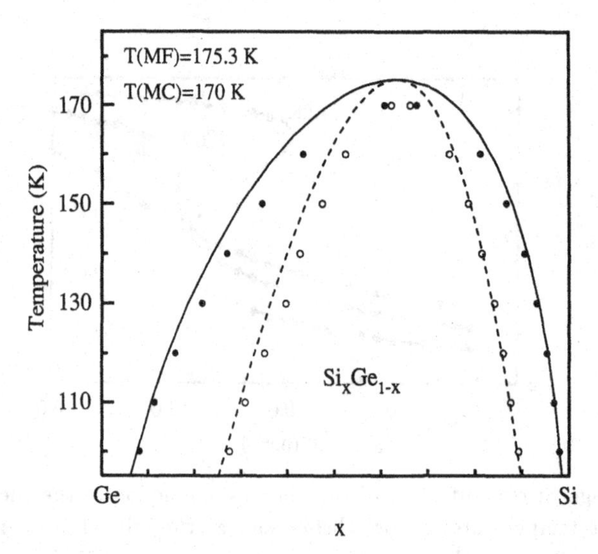

Figure 3. Phase diagram of Si_xGe_{1-x} alloys. Dots: Monte Carlo predictions for the miscibility gap (full) and spinodal line (open). Full line: mean field prediction for the miscibility gap. Dashed line: mean field prediction for the spinodal line.

The resulting phase diagram is drawn in Fig. 3 where solid circles show the miscibility gap and open circles the spinodal line as obtained by our MC simulations. For comparison the line of the miscibility gap and the spinodal line obtained by the mean-field (MF) approximation are also displayed by a continuous and a dashed line respectively. The value of T_c estimated by MC is ≈ 170 K, whereas MF gives 175.3 K. The two results are quite similar, as the thermodynamics of the system is dominated by the (configuration independent) elastic contribution to the formation energy. Our results are at variance with previous cluster expansion calculations,[41] but they agree well with recent MC simulations employing semiempirical interatomic potentials.[42]

MC simulations allow to calculate the bond-length distribution in the alloy as a function of the molar composition. In Fig. 4 we display our results obtained at $T = 300$ K. Similar results have been obtained at different temperatures above T_c. The histogram in the left panel displays the bond-length distribution corresponding to an average composition $x = 0.45$, and shows three well distinct peaks, whose maxima are close to —but not coinciding with— the equilibrium bond lengths of pure Si, pure Ge, and of zincblende SiGe (reported on the right of the figure). In the right panel, we summarize our results for different concentrations. Full dots indicate the maxima of the peaks, error bars their widths, and open circles indicate the average bond length, as obtained by the lattice constant of the alloy. These data do not depend on temperature for any temperature above T_c, within our statistical errors, and they indicate that Vegard's law is followed very closely. Deviations of the average lattice parameter from Vegard's law —as obtained for different temperatures in the range of 200 – 400 K— are displayed on a magnified scale in Fig. 5, along with the predictions of the regular-solution approximation (continuous line). All the data coincide with the predictions of this approximation, thus confirming that Si_xGe_{1-x} is a model random alloy at all temperatures in this range and, *a fortiori*, at higher temperatures.

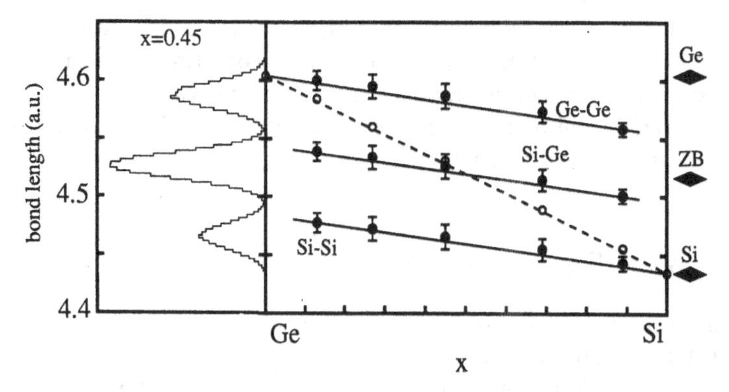

Figure 4. Left panel: bond length distribution in $Si_{0.45}Ge_{0.55}$, measured by MC at T=300 K. Right panel: dependence of maxima of the peaks upon Si concentration (full dots; the error bars indicate the width of the peaks); dependence upon composition of the average bond length, as obtained from the alloy lattice parameter (open circles). The solid and dashed lines are intended as guidelines for the eyes. The diamonds on the right indicate the theoretical equilibrium bond lengths of pure Si and Ge, and of zincblende SiGe.

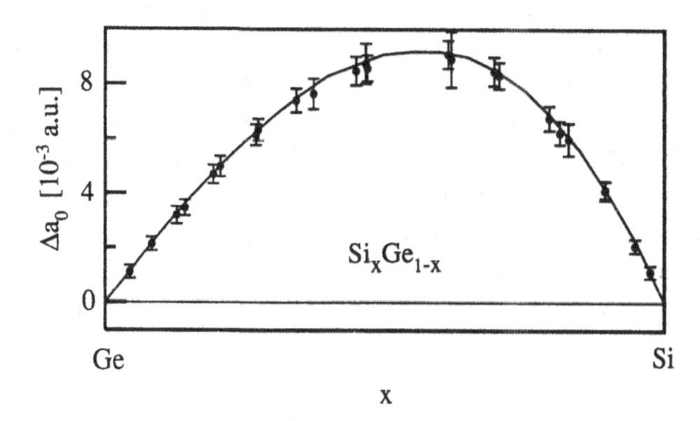

Figure 5. Deviation of the alloy lattice parameter from linearity with respect to concentration (Vegard's law). All the data taken in the temperature range between 200 and 400 K coincide with those predicted by the regular solution approximation (solid line), within statistical errors.

IV. Conclusions

In this paper we have shown that a good description of the energetics of semiconductor alloys can be obtained in an economical and rigorous way, provided that the alloy can be described accurately by perturbation theory. When this is the case, the alloy problem -including lattice relaxation- can be mapped onto a lattice-gas model with two-body interactions only, whose values can be computed from first principles. The lattice-gas model obtained in this way is conveniently studied with conventional Monte Carlo simulations. We have applied this framework to the study of bulk thermodynamics of the Si_xGe_{1-x} alloy. Our results indicate that Si_xGe_{1-x} is a model random alloy with a miscibility gap of about $\approx$ 170 K. No tendency towards ordering has been observed. Our results confirm that bulk thermodynamics should not be responsible for the ordering transition observed in epitaxially grown samples at much higher temperature, thus supporting the kinetic and/or surface origin of such a phenomenon.

Acknowledgements

This work has been cosponsored by the Italian Consiglio Nazionale delle Ricerche under Grant No. 90.00653.PF69, by the Swiss National Science Foundation under Grant No. 20–5446.87, and by the European Research Office of the U.S. Army under Grant No. DAJA 45–89–C–0025.

References

1. P. Hohenberg and W. Kohn, *Phys. Rev.* **136**, 864 (1964).
2. W. Kohn and L. J. Sham, *Phys. Rev.* **140**, 1133 (1965).
3. For a recent review on linearized all-electron methods in DFT, see for instance: O. K. Andersen, O. Jepsen, and M. Sob, in: *Electronic Band Structure and Its Applications*, edited by M. Yussouf, Springer-Verlag, Berlin, 1987. p. 1.
4. For a recent review of the pseudopotential approach to DFT, see for instance: W. E. Pickett, *Computer Phys. Reports* **9**, 115 (1989).
5. O. K. Andersen, *Phys. Rev.* **B12**, 3060 (1975); D. D. Koelling and G. O. Arbman, *J. Phys. F* **5**, 2041 (1975); E. Wimmer, H. Krakauer, M. Weinert, and A. J. Freeman, *Phys. Rev.* **B24**, 864 (1981).
6. D. R. Hamann, M. Schlüter, and C. Chiang, *Phys. Rev. Lett.* **43**, 1494 (1979).
7. E. R. Davidson, in *Methods in Computational Molecular Physics*, edited by G. H. F. Diercksen and S. Wilson, Vol. 113 of NATO Advanced Study Institute, Series C, Plenum PUb. Co., New York, 1983. p. 95.
8. I. Štich, R. Car, M. Parrinello, and S. Baroni, *Phys. Rev.* **B39**, 4997 (1989).
9. R. Car and M. Parrinello, *Phys. Rev. Lett.* **55**, 2471 (1985).
10. G.B. Stringfellow, *J. Phys. Chem. Solids* **34**, 1749 (1973).
11. L. G. Ferreira, S.-H. Wei, and A. Zunger, *Phys. Rev.* **B40**, 3197 (1989); S.-H. Wei, L. G. Ferreira, and A. Zunger, *Phys. Rev.* **B41**, 8240 (1990).
12. M. B. Panish and M. Ilegems, *Prog. Solid State Chem.* **7**, 39 (1972).
13. T. S. Kuan, T. F. Kuech, W. I. Wang, and E. L. Wilkie, *Phys. Rev. Lett.* **54**, 201 (1985); H. R. Jen, M. J. Cherng, and G. B. Stringfellow, *Appl. Phys. Lett.* **48**, 1603 (1986); M. A. Shahid, S. Mahajan, D. E. Laughlin, and H. M.Cox, *Phys. Rev. Lett.* **58**, 2567 (1987); A. Gomyo, T. Suzuki, and S. Iijima, *Phys. Rev. Lett.* **60**, 2645 (1988).
14. A. Ourmazd and J. C. Bean, *Phys. Rev. Lett.* **55**, 765 (1985);
15. D. J. Lockwood, K. Rajan, E. W. Fenton, J.-M. Baribeau, and M. W. Denhoff, *Solid State Commun.* **61**, 465 (1987).
16. F. K. LeGoues, V. P. Kesan, and S. S. Iyer, *Phys. Rev. Lett.* **64**, 40 (1990).
17. F. K. LeGoues, V. P. Kesan, S. S. Iyer, J. Tersoff, and R. Tromp, *Phys. Rev. Lett.* **64**, 2038 (1990).
18. P. Bogusławski, *Phys. Rev.* **B42**, 3737 (1990).

19. S. Froyen and A. Zunger, *Phys. Rev. Lett.* **66**, 2132, 1991.
20. D. de Fontaine, *Solid State Physics*, edited by H. Ehrenreich, F. Seitz, and D. Turnbull. Academic, New York, 1979. Vol. 34, p. 73.
21. J.W. Connoly and A.R. Williams, *Phys. Rev.* **B27**, 5169 (1983).
22. R. Kikuchi, *Phys. Rev.* **81**, 988 (1951).
23. S. Baroni, P. Giannozzi, and A. Testa, *Phys. Rev. Lett.* **58**, 1861 (1987).
24. P. Giannozzi, S. de Gironcoli, P. Pavone, and S. Baroni, *Phys. Rev.* **B43**, 7231 (1991).
25. S. de Gironcoli, P. Giannozzi, and S. Baroni, *Phys. Rev. Lett.* **66**, 2116 (1991).
26. V. Heine and D. Weaire, *Solid State Physics*, edited by H. Ehrenreich, F. Seitz, and D. Turnbull; Academic, New York, 1970. Vol. 24, p. 250.
27. J. Hafner, *From Hamiltonians to Phase Diagrams*, Springer Series in Solid State Science, Vol. 70; Springer-Verlag, Berlin, 1987.
28. The fact that to second order in the perturbation the lattice interactions are only two-body interactions is closely related to a well known similar property obtained for continuous interactions in pseudopotential perturbation theory. See for instance: M. H. Cohen, *J. Phys. Radium* **23**, 643 (1962).
29. H. Hellmann, *Einführung in die Quantenchemie* Deuticke, Leipzig, 1937; R. P. Feynman, *Phys. Rev.* **56**, 340 (1939).
30. X. Gonze and J.P. Vigneron, *Phys. Rev.* **B39**, 13120 (1989).
31. H. Kanzaki, *J. Phys. Chem. Solids* **2**, 24 (1952); H. Kanzaki, *J. Phys. Chem. Solids* **2**, 107 (1952).
32. M. A. Krivoglaz, *Zh. Ekso. Teor Fiz.* **33**, 204 (1958) [*Sov. Phys. JETP* **7**, 139 (1958)]; M. A. Krivoglaz, *Theory of X-ray and Neutron Scattering by Real Crystals*, Plenum Pub. Co., New York, 1969.
33. S. Froyen and C. Herring, *J. Appl. Phys.* **52**, 7165 (1981).
34. S. L. Adler, *Phys. Rev.* **126**, 413 (1962); N. Wiser, *Phys. Rev.* **129**, 62 (1963).
35. C. D. Thourmound, *J. Phys. Chem.* **57**, 827 (1953).
36. M. Hansen, *Constitution of Binary Alloys*, McGraw-Hill, New York, 1958, 2nd ed. p. 774.
37. P. B. Littlewood, *Phys. Rev.* **B34**, 1363 (1986); J. L. Martins and A. Zunger, *Phys. Rev.* **56**, 1400 (1986); S. Ciraci and I. P. Batra, *Phys. Rev. B* **38**, 1835 (1988); B. Koiller and M. O. Robbins, *Phys. Rev.* **B40**, 12554 (1989).
38. H. von Känel, E. Müller, H.-U. Nissen, W. Bacsa, M. Ospelt, K. A. Mäder, R. Stalder, and A. Baldereschi, *J. Cryst. Growth* (1991), in press; E. Müller, H.-U. Nissen, K. A. Mäder, M. Ospelt, H. von Känel, submitted to *Phil. Mag. A.*
39. A. Fleszar and R. Resta, *Phys. Rev.* **B34**, 7140 (1986).
40. The "chemical potential" of the auxiliary lattice gas is indeed the difference between the Si and Ge chemical potentials.
41. A. Qteish and R. Resta, *Phys. Rev.* **B37**, 6983 (1988).
42. P. C. Kelires and J. Tersoff, *Phys. Rev. Lett.* **63**, 1164 (1989).

First Principles Calculation of Phase Diagrams

Juan M. Sanchez

Center for Materials Science and Engineering
The University of Texas
Austin, Texas 78712
USA

Abstract

Phenomenological theories of phase equilibrium have been successfully used to characterize the main contributions to alloy phase stability, the interpretation of complex and extensive experimental data and, in some instances, the prediction of metastable phases. At present, there is increased interest in elucidating the extent to which microscopic quantum theory can be combined with statistical mechanics to produce a reliable description of phase equilibrium and, in particular, phase diagrams. This question will be explored using a first-principles statistical mechanics theory which incorporates the calculation of electronic total energies in the local density approximation, configurational entropies, vibrational modes and local volume relaxations in disordered systems. Applications of the theory to the binary Ni-Al and ternary Ni-Al-Ti systems are given using the Linear Muffin-Tin Orbital method for the total energy calculations, the Cluster Variation method for the description of the configurational entropy, and the Debye-Grüneisen approximation for the vibrational modes.

I. Introduction

The limited information usually available for equilibrium and metastable phase diagrams represents a major hurdle in the design and development of new materials systems. Engineering alloy systems for structural applications often exhibit a great variety of equilibrium and metastable phases which compete closely for stability within given temperature and composition ranges. As such, the accurate experimental determination of the equilibrium phase diagram, especially for multicomponent alloys, is generally a time consuming undertaking capable of significantly slowing down any alloy development program.

During the last few years, however, the computation of phase diagrams has emerged as a distinct alternative to the experimental route. At present, the computational approach to alloy phase diagram determination offers a potentially valuable tool in the design and development of new materials.

The stage for the current state of development was set by early studies of simple phenomenological models, which were shown capable of reproducing quite well the most important features of alloy phase diagrams.[1–5] These models were based, essentially, on generalizations of the Ising Hamiltonian and included pair and many-body interactions. In these developments, the Cluster Variation Method (CVM) of Statistical Mechanics[6] played key role since it provided an efficient and computationally economical way of describing the configurational thermodynamic of alloys.

The early success of the CVM calculations revived the old dream of computing alloy phase diagrams from first principles; i.e. from the knowledge of the electronic structure of the alloy. Indeed, one of the most significant recent developments in alloy theory, density functional theory and its computational version, the local-density approximation (LDA),[7] was fully developed and ready to be applied together with the statistical models to tackle the delicate problem of alloy stability at finite temperatures.

The implementation of the local-density approximation by several investigators, together with the development of efficient linear methods to study the electronic structure of solids, has led to fully *ab-initio* calculations of the total energy at zero temperature of pure solids,[8–10] relatively simple compounds[11,12] and disordered alloys.[13–15] By reproducing a wide range of physical properties within a few percent of the experimental values, these quantum mechanical total energy calculations have provided conclusive evidence in favor of the local-density approximation.

In the language of the CVM, the statistical thermodynamics of the alloy is described in terms of atomic configurations of clusters of lattice sites. This localized description of the state of partial order in the alloy can be conveniently accomplished using multisite correlation functions.[1,16] Specifically, it has been shown by Sanchez et al.[16] that any function of configuration, of which the energy is only a particular case, can be expanded in terms of multisite characteristic functions which form an orthonormal basis in configuration space. For example, the CVM, which was originally proposed in the context of the variational principle of statistical mechanics, can also be shown to follow from a similar cluster expansion. Although the cluster expansion can be applied quite generally to any function that depends upon the configuration of the system, its usefulness rests heavily on the rate of convergence in terms of the size and complexity of the clusters.

The cluster expansion of the expectation value of the configurational energy results in the commonly used bi-linear expression in terms of effective multisite interactions and correlation functions, the latter being expectation values of the orthogonal characteristic functions.[16] The effective pair and many-body interactions fully describe the statistical thermodynamics of the alloy. As mentioned, numerous studies of binary and multicomponent alloys, whereby the multisite interactions were determined using ground state analysis and available thermochemical data such as energies of formation, served to establish the usefulness of the cluster expansion for the description of states of partial order. The attractiveness of this cluster representation for the configurational energy rests on the fact that the treatment of the configurational thermodynamics of partially ordered alloys, i.e. alloys displaying both short- and/or long-range order, becomes essentially isomorphic to a generalized Ising model where the chemical interactions are of relatively short range. Although these interactions will in general include many-body terms, as well as temperature and volume dependence, they can be easily treated using the CVM to calculate configurational free energies and, from them, the solid state portion of phase diagrams.

The success of this cluster representation for the energy of alloys, proven in the context of phenomenological models, led Connolly and Williams[17] to propose the use *ab-initio* total energy calculations of ordered compounds, in lieu of experimental data, as a mean of obtaining the set of effective pair and multisite chemical interactions. Although a simple extension of the phenomenological approach, the proposal of Connolly and Williams reflected the degree of confidence with which total energies could be calculated using the local-density approximation, and opened the door to first principles calculation of phase diagrams.

Among the first applications of this first-principles approach, were studies of temperature-composition binary phase diagrams of noble-metal alloys[18] and semiconductor alloys.[19] Subsequently, numerous other cases have been investigated with relatively good results.[20–27] In general, the cluster expansion of the configurational energy converges relatively fast, thus offering a practical method for the determination of the chemical interactions in a disordered alloy system from total energy band calculations for a relatively small set of ordered compounds.

The most serious shortcomings of the first principles phase diagram calculations carried out to date[18–25] are the inadequate treatment of local volume and elastic relaxation and the neglect of vibrational modes. The treatment of local relaxations present such a formidable task that few attempts have been made to include this effect in first principles calculation. The contributions due to local volume relaxations, as distinguished from global volume relaxations which are routinely and easily incorporated, and of vibrational entropies are expected to be more tractable. In particular, local volume relaxations should play an important role for systems where there is a significant difference in the molar volumes of the constituents elements. This point is well illustrated, for example, by the recent calculation of the Ag-Cu phase diagram, where the effect of relaxation and of vibrational modes are investigated.[27]

In the next section we present a cluster theory of the configurational thermodynamic of alloys. The description provides the formal framework for the treatment of short-range order (SRO) effects in the configurational energy and in the configurational entropy. The contributions to the free energy due to local volume relaxations and vibrational modes are also discussed. We make contact with microscopic electronic theories via the Linear Muffin-Tin Orbital (LMTO) approximation, which is used to calculate the total energies of selected compounds in the Ni-Al-Ti systems. The first principles statistical thermodynamic theory, which includes configurational, vibrational and local volume relaxation effects, is applied to the computation of the equilibrium solid state phase diagrams for the Ni-Al system and for the ternary Ni-Al-Ti system. The Ni-Al system is chosen as an interesting example in which *bcc*- and *fcc*-based phases coexist giving rise to a relatively complex binary phase diagram. The Ni-Al-Ti system, on the other hand, exemplifies and represents the first application of the method to ternary alloys.

II. Configurational Thermodynamics

In this section we discuss briefly the general formalism for the description of the configurational thermodynamics of alloys. The main result is the cluster expansion which provides the basis for the description of disordered alloys from the knowledge of the energy (binding curves) of ordered compounds. For the sake of simplicity we consider only binary systems although the theory can be easily extended to multicomponent alloys.[16]

II.1. Cluster expansion

The configuration of a crystalline binary alloy is described in terms of spin or occupation numbers σ_i at each lattice site i which take, respectively, values $+1$ and -1 for components A and B. Any configuration of the system is then fully specified by the N-dimensional vector $\sigma = \{\sigma_1, \sigma_2, ..., \sigma_N\}$, where N is the number of lattice points. In general, one is confronted with the problem of describing functions that depend explicitly on the occupation variables σ_i, such as the energy of formation of the alloy. In order to provide an unambiguous description of such functions, it is convenient to introduce an orthogonal functional basis in configurational space. Although in the thermodynamic limit the dimension of the complete orthogonal basis is infinite, judicious choice of the basis functions allow us to obtain accurate approximations to the thermodynamic potentials in terms of a subset of finite dimension.

For a single site i, the set of two polynomials in the discrete variable σ_i, namely the polynomial of order 0, $\phi_0(\sigma_i) = 1$, and the polynomial of order 1, $\phi_1(\sigma_i) = \sigma_i$, form a complete and orthonormal set, with the inner product between two functions of configuration f(σ_i) and g(σ_i) in the one-dimensional discrete space spanned by σ_i defined as:

$$\langle f(\sigma_i) \cdot g(\sigma_i) \rangle = \frac{1}{2} \sum_{\sigma_1 = \pm 1} f(\sigma_i) g(\sigma_i). \tag{1}$$

The set of orthonormal characteristic functions in the N-dimensional discrete space spanned by the vector σ is obtained from the direct product of the $\{\phi_0(\sigma_i), \phi_1(\sigma_i)\}$, where i spans all crystal sites (i =1,2,...,N). For a binary system, the resulting characteristic functions, $\Phi_\alpha(\sigma)$, are given by products of the spin operator σ_i over the sites of all possible clusters $\alpha = \{i_1, i_2, ..., i_n\}$ in the crystal:[16]

$$\Phi_\alpha(\sigma) = \prod_{i \in \alpha} \sigma_i = \sigma_{i_1} \sigma_{i_2} ... \sigma_{i_n}. \tag{2}$$

Accordingly, there is a one to one correspondence between the set of orthogonal functions $\Phi_\alpha(\sigma)$ and the set of all clusters α in the crystal, including the empty cluster for which $\Phi_0(\sigma) = 1$.

The orthogonality of the characteristic functions $\Phi_\alpha(\sigma)$ is expressed by:[16]

$$\frac{1}{2^N} \sum_\sigma \Phi_\alpha(\sigma) \Phi_\beta(\sigma) = \delta_{\alpha,\beta}. \tag{3}$$

In view of Eq. (3), any function of configuration, F(σ), may be written as,

$$F(\sigma) = \sum_\alpha F_\alpha \Phi_\alpha(\sigma), \tag{4}$$

where the sum extends over all clusters in the crystal, including the empty cluster, and where F_α is given by,

$$F_\alpha = \langle F(\sigma) \cdot \Phi_\alpha(\sigma)\rangle = \frac{1}{2^N}\sum_\sigma F(\sigma)\Phi_\alpha(\sigma). \tag{5}$$

Thus, the terms F_α are the projections of $F(\sigma)$ on the orthogonal cluster basis.

It should be noted that the space group symmetry of the crystal requires that the cluster projections F_α of the function $F(\sigma)$ be the same for all clusters α which are related by a symmetry operation (translation or point group). Accordingly, the cluster expansion in Eq. (4) becomes:

$$F(\sigma) = \sum_{n=0}^{N} F_n\Theta_n(\sigma), \tag{6}$$

where n labels the set of inequivalent clusters in the crystal. In the case of a disordered lattice, these cluster are only distinguished by their number of points and their geometry. In Eq. (6), the $\Theta_n(\sigma)$ are given by:

$$\Theta_n(\sigma) = \sum_{\alpha\in n} \Phi_\alpha(\sigma). \tag{7}$$

In view of the orthogonality of the $\Phi_\alpha(\sigma)$, we also have:

$$\frac{1}{2^N}\sum_\sigma \Theta_n(\sigma)\Theta_m(\sigma) = z_n N\delta_{n,m} \tag{8}$$

where z_nN is the total number of n-type clusters in the crystal.

The most common applications of Eq. (6) are for the cluster expansion of expectation values of functions of configurations, such as the average of the configurational energy. With the notation $\xi_n = \langle\Phi_\alpha(\sigma)\rangle$ for the expectation value of the characteristic functions, where α is any cluster belonging to the equivalent set n, we obtain:

$$\bar{F} = \langle F(\sigma)\rangle = N\sum_{n=0}^{N} z_nF_n\xi_n. \tag{9}$$

The usefulness of this cluster expansion rests on the fast convergence of the projections F_n. In Section III, the cluster expansion given by Eq. (9) is used to obtain the renormalized contributions to the configurational energy arising from chemical interactions, local volume relaxations and vibrational modes.

II.2. Electronic structure calculations

As mentioned, the development of local-density functional theory[7] has been instrumental in our ability to calculate the total energy of ordered compounds from the knowledge of their electronic structure. These calculations, which use only atomic numbers as input, correctly reproduce 0 K ground state properties of the elements and of ordered compounds.

Here, the electronic structure results for the binary Ni-Al system[20,28] were obtained using the augmented spherical wave (ASW) approximation.[8,11,12] For the ternary Ni-Al-Ti system, the calculations were carried out using the linear muffin-tin orbital (LMTO) method,[9] with exchange and correlation treated in the local-density approximation. Total energy calculations for the pure elements and for each compound are performed in the atomic sphere approximation, including corrections due to the overlap of atomic spheres, for approximately 15 values of the average Wigner-Seitz radius centered around the minimum of the electronic binding curve. In all cases, the energy is minimized with respect to the ratio of atomic radii for each element in the compound.

A convenient representation of the calculated total energy curves for a given compound is provided by a Morse function of the form:[29]

$$E(r) = A - 2Ce^{-\lambda(r-r_0)} + Ce^{-2\lambda(r-r_0)}, \tag{10}$$

where $E(r)$ is the calculated electronic total energy of the rigid lattice and A, C, λ, and r_0 are fitting parameters. Here, the variable r, is the Wigner-Seitz atomic radius related to the volume per atom by the relation $\Omega = (4\pi/3)r^3$. For the compounds, r is the *effective* Wigner-Seitz radius obtained from the average of the constituent atomic volumes. It follows from Eq. (10) that r_0 is the Wigner-Seitz radius corresponding to the minimum in the binding curve and that C is the cohesive energy of the rigid lattice.

The results of the electronic structure calculations can be extended to include the vibrational modes of configurationally ordered systems, i.e. pure elements and ordered compounds, by means of a Debye-Grüneisen analysis[29] of the calculated binding energies.[27] In particular, the analysis yields theoretical bulk moduli, Debye temperatures, and Grüneisen constants from which the vibrational free energies of pure metals and hypothetical, chemically ordered compounds, are obtained. For simple elemental systems[29] and alloys,[27] the theoretical thermal properties are in reasonable agreement with experiment.

In terms of properties of the rigid lattice, the vibrational free energy, $F(r,T)$, is given by,

$$F(r,T) = \frac{9}{8}k_B\theta + E(r) - k_BT[D(\theta/T) - 3ln(1 - e^{-\theta/T})], \tag{11}$$

where k_B is Boltzmann's constant, $D(x)$ is the Debye function, $E(r)$ is the electronic binding energy, and where the volume dependence of the Debye temperature, θ, is given by:

$$\theta = \theta_0(\frac{r_0}{r})^{3\gamma}, \tag{12}$$

with θ_0 the Debye temperature corresponding to r_0, and with γ the Grüneisen constant.

For the ordered compounds, the free energy given by Eq. (11) represents the volume and temperature dependent binding energy in the absence of configurational disorder.

Thus, the results of the electronic structure calculations are cast in the form of a set of Morse parameters (see Eq. 10), the Debye temperature θ_0 and the Grüneisen constant γ for several compounds and for the pure elements. For the purpose of proceeding with the computation of equilibrium phase diagrams, we present next the determination of the effective interactions, using the cluster expansion developed in Section II.1, followed by the calculation of the configurational entropy in the CVM approximation. We defer until Section III further discussion of the electronic structure parameters obtained for the binary Ni-Al and ternary Al-Ni-Ti systems.

II.3. The effective interactions

The cluster expansion developed in Section II.1, and in particular Eq. (9), may be applied to the energy of a set of ordered compounds, for which the correlation functions are known *a-priori*, in order to obtain effective chemical interactions applicable to the disordered system. Here we apply the procedure to the vibrational free energies of the ordered compounds calculated in the Debye-Grüneisen approximation. The resulting temperature and volume dependent effective interactions are then used in a CVM treatment of the configurational entropy in order to include contributions due to configurational disorder into the total free energy. Phase diagrams calculations that include vibrational entropies have been carried out previously for Cu-Ag[27] and Ru-Nb[26] alloys.

In practice, the applicability of the method depends upon the convergence of the cluster expansion for relatively small clusters, and on the availability of a set of total energies for which the required inversion of Eq. (9) is defined. At present, selection criteria for clusters giving a converged cluster expansion are not available. Thus, a maximum interaction range is assumed *a-priori* , much as is done with the correlation range in the CVM. In some instances, attempts to ascertain the accuracy of the approximation have been made by comparing the total energy of compounds not included in the inversion procedure, with the values obtained using the assumed cluster expansion.[23,24]

For the *fcc* lattice in the tetrahedron approximation, the inversion of Eq. (9) is straightforward, requiring the calculation of total energies for only five high symmetry structures. For larger cluster approximations, this inversion is not immediately apparent and *ad-hoc* approaches, such as least-square fitting of the calculated total energies, have been proposed. It can be shown, however, that within a given maximum cluster approximation, there is always a natural set of relevant structures, given by the vertices of a convex configurational polyhedron, for which the inversion of Eq. (9) is unique.[30]

Defining the correlation functions $\{\xi_{k,n}\}$, where k labels a set of ordered compounds and n labels the interactions included in the the cluster expansion, the vibrational free energy (per atom) for each of the ordered structures, $F_k(r,T)$, takes the form:

$$F_k(r,T) = \sum_{n=0}^{m} z_n V_n(r,T)\xi_{k,n}, \tag{13}$$

where $V_n(r,T)$ are the volume and temperature dependent effective interactions.

By properly choosing the set of ordered compounds ($k = 0,1,...,m$) and of the cluster interactions ($n = 0,1,...,m$), Eq. (13) can be inverted. At a fixed Wigner-Seitz radius and temperature, we have:

$$V_n(r,T) = \sum_{k=0}^{m} \omega_{n,k} F_k(r,T), \tag{14}$$

where $\omega_{n,k}$ are obtained by inversion of the matrix with elements $z_n\xi_{k,n}$.

If interactions are obtained using the same volume for all five compounds, the approach yields, once configurational effects are included, a total free energy functional that depends on the configurational variables ξ_n and volume (or r). Volume relaxation can then incorporated globally by minimizing this functional with respect to r, in addition to the usual minimization with respect to the correlation functions ξ_n. This

Table I

Morse parameters for the Ni-Al system from total energies calculated in the ASW approximation. The energies of formation are referred to Ni(*fcc*) and Al(*fcc*) at their $T = 0$ K equilibrium volume.

System	r_0(a.u.)	λ(a.u.$^{-1}$)	C (Ryd)	A (Ryd)	B (Kbar)	θ_0 (K)	γ
Ni (*fcc*)	2.5853	1.4047	0.3645	0.3645	2171	407	1.8158
Ni_3Al ($L1_2$)	2.6170	1.3217	0.3828	0.3481	1994	422	1.7294
NiAl ($L1_0$)	2.6885	1.1984	0.3973	0.3568	1656	424	1.6109
$NiAl_3$ ($L1_2$)	2.8167	1.1308	0.3436	0.3277	1217	413	1.5926
Al (*fcc*)	2.9564	1.1184	0.2665	0.2665	880	409	1.6532
Ni (*bcc*)	2.5835	1.2006	0.4577	0.4631	1993	390	1.5509
Ni_3Al (DO_3)	2.6237	1.2906	0.3959	0.3624	1961	419	1.6931
NiAl (B2)	2.6775	1.1309	0.4549	0.4003	1696	429	1.5140
NiAl(B32)	2.7043	1.1308	0.4270	0.3967	1576	415	1.5291
$NiAl_3$ (DO_3)	2.8231	1.1892	0.2912	0.2804	1139	399	1.6786
Al (*bcc*)	2.9780	1.0983	0.2483	0.2524	785	387	1.6354

global volume relaxation is based on the assumption that the effective local volumes occupied by clusters in the disordered alloy are independent of their configuration. Although this assumption has been applied to most of the first-principles calculations done to date, it appears physically implausible in cases where there is a significant difference between the atomic volumes of the constituents elements.

Other schemes aimed to account for the effect of local volume relaxations have also been proposed. For example, the cluster interactions may be obtained from the total energies $F_k(r_k, T)$ of each compound calculated at different Wigner-Seitz radii r_k. The Wigner-Seitz radius, r_k, corresponds to the equivalent volume per atom in the disordered alloy occupied by clusters with configurations characteristic of compound k. In this local volume relaxation scheme, Eq. (14) becomes:

$$V_n(r,T) = \sum_{k=0}^{m} \omega_{n,k} F_k(r_k, T), \tag{15}$$

where r stands for the set of Wigner-Seitz radii.

The set r is chosen by minimizing the total free energy with respect to each of the r_k subject to appropriate constraints. In the absence of any constraint, the approach is equivalent to the minimization of the vibrational free energy for each compound independently, and it is commonly referred to as total volume relaxation. Thus, in this scheme, the local volume of each cluster in the alloy is allowed to relax fully to the value found in the ordered state implying the existence of well define interatomic distance which are approximately independent of the sourrounding chemical environment.

An approach intermediate to global and total volume relaxations, both of which appear physically implausible, is to define atomic radii for each constituent, r_A and r_B, such that the Wigner-Seitz radius r_k for each ordered structure k is given by:

$$r_k^3 = (1 - c_k)(r_A)^3 + c_k(r_B)^3, \tag{16}$$

where c_k is the concentration of component B in the ordered compound k. The local atomic radii, which are clearly dependent on the state of ordered and temperature,

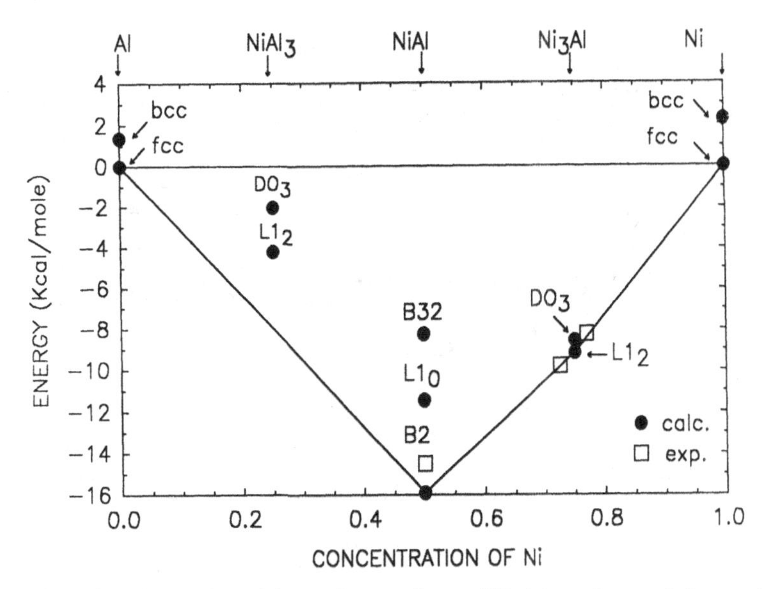

Figure 1. Calculated energies of formation at the equilibrium volume of *fcc*- and *bcc*-based compounds in the Ni-Al system.

are then obtained variationally by minimization of the total free energy with respect to both r_A and r_B. This approach, which results in partial relaxation of the local volumes in the alloy, has been used to calculate the phase diagram of the Ru-Nb[26] and Ag-Cu systems.[27]

II.4. Configurational Entropy

The CVM, originally proposed by Kikuchi[2] in 1951, is formulated here in terms of a cluster expansion of the configurational entropy.[16] For a given probability distribution $X(\sigma)$, the configurational entropy is given exactly by:

$$S = -k_B \sum_{\sigma} X(\sigma) ln X(\sigma) \tag{17}$$

where the sum is carried over all 2^N configurations in the crystal.

A tractable representation of Eq. (17) may be obtained by considering an infinite series of clusters with entropies defined by:

$$S_\alpha = -k_B \sum_{\sigma_\alpha} X_\alpha(\sigma_\alpha) ln X_\alpha(\sigma_\alpha), \tag{18}$$

where the cluster probability distribution $X_\alpha(\sigma_\alpha)$ is given by the sum of the $X(\sigma)$ over all configurational variables σ_i outside cluster α. This series of cluster entropies trivially converges to the exact configurational entropy as the size of the cluster α increases to include all points in the crystal. Using an exact Möbius transformation, we may also write the cluster entropies, S_α, in terms of a set of *irreducible* cluster contributions, $\hat{S}_\alpha$, as:[10]

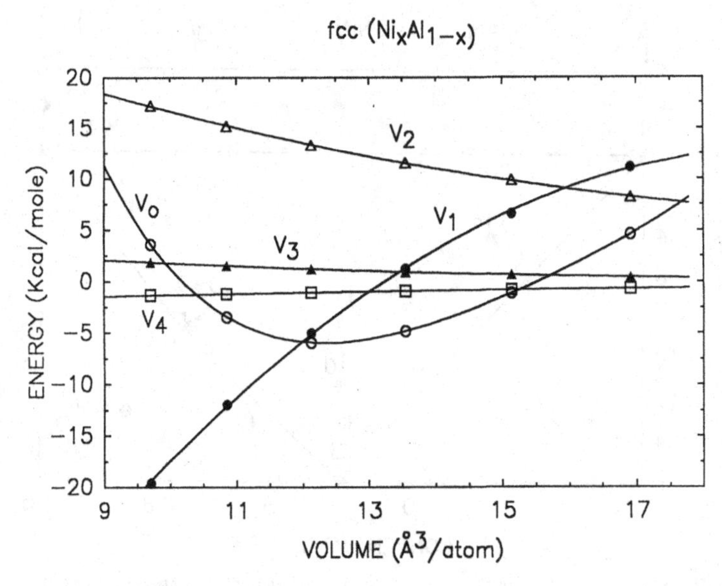

Figure 2. Calculated effective interactions as a function of average atomic volume for the empty (V_0), point (V_1), nearest-neighbor pair (V_2), triangle (V_3) and tetrahedron (V_4) clusters of the *fcc* lattice with global volume relaxations.

$$S_\alpha = \sum_{\beta \subseteq \alpha} \hat{S}_\beta \tag{19}$$

where the sum runs over all the subclusters of α, including α, and excludes the empty cluster.

The key approximation made in the CVM consists of neglecting the irreducible entropy contributions $\hat{S}_\alpha$ for clusters larger than a given maximum cluster. This closure condition allow us to express the total configurational entropy, S, in terms of a *finite* sum of irreducible contributions. Using the space group symmetry of the crystal, Eq. (19) for $\alpha \to N$ becomes:[10]

$$S = N \sum_{n=1}^{m} z_n \hat{S}_n, \tag{20}$$

or, in terms of the cluster entropies:

$$S = N \sum_{n=1}^{m} z_n a_n S_n = -N k_B \sum_{n=1}^{m} z_n a_n \sum_{\sigma_n} X_n(\sigma_n) ln X_n(\sigma_n), \tag{21}$$

where, as defined previously, n labels inequivalent clusters and m labels the maximum cluster. The coefficients a_n, obtained by inverting Eq. (19), are given by,[10]

$$\sum_{\beta \supseteq \alpha}' a_\beta = 1, \tag{22}$$

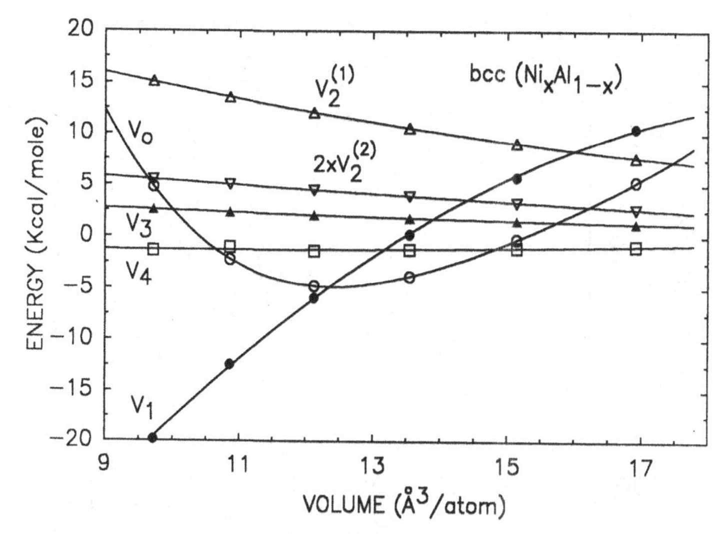

Figure 3. Calculated effective interactions as a function of average atomic volume for the empty (V_0), point (V_1), nearest-neighbor pair ($V_2^{(1)}$), next-nearest-neighbors ($V_2^{(2)}$), triangle (V_3) and tetrahedron (V_4) clusters of the *bcc* lattice with global volume relaxations.

where the equation is valid for each subcluster α of the maximum cluster, and where the sum runs over all subclusters β of the maximum clusters that contain or equal α. For example, in the tetrahedron approximation of the *fcc* lattice, the coefficients a_n are equal to 5, −1, 0 and 1 for the point, pair, triangle and tetrahedron clusters, respectively.

The total free energy functional (per atom) of the disordered alloy, including local volume relaxations and vibrational modes, is given by:

$$F_{tot} = \sum_{n=0}^{m} z_n V(r,T)\xi_n + k_B T \sum_{n=1}^{m} z_n a_n \sum_{\sigma_n} X_n(\sigma_n) ln X_n(\sigma_n). \tag{23}$$

Here the effective interactions $V(r,T)$ are given by Eq. (15). Using the cluster expansion described in Section II.1, the cluster probability distributions can be written in terms of the multisite correlation functions:[10]

$$X_n(\sigma_n) = \frac{1}{2^n}[1 + \sum_{n=1}^{4} \Theta_n(\sigma_n)\xi_n] \tag{24}$$

where the $\Theta_n(\sigma_n)$ are the characteristic functions defined as sums of products of the configurational variables σ_i for lattice sites i belonging to cluster n (see Eqs. (2) and (7)). Thus the free energy functional given by Eq. (23) is function only of the set of Wigner-Seitz radii r and of the correlation functions ξ_n.

At a given temperature and concentration, the latter being given by the point correlation ξ_1, the equilibrium free energy is obtained by minimizing the free energy functional with respect to the remaining correlation functions, ξ_n, and the set of Wigner-Seitz radii r. As mentioned in Section II.3, the minimization with respect to r may be carried out using three different schemes: i) constraining r_k to be all equal,

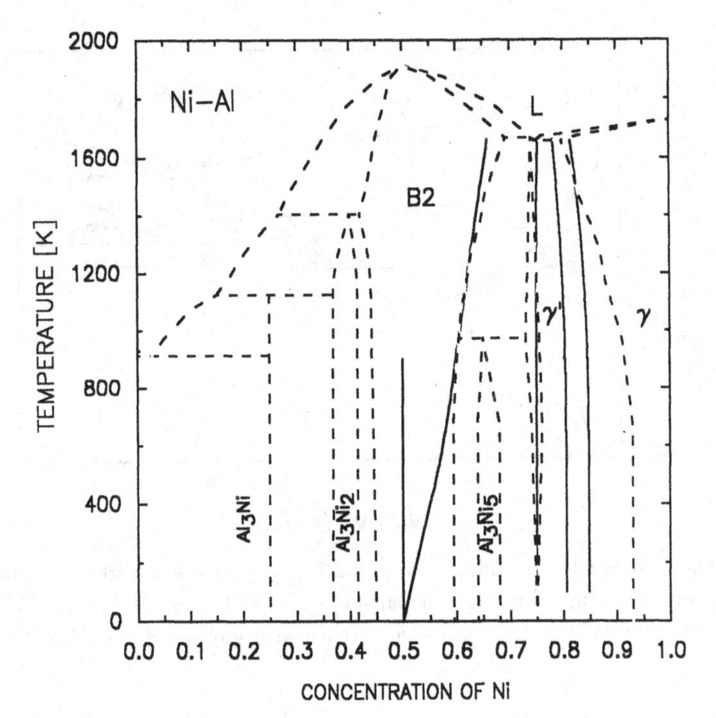

Figure 4. Calculated phase boundaries (solid lines) for the Ni-Al system compared to the experimental phase diagram (broken lines).

which results in global volume relaxation without allowing relaxation of local volumes; ii) varying the r_k independently which results in total relaxation of local volumes; and iii) subjecting the r_k to the external constraints, such as that of Eq. (16), which gives partial relaxation of the local volumes.

III. Applications of the theory

The first-principles statistical theory of alloy phase equilibrium developed in Sec.II is applied here to two alloy systems of significant technological interest: the binary Ni-Al and the ternary Al-Ni-Ti systems. The first example represents a relatively complex alloy system involving *bcc*- and *fcc*-phases while the second example stands out as the first application of the method to ternary alloys.

Total energy electronic structure calculations, as described in Sec. II.2, were carried out as function of volume for six *bcc*-based and five *fcc*-based structures. The structures of the different compounds together with the Morse parameters, Bulk moduli, Debye temperatures and Grüneisen constants are give in Table I.

The energy of formation at the equilibrium volume for each compound, calculated using the ASW and LDA approximations, is shown in Fig. 1. For comparison, available experimental data is also shown in the figure. The ground state diagram of Fig. 1 correctly reproduces, both qualitatively and quantitatively, the observed *fcc*- and *bcc*-based phases experimentally observed in Ni-Al.

Following the global relaxation procedure described previously, effective chemical interactions were obtained for both structures, *fcc* and *bcc*, and they are shown in

Table II

Morse parameters for *fcc*-based phases of the Ni-Al-Ti system from total energies calculated in the LMTO approximation. The energies of formation are referred to Ni(*fcc*), Al(*fcc*) and Ti(*fcc*) at their T = 0 K calculated equilibrium volume. Binary phases are *fcc* (pure elements), $L1_2$ (A_3B and AB_3), $L1_0$ (AB) and tetragonal (A_2BC).

System	r_0(a.u.)	λ(a.u.$^{-1}$)	C (Ryd)	A (Ryd)	B (Kbar)	θ_0 (K)	γ
Al	2.9470	1.2389	0.2142	0.2142	871	406	1.8255
Ni	2.5829	1.4800	0.3923	0.3923	2596	445	1.9113
Ti	3.0668	0.9583	0.5705	0.5705	1333	385	1.4695
Ni_3Al	2.6191	1.3722	0.4130	0.3745	2317	455	1.7969
NiAl	2.6829	1.3410	0.3747	0.3298	1960	461	1.7989
$NiAl_3$	2.7995	1.2748	0.2988	0.2765	1354	434	1.7844
Al_3Ti	2.9272	1.1525	0.3510	0.3199	1243	442	1.6868
AlTi	2.9709	1.0630	0.4613	0.4283	1369	434	1.5790
Al_3Ti	3.0097	1.0213	0.5117	0.4879	1384	411	1.5369
Ni_3Ti	2.6703	1.4454	0.3953	0.3540	2414	447	1.9298
NiTi	2.7967	1.2999	0.4171	0.3862	1967	423	1.8177
Ni_3Ti	2.9307	1.1048	0.4961	0.4813	1612	402	1.6189
Al_2NiTi	2.8218	1.1904	0.4282	0.3687	1678	452	1.6796
$AlNi_2Ti$	2.7734	1.2584	0.4202	0.3960	1873	433	1.7450
$AlNiTi_2$	2.8934	1.1245	0.4609	0.4391	1572	417	1.6268

Figs. 2 and 3 as a function of average atomic volume. The interactions are for pair, triangle and tetrahedron clusters in both lattices. In the *fcc* lattice the range of interactions extend only to nearest-neighbors, whereas the *bcc* lattice also includes second-neighbors.

The calculated solid state portion of the phase diagram is compared in Fig. 4 with the experimentally determined diagram. The phase diagram calculations were carried out using the CVM in the tetrahedron-octahedron approximation for *fcc*-based and the tetrahedron approximation for *bcc*-based structures. Furthermore, vibrational modes and local volume relaxations are included. The solubility limits for the NiAl phase (B2) are reproduce quite well by the calculations. The calculated phase boundaries, although qualitatively correct, are less accurate for the Ni_3Al ($L1_2$) and *fcc* disordered phases. A likely reason for the disagreement is the fact that only nearest-neighbor interactions are included in the *fcc*-based phases. Nevertheless, the results are encouraging, particularly since the method is parameter free and uses only atomic numbers as input.

As an example of the application of the method to ternary systems, we consider here the *fcc*(γ)-$L1_2$(γ') phase equilibrium in the Ni-Al-Ti system. Allowing only nearest-neighbor interactions (pair and many-many body) in the *fcc* lattice, it is necessary to calculate the total energy as a function of volume for 15 structures. The corresponding Morse potentials obtained from LMTO total energy calculations are shown in Table II. It should be noted that, for the NiAl system, the results obtained with the ASW and LMTO calculations are in general agreement (within 10%). The largest discrepancy is seen for the energy of formation of the metastable $NiAl_3$ compound (30%). The calculated $\gamma - \gamma'$ two phase boundaries at 1023 K are shown in Fig. 5. Although the two-phase region is considerably narrower than observed ex-

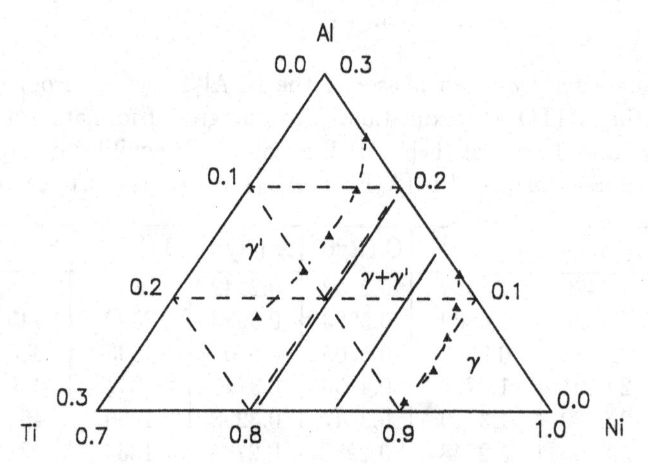

Figure 5. Calculated isothermal section at 1023 K showing the $\gamma-\gamma'$ phases of the Ni-Al-Ti, compared to experimental results. The phase boundaries were calculated in the tetrahedron approximation of the CVM using a totally relaxed local volumes and vibrational modes are not included.

perimentally, the general trends and tie-line directions are well reproduced by the calculations.

IV. Conclusions

The calculation of phase diagrams from the knowledge of the electronic structure of compounds has, over the last few years, emerged as a potentially useful tool in alloy design. We have attempted to demonstrate that the basic theoretical principles, as well as the limitations of the method, are presently well understood. We have seen with two examples, Ni-Al and Ni-Al-Ti, that the theory is capable of giving a semiquantitative description using only atomic numbers as input. Thus, although much work remains to be done, it appears that the we are on the way to a truly predictive first-principles theory of alloy phase stability. Among the problems that are likely to be the focus of future work are the effect of elastic relaxations and applications to multicomponent systems.

Acknowledgement

This work was supported in part by NSF Grant No. DMR-89-96244 and by a Grant for International Joint Research Project from the NEDO, Japan.

References

1. J. M. Sanchez and D. de Fontaine, *Phys. Rev.* **B17**, 2926 (1978).
2. R. Kikuchi, J.M. Sanchez, D. de Fontaine, and H. Yamauchi, *Acta Metall.* **28**, 651 (1980).

3. C. Sigli and J.M. Sanchez, *Acta Metall.* **33**, 1097 (1985).
4. J.M. Sanchez, J.R. Barefoot, R.N. Jarret, and J.K. Tien, *Acta Metall.* **32**, 1519 (1984).
5. C.E. Dahmani, M.C. Cadeville, J.M. Sanchez, and J.L. Morán-López, *Phys. Rev. Lett.* **55**, 1208 (1985).
6. R. Kikuchi, *Phys. Rev.* **81**, 988 (1951); J. Chem. Phys. **60**, 1071 (1974).
7. W. Kohn and L. J. Sham, *Phys. Rev.* **140**, A1133 (1965); P. Hohenberg and W. Kohn, *ibid.* **136**, B864 (1964).
8. V. L. Moruzzi, J. F. Janak and A. R. Williams, *Calculated Electronic Properties of Metals* (Pergamon, New York 1978).
9. O.K. Andersen, O. Jepsen, and D. Glötzel, *Highlights of Condensed Matter Theory, Proceedings of the International School of Physics Enrico Fermi*, North-Holland, Amsterdam, 1985.
10. M.T. Yin and M.L. Cohen, *Phys. Rev. Lett* **45**, 1004 (1980).
11. A.R. Wiliams, C.D. Gelatt, and V.L. Moruzzi, *Phys. Rev. Lett.* **44**, 429 (1980).
12. C.D. Gelatt, A.R. Wiliams, and V.L. Moruzzi, *Phys. Rev.* **B27**, 2005 (1985).
13. J.S. Faulkner, *Prog. Mater. Sci. 27, pp. 1-187, 1982 (Pergamon Press).*
14. H. Winter and G.M. Stocks, *Phys. Rev.* **B27**, 882 (1982).
15. H. Winter, P.J. Durham, and G.M. Stocks, *J. Phys. F* **14**, 1047 (1984).
16. J.M. Sanchez F. Ducastelle and D. Gratias, *Physica* **128A**, 334 (1984).
17. J.W.D. Connolly and A.R. Williams, *Phys. Rev.* **B27**, 5169 (1983).
18. K. Terakura, T. Oguchi, T. Mohri, and K. Watanabe, *Phys. Rev.* **B35**, 2169 (1987).
19. A.A. Mbaye, L.G. Ferreira, and A. Zunger, *Phys. Rev. Lett.* **58**, 49 (1987).
20. A.E. Carlsson and J.M. Sanchez, *Solid State Commun.* **65**, 527 (1988).
21. T. Mohri, K. Terakura, T. Oguchi and K. Watanabe, *Acta Metall.* **36**, 547 (1988).
22. A. Zunger, S.-H. Wei, A.A. Mbaye and G.L. Ferreira, *Acta Metall.* **36**, 2239 (1988).
23. S. Takizawa, K. Terakura and T. Mohri, *Phys. Rev.* **B39**, 5792 (1989).
24. L.G. Ferreira, S.-H. Wei and A. Zunger, *Phys. Rev.* **B40**, 3197 (1989); *ibid.* **B41**, 8240 (1990).
25. M. Sluiter, D. de Fontaine X.Q. Guo R. Podloucky and A.J. Freeman, *Physical Rev.* **B42**, 10460 (1990).
26. J.D. Becker, J.M. Sanchez and J.K. Tien, *Mat. Res. Soc. Symp. Proc., Vol. 213*, pp. 113-118, 1991.
27. J.M. Sanchez, J.P. Stark and V.L. Moruzzi, *Phys. Rev. B, in press.*
28. J.M. Sanchez, J.D. Becker and A.E. Carlsson, in *Computer Aided Innovation of New Materials*, M. Doyama, T.Suzuki, J. Kihara and R. Yamamoto (Editors), pp. 791-794, Elsevier Science Publishers, 1991.
29. V. L. Moruzzi, J. F. Janak and K. Schwarz, *Phys. Rev.* **B37**, 790 (1988).
30. J.M. Sanchez and D. de Fontaine, in *Structure and Bonding in Crystals*, M. O'Keeffe and A. Navrotsky (Editors), Vol. II, p. 117, Academic Press, 1981.

On the Causes of Compositional Order in NiPt Alloys

B. L. Györffy[1], F. J. Pinski[2], B. Ginatempo[3], D. D. Johnson [4], J. B. Staunton[5], W. A. Shelton*[6], D. M. Nicholson[6], and G. M. Stocks[6]

[1] *H.H. Wills Physics Laboratory*
University of Bristol
Bristol BS8 1TL, U.K.

[2] *Department of Physics*
University of Cincinnati
Cincinnati, Ohio 45221, U.S.A.

[3] *Istituto di Fisica Teorica*
Universta di Messina
Messina, Italy

[4] *Sandia National Laboratories*
Livermore, California 94551-0969, U.S.A.

[5] *Department of Physics*
University of Warwick
Coventry CV4 7AL, U.K.

[6] *Metals and Ceramics Division*
Oak Ridge National Laboratory
Oak Ridge, Tennessee 37831, U.S.A.

Abstract

We review, briefly, the arguments which gave rise to the current controversy concerning the origin of compositional order in $Ni_cPt_{(1-c)}$ alloys. We note that strain fluctuations

* Present address: Computational Physics Inc., P.O. Box 788, Annandale, Virginia, U.S.A.

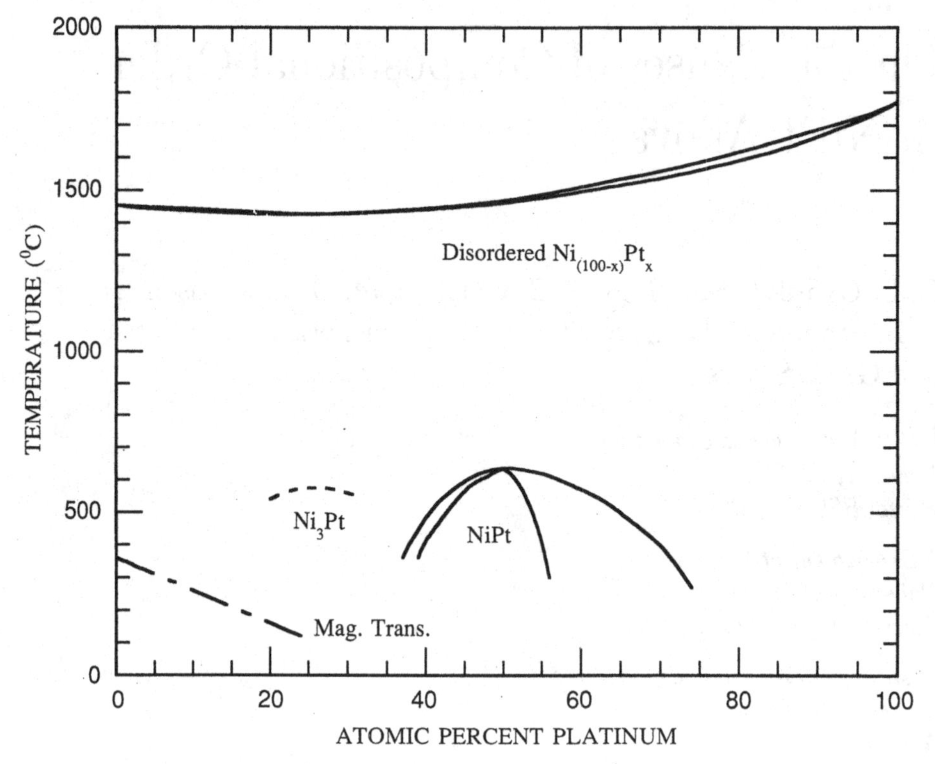

Figure 1. NiPt phase diagram.[1]

play an important role in determining the state of compositional order in this system and outline a theoretical framework that takes account of them.

I. The Nature of the Dilemma

As shown in fig. 1 at high temperatures (T > 1000 K) $Ni_cPt_{(1-c)}$ alloys form *fcc* solid solutions for all concentrations.[1] Moreover, they display a robust tendency for compositional order. Experimentally this manifests itself in significant ordering-type short range order in the disordered, solid-solution phase and in formation of various ordered structures ($L1_2$, $L1_0$ etc.) upon slow cooling to low temperatures. The former is observed in neutron diffuse-scattering experiments[2] and the evidence for the latter is provided by the well studied super-lattice Bragg peaks. Concentration oscillations at the surfaces of these alloys[3] also suggest a strong thermodynamic preference for Ni-Pt rather than Pt-Pt or Ni-Ni neighbours. Phonons, measured and calculated also tell the same tale.[4] The dilemma we shall discuss here concerns not the fact but the physical origin of the above well established phenomena.

There are two microscopic approaches to the problem of explaining the occurrence of compositional short and long range order. The first is based on effective lattice Hamiltonians which describe the system in terms of the occupational variable ξ_i, which takes the value 1 if there is an A atom at the i-th site and 0 if the atom at the i-th site is of the B type, and semi-phenomenological interaction parameters. For pairwise interactions only, such a Hamiltonian takes the form

$$H = \frac{1}{2}\sum_{ij}[v_{ij}^{AA}\xi_i\xi_j + v_{ij}^{AB}\xi_i(1-\xi_j) + v_{ij}^{BA}(1-\xi_i)\xi_j + v_{ij}^{BB}(1-\xi_i)(1-\xi_j)] \quad (1)$$

where $v_{ij}^{\alpha\beta}$ is the interaction energy between an α (=A or B) and a β (=A or B) type atom at the sites i and j respectively. In this language an ordering tendency is due to AB bonds being more attractive than AA or BB bonds, namely, $v_{ij} \equiv v_{ij}^{AA}+v_{ij}^{BB}-2v_{ij}^{AB} > 0$ By contrast $v_{ij} < 0$ implies clustering. If the parameters are not known apriori but are determined by fitting to experimental data, (which is the usual *modus operandii* when applying this approach[5,6]) no conflict arises between theory and experiments as to the cause of ordering. Indeed, Dahmani *et al.*[2] found, by fitting to the results of their neutron scattering experiments in the $Ni_cPt_{(1-c)}$ alloys, that v_{ij} is large and positive for nearest neighbour sites and very small for i and j further apart. So at this level of description there is, again, no controversy. However, the matter becomes more complicated if we follow the alternative approach and attempt to answer the question of what drives the ordering process on the basis of a, more or less, first principles, electronic, description of the concentration fluctuations. It is at this level that the observed behaviour of the $Ni_cPt_{(1-c)}$ system becomes problematic.

In short, fairly general, theoretical arguments suggest that alloys of transition metals with roughly half-filled *d*-bands order while those with almost full or almost empty *d*-bands cluster in the disordered state and, therefore, phase separate at low temperatures.[7–11] Moreover, this prediction is borne out by a considerable body of experimental data.[10,11] However, there are many exceptions to the above *rule*. Of these $Ni_cPt_{(1-c)}$ is one of the best known examples.[12]

Treglia and Ducastelle[12] studied this case with some care. However, using an otherwise fairly reliable theoretical scheme, based on a tight-binding model Hamiltonian for describing the electrons, they concluded that there was no simple way of avoiding the prediction that $Ni_cPt_{(1-c)}$ should cluster in the disordered phase and phase separate at low temperatures. In the end, they suggested that the spin-orbit coupling, which was neglected in their non-relativistic treatment of the problem, might give rise to a repulsion between the Pt atoms and hence override the usual band filling argument. To substantiate or reject this interesting hypothesis was the purpose of two recent reconsiderations of the problem by Pinski *et al.*[13] and Lu *et al.*[14,15] Unfortunately, these calculations resulted in apparently conflicting results. The first suggested that the ordering tendency in the $Ni_cPt_{(1-c)}$ system is due to a size effect overriding the band filling mechanism whilst Lu *et al.*[14] concluded that it is a relativistic effect but not specifically to do with the spin-orbit coupling. In this note we wish to examine the implications of these seemingly conflicting results.

The calculations themselves are quite straightforward in both cases and as such are not contoversial. Pinski et al.[13] calculated the Warren-Cowley short-ranged order parameter $\alpha(q)$ and found that it increases from the centre of the Brillouin zone (Γ-point) to the zone boundary (X-point) indicating an ordering tendency (see fig. 2). In addition to self-consistency, these calculations differed from those of Treglia and Ducastelle[12] in their treatment of what, in tight binding language is called *off diagonal* randomness. It was then argued that this feature, which is the consequence of the size difference between the Ni and Pt atoms, was responsible for suppressing the normally powerful band filling mechanism. On the other hand, Lu et al.,[14,15] calculated the formation enthalpy

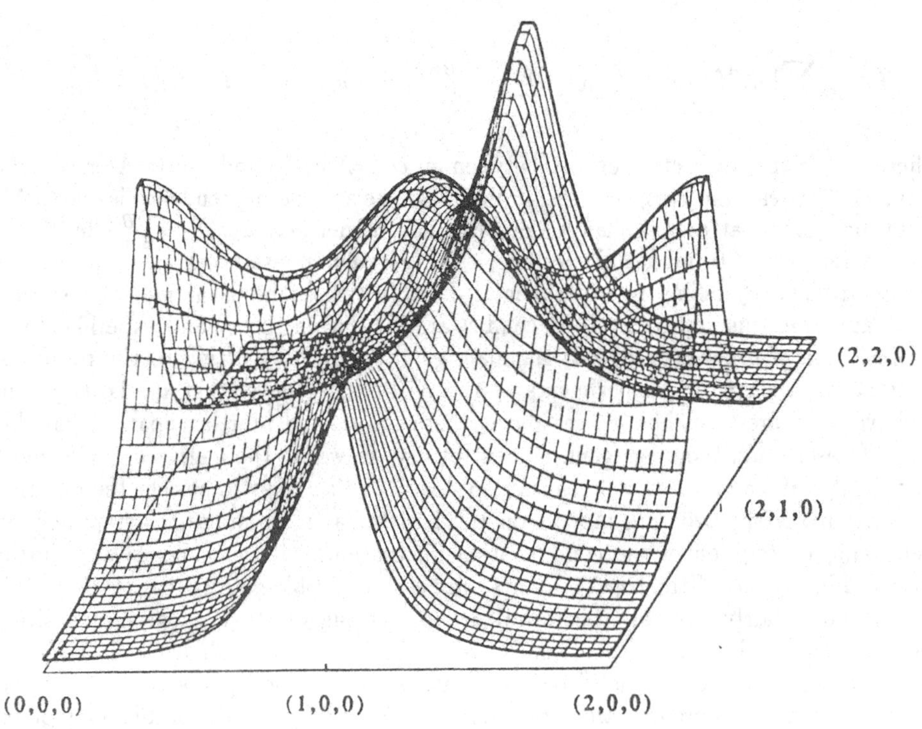

Figure 2. $\alpha_{\vec{q}}$ in the $q_z = 0$ plane for $Ni_{0.5}Pt_{0.5}$ for a temperature about 10% greater than the theoretical ordering temperature. The peaks are at the X-points of the *fcc* Brillouin zone, indicating the tendency for the alloy to order along the (100) direction, *i.e*, into the $L1_0$ structure at low temperatures.[13]

$$\Delta E(V_{L1_0}, V_{Ni}, V_{Pt}) = E(NiPt, V_{L1_0}) - \frac{1}{2}[E(Ni, V_{Ni}) + E(Pt, V_{Pt})], \qquad (2)$$

where V_α is the equilibrium volume of the system indicated by the suffix $\alpha = L1_0$, Ni and Pt respectively. The calculations were performed using state-of-the-art self-consistent, scalar-relativistic and non-relativistic FLAPW calculations.[14,15] As shown in fig. 3 they found that non-relativistically ΔE is positive implying phase separation at low temperatures, but scalar-relativistically it is negative and, therefore, consistent with the observed ordering tendency. Thus, they concluded that ordering, in these alloys, is a relativistic effect, albeit they found the mass-velocity and Darwin corrections to be the decisive factor rather than the spin-orbit coupling mechanism suggested by Treglia and Ducastelle.[12] In fact they argue that spin-orbit coupling favours clustering.

Before attempting to reconcile the apparently conflicting views we wish to digress briefly in order to discuss the theories of compositional order relevant to the above controversy.

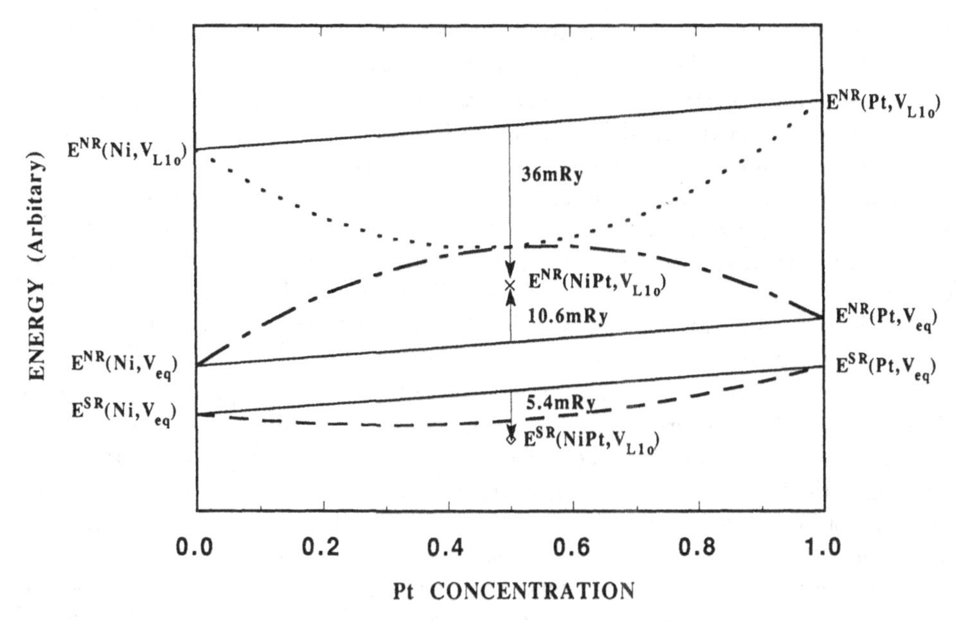

Figure 3. The local density approximation (LDA) ground state energies and configurationally averaged energies of various ordered and disordered NiPt alloys. The superscripts NR and SR denote non-relativistic and scalar-relativistic respectivly. V_{eq} denotes the equilibrium volume for the system specified. V_{L1_0} is the volume of the $L1_0$ ordered compound and $V_{Ni_{0.5}Pt_{0.5}}$ is that for the corresponding solid solution. The points marked by $\times$ and $\diamond$ are due to Lu *et al.*[14,15] The dotted line, dot-dashed line and dashed line are the *conjectured* configurationally averaged energies for solid solution alloys as a function of the Ni concentration c. The dotted line is for a rigid lattice having the volume of corresponding to the 50/50 alloy. Whilst, the dot-dashed and dashed lines denote non-relativistic and scalar-relativistic calculations performed at the equilibrium volumes of the alloys in question.

II. First Principles Theories of Compositional Order

Currently three ways of proceeding are being actively persued: the first is the effective Hamiltonian approach advocated by Connolly and Williams,[16] the second is the electronically based mean field theories of Ducastelle and Gautier[17] and Gyorffy and Stocks,[18,19] and the third is the hybrid, quasi-random structures scheme of Zunger *et al.*[20] As the latter has not, as yet, been brought to bear on the problem of ordering in metallic alloys we shall not consider it here any further.

The first strategy consists of determining the energy parameters in lattice Hamiltonians, like the one in eq. 1, by requiring that they reproduce the results of first principles total energy calculations for various ordered configurations. The effective Hamiltonian so obtained is then used in full lattice statistical mechanics calculations based on either cluster variational (CVM) or Monte Carlo methods. Clearly, the virtues of this procedure are conceptual simplicity and generality. Its principle draw-back is its lack of foundation in theoretical physics. Namely, there is no guarantee that a suitable lattice Hamiltonian exists which reproduces the complex, non-local, long-ranged energetics of the electron *glue* which holds solids together. For a discussion of the non-covergence of the many-atom-interaction series in solids the reader is referred to the recent paper by

Heine *et al.*[21] Nevertheless, it often works, although in what follows, its applicability to the $Ni_cPt_{(1-c)}$ case will be the subject of some reservations.

The alternative, mean-field theory, approach does not eliminate the electronic variables to produce a lattice model Hamiltonian but attempts to carry out the required sums over the compositional ensemble in terms of electronic free energies. Of course, this cannot be done without serious approximations. In fact, the only tractable scheme, at the moment, is the mean field theory based on the Coherent Potential Approximation (CPA) treatment of the electrons in the compositionally disordered crystal potential. As a simplification the electrons may be described by a tight-binding model,[11,12,17] but frequently a fully first-principles multiple scattering method, based on a density functional theory description of the crystal potential, is used,[18,19] (SCF-KKR-CPA).

Clearly, the main shortcoming of the latter approach is the limitation imposed on it by the mean-field theory assumption. On the other hand, it is a very generally applicable procedure since it makes no assumption about the functional form of the variation of the free energy from configuration to configuration. In other words, the free energy is not broken down into 2,3,..., $n-$body contributions.[21] In particular, the above mean field theory deals readily with the cases where the interactions are long ranged and oscillatory.[18,19] Furthermore, unlike the Connolly-Williams method, but like the method of quasi-random structures, it deals directly with the disordered state whose instability frequently signals the onset of compositional order. Thus, near the transition temperature, T_c, it can be regarded as a first principles Landau theory.[6,19]

Let us now focus attention on the correlation function $\alpha(ij)$ defined by

$$c_i(1-c_j)\alpha_{ij} = \langle \xi_i \xi_j \rangle - \langle \xi_i \rangle \langle \xi_j \rangle. \tag{3}$$

In the disordered state where $c_i = \bar{c}$ for all i and $\alpha(ij)$ depends only on the vector $\vec{\mathbf{R}}_{ij}$ which connects the sites i and j, the lattice Fourier transform of $\alpha(ij)$ can be written in terms of the Ornstein-Zernike direct correlation function $\alpha_{\vec{\mathbf{q}}}$ as follows[18]

$$\alpha_{\vec{\mathbf{q}}} = \frac{1}{1-\beta\bar{c}(1-\bar{c})S^{(2)}(\vec{\mathbf{q}})}, \tag{5}$$

where β is the inverse temperature $1/k_BT$. Note that $\alpha(\vec{\mathbf{q}})$ is the Warren Cowley short-range order parameter mentioned earlier and is measured in diffuse scattering experiments. The above formula is exact and can be taken as the definition of the direct correlation function $S^{(2)}(\vec{\mathbf{q}})$. The reason for an interest in $S^{(2)}(\vec{\mathbf{q}})$ is that it is more directly related to the interactions between atoms than is α_{ij} which, like the radial distribution function in liquids, is dominated by simple packing considerations. Thus a useful way of comparing different theories and experiments is to compare the direct correlation functions they imply. For instance, the usual mean field approximation[6,22,23] for the model Hamiltonian given in eq. 1, results in $S^{(2)}(\vec{\mathbf{q}}) = v(\vec{\mathbf{q}})$ which is the lattice Fourier transform of v_{ij}. Of course, when the CVM or Monte Carlo calculations are used the relationship between the potential parameters, $v(\vec{\mathbf{q}})$, and the direct correlation function is more complicated. Nevertheless, its calculation in either case is straightforward.

In the case of the first principles mean field theory we have referred to above, $S^{(2)}(\vec{\mathbf{q}})$ works out to be the lattice Fourier transform of

$$S^{(2)}_{ij} = \left(\frac{\partial^2 \Omega^{CPA}}{\partial c_i \partial c_j} \right)_{c_i, c_j = \bar{c}\ \forall i}, \tag{5}$$

where $\Omega^{CPA}(\{c_i\})$ is the electronic grand potential averaged, using the coherent potential approximation, with respect to the inhomogeneous product distribution function

$$P(\{\xi_i\}) = \prod_i p_i(\xi_i), \tag{6}$$

with each factor being parameterized by the local concentration c_i as follows

$$p_i(\xi_i) = c_i\xi_i + (1 - c_i)(1 - \xi_i). \tag{7}$$

Starting with the basic equation of the inhomogeneous KKR-CPA, which corresponds to the distribution in eq. 6, one can derive an explicit expression for $S^{(2)}(\vec{\mathbf{q}})$ in terms of quantities which are readily available at the end of an SCF-KKR-CPA calculation. The results shown in fig. 2 were calculated using this theory retaining only the band structure contributions to the electronic grand potential *i.e.* contributions resulting from double counting corrections have been neglected, an approximation that is expected to be valid in alloys where charge transfer is small. To make it explict that $S^{(2)}(\vec{\mathbf{q}})$ has the form of a susceptibility it is possible to approximate the full expression by

$$S^{(2)}(\vec{\mathbf{q}}) = |M|^2 \sum_{\vec{\mathbf{k}}} \int d\epsilon \int d\epsilon' \frac{f(\epsilon) - f(\epsilon')}{\epsilon - \epsilon'} A^B(\vec{\mathbf{k}}, \epsilon) A^b(\vec{\mathbf{k}} + \vec{\mathbf{q}}; \epsilon') \tag{8}$$

where $|M|^2$ indicates matrix elements, $f(\epsilon)$ is the usual Fermi function and $A^B(\vec{\mathbf{k}}, \epsilon)$ is the Bloch spectral function which describes the electronic structure of the random alloy.

As is well known if, in the solid solutiuon, the compositional fluctuations are such that like atoms cluster together $\alpha(\vec{\mathbf{q}})$ has a peak at $\vec{\mathbf{q}} = 0$. In the opposite case of ordering type fluctuations $\alpha(\vec{\mathbf{q}})$ peaks at $\vec{\mathbf{q}}_0 \neq 0$ where $\vec{\mathbf{q}}_0$ is the wave vector of the compositional modulation. When the fluctuations are precursors to the $L1_0$ ordered state, $\vec{\mathbf{q}}_0$ is one of the X-points in the Brillouin zone. It will be of interest later that, in principle, $\alpha(\vec{\mathbf{q}})$ may peak at both the $\Gamma-$ and X-points. This is a very interesting circumstance because it may give rise to competition between phase separation and ordering. In particular, for $c = 0.5$ on a *fcc* lattice, both phase separation and ordering into the $L1_0$ structure are second order phase transitions and hence are due to instabilities of the disordered phase to $\vec{\mathbf{q}} = 0$ and $\vec{\mathbf{q}} = \vec{\mathbf{q}}_0$ fluctuations respectively. These instabilities are signalled by the divergence of $\alpha(\vec{\mathbf{q}})$ at the appropriate wave vectors. In the case of the two peaks envisioned above, as the temperature is lowered, one of them wins by diverging first with a transition temperature T_c. The state corresponding to the other becomes a metastable state for $T < T_c$.

Returning to our main concern, we note that the $\vec{\mathbf{q}}$ dependence of $\alpha(\vec{\mathbf{q}})$ for $Ni_{0.5}Pt_{0.5}$ that is shown in fig. 2 implies ordering type short range order in the disordered state.[13] Pinski *et al.*[13] found that the structure in $\alpha(\vec{\mathbf{q}})$ arose predominantly from hybridization of t_{2g} *d*-states produced by a balance between diagonal and off-diagonal disorder. Namely, the random overlap of atoms drives ordering and overpowers the usual band-filling effect which, for $Ni_cPt_{(1-c)}$, should result in phase separation. The differences between the charge overlaps of Pt-Pt, Pt-Ni and Ni-Ni nearest neighbour pairs may be described as a consequence of Pt being a big atom and Ni being a small atom. Consequently, the above mechanism of ordering was regarded as the quantum mechanical description of the empirical rule, refered to by Hume-Rothery,[24] that big atoms and small atoms order if they mix at all.

Whilst these results are in good qualitative agreement with the experimental facts they are, clearly, in conflict with the unexceptionable non-relativistic calculations of Lu *et al.*[14] which, as will be recalled, yield a positive enthalpy of mixing, ΔE, and hence imply, on the basis of the third law of thermodynamics, phase separation. Although, relativistically they find $\Delta E < 0$ and, hence, conclude that ordering is a relativistic effect, the matter needs further attention for two reasons. Firstly, the inconsistency of the non-relativistic results may hid some conceptual flaw in our understanding of ordering and phase separation in alloys. Secondly, whilst the relativistic calculations conclusively indentified the ground state, they do not conclusively indentify the ordering mechanism. Fortunately, some recent calculations[15,25] clarify, to some extent, both aspects of the problem.

Recently, Lu *et al.*[15] found that, if the ground state energies of the pure metals in eq. 2 are calculated at the volume of the ordered $L1_0$ intermetallic compound, rather than at their respective equilibrium volumes, ΔE turns out to be negative even in a non-relativistic calculation. Thus as shown in fig. 3 $E(L1_0)$ is above the $cE^{NR}(Ni, V_{NiPt})+(1-c)E^{NR}(Pt, V_{NiPt})$ line because pure Ni contracts and pure Pt expands, compared with the volume of the intermetallic compound, in order to minimize the alloy ground state energy.

Also, we have calculated $\bar{E}(Ni_{0.5}Pt_{0.5}, V_{Ni_{0.5}Pt_{0.5}})$, the configurationally averaged energy for the disordered $Ni_{0.5}Pt_{0.5}$ alloy, using the same SCF-KKR-CPA method that was used in calculating $\alpha(\vec{q})$ in our previous paper.[13] As expected, on the basis of the strong ordering tendency we found earlier, $\bar{E}(Ni_{0.5}Pt_{0.5}, V_{Ni_{0.5}Pt_{0.5}})$ lies above $E(NiPt, V_{L1_0})$ [25]. The SCF-KKR-CPA is exact in the impurity limit, for a fixed crystal potential function which, in the present case, is the local density approximation (LDA) functional,[19] and therefore reduces, smoothly, to the pure metal calculation for $c = 0$ and 1. Therefore, on the basis of the above results, we conjecture that when the average energy, $\bar{E}(c)$, is calculated as a function of the concentration it will be given by either the dotted or the dot-dashed lines of fig. 3 depending on whether the lattice parameter is kept at its value for the 50/50 disordered alloy or is relaxed in order to minimise the average energy at each concentration. These conjectures together with fig. 2 provide the following simple explanation of the inconsistancy in the non-relativistic theory.

As noted before phase separation is a $\vec{q} = 0$ instability described by

$$\alpha(0) = \frac{1}{1 - \beta\bar{c}(1-\bar{c})S^{(2)}(0)}, \tag{9}$$

where

$$\begin{aligned} S^{(2)}(0) &= \frac{1}{N}\sum_{ij} S^{(2)}_{ij}, \\ &= \frac{1}{N}\sum_{ij}\left(\frac{\partial^2\Omega^{CPA}}{\partial c_i \partial c_j}\right) \\ &= \frac{1}{N}\frac{\partial^2\Omega^{CPA}(c)}{\partial c^2}, \end{aligned} \tag{10}$$

and $\Omega^{CPA}(c)$ is the configurationally averaged electronic grand potential for the homogeneous alloy as calculated within the SCF-KKR-CPA. Evidently, $\Omega^{CPA}(c)$ should be used for $\bar{E}(Ni_c Pt_{(1-c)}, V_\alpha)$ in fig. 3. Thus, on a rigid lattice $\Omega^{CPA}(c)$ is given by the dotted line in fig. 3 and hence $S^{(2)}(0) > 0$ on account of eq. 10. Consequently, as is clear from eq. 9, $\alpha(0)$ will not diverge at any temperature. This is consistent with the $\alpha(\vec{q})$ calculated by Pinski *et al.*[13] non-relativistically and on a rigid lattice. As shown in fig. 2 $\alpha(\vec{\mathbf{q}})$ is peaked at the zone boundary and only its value at the X-point can diverge. In short, both the conjectured shape of $\Omega^{CPA}(c)$ on a rigid lattice and the zone boundary value of $\alpha(\vec{\mathbf{q}})$ indicate ordering.

On the other hand if the lattice is allowed to relax at each concentration, $\Omega^{CPA}(c)$ is conjectured to be given by the dot-dashed line in fig. 3, thus $(\partial^2\Omega^{CPA}(c)/\partial c^2) < 0$ and therefore $S^{(2)}(0) > 0$. Consequently, $\alpha(0)$ will diverge at some temperature T_c^{ps} signaling phase separation in agreement with the arguments of Lu *et al.*[14] Unfortunately, the function $\Omega^{CPA}(c)$ is not sufficient to determine what will happen to $\alpha(\vec{\mathbf{q}})$ at finite wave vectors. It can only predict that the solid solution will phase separate unless some other instability, at finite $\vec{\mathbf{q}} = \vec{\mathbf{q}}_0$, interfers at some $T_c^{\vec{\mathbf{q}}_0} > T_c^{ps}$. Since the mean-field theory used by Pinski *et al.*[13] was designed to describe compositional fluctuations on a rigid lattice, it is clear that it cannot describe $\alpha(\vec{\mathbf{q}})$ in the present circumstance dominated by lattice relaxation effects.

One way forward is to follow the Connolly-Williams argument[16] and fit effective Hamiltonians to fully relaxed total energy calculations such as those by Lu *et al.*[14] Whilst $\alpha(\vec{\mathbf{q}})$ obtained in this way may lead to useful insights it is frought with difficulties because the potential parameters representing elastic forces are likely to be long ranged.

For these reasons, as well as for wishing to describe the ordering process in explicitly electronic terms, in the next section we shall outline a generalization of the mean-field theory that includes an account of stain fluctuations on an equal footing with compositional fluctuations.

III. Strain Fluctuations and the State of Compositional Order

Even in crystalline solids atoms do move about. At finite temperature they move from lattice site to lattice site by diffusion, as well as vibrating about their equilibrium positions. In a multicomponent system this means that a given site is occupied by different atoms at different times. Evidently, when a big atom replaces a small one, the environment of the site in question responds by expanding. Alternatively, when a large atom is interchanged with a small atom the neighbouring atoms relax towards the site. Here, we are interested in describing these strain fluctuations whithin the same kind of first principles framework that we applied to concentration fluctuations. Although these strain fluctuations are ubiquitous in metallic alloys and have received considerable experimental and theoretical attention from physicists, metallurgists and material scientists for the best part of a hundred years, very little effort has been directed towards obtaining a fully microscopic, electronic, theory of them. For a careful discussion of this point see Zunger *et al.*[26] Consequently, what follows can only be some very preliminary remarks.

As a straightforward generalization of the mean field theory encapsulated in eqs. 3–7 we may consider the electronic grand potential for the inhomogenous concentration configuration $\{c_i\}$ on a relaxed lattice. That is to say each lattice point $\vec{\mathbf{R}}_i^0$ is assumed to have moved to the relaxed position $\vec{\mathbf{R}}_i = \vec{\mathbf{R}}_i^0 + \vec{\mathbf{u}}_i$ under the influence of the compositional (Kanzaki) forces in order to minimize the free-energy $\Omega^{CPA}(\{c_i\}, \{\vec{\mathbf{u}}_i\})$. Describing the total displacement field by the set $\{\vec{\mathbf{u}}_i\}$, going through the arguments of Gyorffy and Stocks[18] and using the chain rule for differentiating with respect to the local concentrations, we find the following expression for the direct correlation function

$$S_{ij}^{(2)} = S_{ij}^{c,c} + \sum_{l,\alpha} S_{il}^{c,u_l^\alpha} \gamma_{lj}^{\alpha}, \tag{11}$$

where $\alpha = x, y$ and z and

$$S_{ij}^{c,c} = \left(\frac{\partial^2 \bar{\Omega}^{CPA}(\{c_i\}, \{\vec{\mathbf{u}}_i\})}{\partial \mathbf{c}_i \partial \mathbf{c}_j} \right)_{c_i, c_j = \bar{c}\ \forall i}, \tag{12}$$

$$S_{ij}^{c,u_j^\alpha} = \left(\frac{\partial^2 \bar{\Omega}^{CPA}(\{c_i\}, \{\vec{\mathbf{u}}_i\})}{\partial \mathbf{c}_i \partial \mathbf{u}_j^\alpha} \right)_{c_i, c_j = \bar{c}\ \forall i}, \tag{13}$$

$$\gamma_{ij}^\alpha = \left(\frac{\partial u_i^\alpha}{\partial c_i} \right)_{c_i, c_j = \bar{c}\ \forall i}. \tag{14}$$

Evidently, the effects of the strain fluctuations on the short range order are described by the new response functions S_{ij}^{c,u_j^α} and γ_{ij}^α. A theory for the latter, very interesting, quantity (note $\delta u_i^\alpha = \sum_j \gamma_{ij}^\alpha \delta c_j$) can be readily developed in the limit where the displacements, $\{\vec{\mathbf{u}}_i\}$ are regarded as small and their effect on Ω^{CPA} is taken into account only in the Harmonic approximation. Under these assumptions we find

$$\gamma_{ij}^\alpha = \sum_{l,\beta} \Phi_{\alpha\beta}^{-1}(i,l) S_{li}^{c,u_i^\alpha}, \tag{15}$$

where the static force constants $\Phi_{\alpha\beta}(i,j)$ are given by

$$\Phi_{\alpha\beta}(i,j) = \left(\frac{\partial^2 \bar{\Omega}^{CPA}(\{c_i\}, \{\vec{\mathbf{u}}_i\})}{\partial u_i^\alpha \partial u_j^\beta} \right)_{c_i = \bar{c}; \forall i : u_i^\alpha ; u_j^\beta = 0; \forall i} \tag{16}$$

The rest of the theory proceeds along lines entirely analogous to that for $S_{ij}^{c,c}$ [18,19] and the end results are complicated response function formulas (for S_{ij}^{c,u_j^α} and $\Phi_{\alpha\beta}(i,j)$) which have to be evaluated using the results of a SCF-KKR-CPA calculations for a homogenous solid solution with concentration $\bar{c}$. Although the implementation of the theory is a major computational task, the above formulas provide a fairly concise conceptual framework for contemplating the influence of strain fluctuations on the compositional short ranged order. The extension of these ideas to include long range order is also straightforward.

From the point of view of our present concern there are a couple of simple comments which follow from the above considerations.

i) Strain fluctuations induce an elastic contribution to the direct correlation function $S_{ij}^{(2)}$ or, in the mean field theory to the interchange energy $v(\vec{\mathbf{q}})$. As was stressed by Khachaturyan,[6] for alloys of large and small atoms this can be a surprisingly large effect and hence it may play an important role in the case of the $Ni_cPt_{(1-c)}$ system.

ii) Given the way elastic forces propagate in solids the strain fluctuation contribution to $S_{ij}^{(2)}$ is bound to be long ranged. Indeed, in the elastic dipole limit, we may expect contributions which fall off as $S_{ij}^{(2)} = |\vec{\mathbf{R}}_{ij}|^{-3} \exp(-\zeta/\vec{\mathbf{R}}_{ij})$. Thus the problem at hand is like that of dislocations where both the short-ranged bonds, in the core, and the long ranged, elastic, forces are equally important. From this point of view, the first principles mean-field theory method advocated here is particularly promising since it requires calculations in $\vec{\mathbf{q}}$-space and treats the small and large $\vec{\mathbf{q}}$ limits on equal footing.

In this respect, it is quite different from supercell or finite cluster based methods.[20] These considerations are also relevant to the applicability of the Connolly-Williams type of approaches. Clearly, to account for large elastic interactions, the effective Hamiltonian must include long-ranged forces for which both the CVM and the Monte Carlo method become difficult to do. Fortunately, as was recently demonstrated by Marias *et al.*,[27] under these circumstances the mean-field theory becomes a better and better approximation.

Having sketched out a framework for treating the effects of strain fluctuations on the ordering process let us return to the dilema posed by the results of the non-relativistic theory. To begin with we note that from eq. 11 we have

$$\begin{aligned} S^{(2)}(0) &\equiv \frac{1}{N}\sum_{ij} S^{(2)}_{ij}, \\ &= \frac{1}{N}\sum_{ij}\left(S^{cc}_{ij} + \sum_{l,\alpha} S^{c,u^\alpha_l}_{il}\gamma^\alpha_{i,l}\right), \\ &= \frac{1}{N}\left(\frac{\partial^2\Omega^{CPA}(c,a)}{\partial c^2} + \frac{\partial^2\Omega^{CPA}(c,a)}{\partial c\partial a}\frac{\partial a}{\partial c}\right), \\ &= \frac{1}{N}\left(\frac{d^2\Omega^{CPA}(c,a)}{dc^2}\right)_{c=0.5}. \end{aligned} \tag{17}$$

Where a is the lattice constant which is assumed to be a function of the concentration c. Thus, as anticipated at the end of the previous section, the direct correlation function at the Γ-point is given by the second derivative of $\Omega^{CPA}(c,a)$, the electonic grand potential for the relaxed lattice. The dot-dashed line in fig. 3 is our conjecture for $\Omega^{CPA}(c,a)$. Hence, as before, we conclude that the disordered phase will be unstable to phase separation at T_c^{ps} unless an ordering instability intervens.

Interestingly, such an unusual circumstance has a good chance of occuring in this system. If we assume that the contribution to $S^{(2)}(\vec{q})$ by the strain fluctuations are of the clustering type for all $\vec{q}$ then, by taking the Fourier transform of eq. 11 and writing

$$S^{(2)}(\vec{q}) = S^{c,c}(\vec{q}) + \sum_\alpha S^{c,u^\alpha}(\vec{q})\gamma^\alpha(\vec{q}), \tag{18}$$

we arrive at the conclusion that $S^{(2)}(\vec{q})$ is given by a sum of two sets of peaks. The members of one set are centered on the Γ-points and are the contributions from the second term in eq. 18, whilst the second set are located at the X-points and come from the first term. Although possible, it is unlikely that the superposition of such peaks will result in monotonic behaviour of $S^{(2)}(\vec{q})$ and $\alpha(\vec{q})$. In short, along the Γ-X line we conjecture a two peaked structure for $\alpha(\vec{q})$ in a non-relativistic theory that includes strain fluctuations. As the temperature is lowered, which of the two gives rise to the first instability will be determined only when full calculations of the new response functions in eq. 18 have been calculated.

Nevertheless, it is instructive to analyse the above hypothetical situation a bit further. Of course, at $T = 0$, the absolute free energy minimum is that of the ground state which, given the ground state search of Lu *et al.*,[14] may be safely assumed to be a phase separated state. However, reaching that state from the homogeneous solid solution phase will be difficult if there is a sizable ordering peak, even if it is not larger

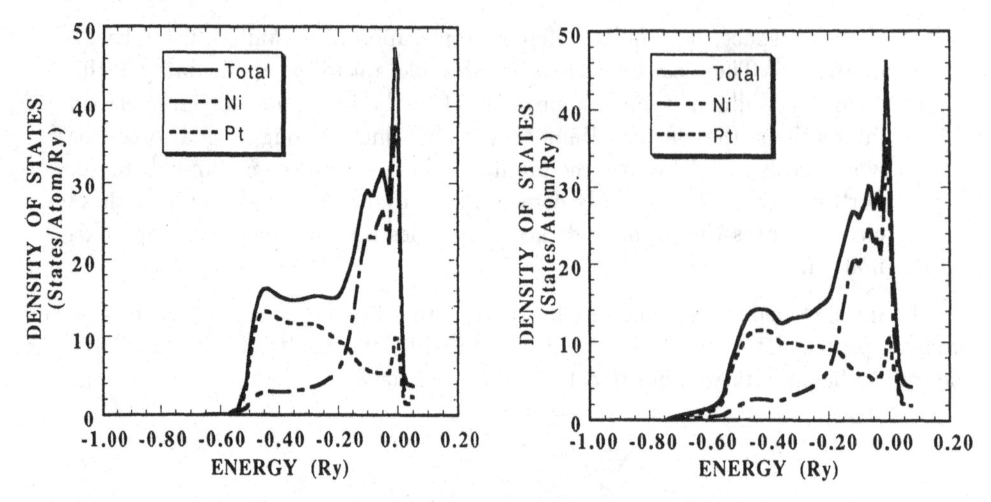

Figure 4. Non relativistic (left) and relativistic (right) SCF-KKR-CPA densities of states (DOS) for $Ni_{0.5}Pt_{0.5}$. Solid line: total DOS. Dashed line: Ni-contribution. Dotted line: Pt-contribution. Charge self consistency reduces the effects of the mass-velocity and Darwin terms, and the relativistic and non relativistic DOS's look surprisingly similar, confirming, indirectly, the validity of the nonrelativistic approach to NiPt.

than the peak at the zone center. The point is that if nearest neighbour exchange of atoms favours unlike neighbour pairs the system will have to wait for long, perhaps astronomically long, times for a small $\vec{q}$ clustering type fluctuation which could ultimately lead to phase separation. Thus a likely outcome of cooling at moderate rates would be a metastable ordered state. Clearly, finding real systems, as opposed to the above hypothetical non-relativistic Ni_cPt_{1-c} alloys, with competing instabilities would be of great physical interest.[28]

Finally, let us speculate on the outcome of a relativistic theory that would describe the experiments on the Ni_cPt_{1-c} alloy system. The relativistic calculations of Lu *et al.*,[14] displayed in fig. 3, suggest that the contribution from strain fluctuations are reduced due to a relativistic contraction of the Pt atom. Indeed from our conjectured plot of $\Omega^{CPA}(c,a)$ for the relativistic theory shown in fig. 3 we expect $\left(\partial\Omega^{CPA}/\partial c^2\right) < 0$ and hence the phase separation instability to be supressed. At the same time we do not expect $S^{c,c}(\vec{q})$ to change very much. In support of this expectation we show in fig. 4 the densities of states for the $Ni_{0.5}Pt_{0.5}$ random alloy from non-relativistic and fully (Dirac equation[29]) relativistic SCF-KKR-CPA calculations. Suprisingly the differences are small, particularly in the hybridized *d*-band complex which, it will be recalled, is the region in the densities of states identified by Pinski *et al.*[13] as giving rise to the X-point peak in $\alpha(\vec{q})$. Thus, we may assume that $S^{c,c}(\vec{q})$ obtained on the basis of a fully relativistic theory will have a negative peak at the zone boundary as before. This suggests that $\alpha(\vec{q})$ will have the same shape as in the non-relativistic calculation whose reults are shown in fig. 2.

In summary thare are two size effects operating in the proplem: one works through strain fluctuations and leads to reduced solubility, the other is a manifestation of the random overlaps of atoms and can cause ordering even when the band filling argument favours clustering. Evidently, in the context of first principles quantum mechanical calculations the first was discovered by Lu *et al.*[14] and the second has been identified by Pinski *et al.*[13] The cause of order in the $Ni_{0.5}Pt_{0.5}$ alloy may then be described as follows. The relativistic contraction of the Pt *s*-orbital[14] eliminates the $\vec{q} = 0$ instability and thereby suppresses immiscibility and therefore allows the second size effect to drive the $L1_0$ ordering process. If substantiated by explicit calculations this explanation will fit nicely with Hume-Rothery's general conclusion[24] that:

It is therefore natural that, provided the solvent and the solute atoms are of sufficiently similar size to permit the formation of a wide solid solution, the tendency to form super-lattices increases with increasing difference in atomic diameters, since the greater this difference, the greater the strain to be relieved. Thus, superlattices are found in the system copper-gold, where the atomic diameters are Cu=2.54 Å, Au =2.88 Å but not in the system silver-gold, where the sizes of the two atoms are nearly the same.

Acknowledgements

Work partially supported by Cray Research, Inc., and the Ohio Supercomputer Center; by Consiglio Nazionale Delle Ricerche (Italy); by Department of Energy, Basic Energy Sciences, Division of Material Sciences; by Science and Engineering Research Council, U.K. and NATO; by U.S. Department of Energy Assistant Secretary of Conservation and Renewable Energy, Office of Industrial Technologies, Advanced Industrial Concepts Materials Program, under subcontract DEAC05-84OR21400 with Martin-Marietta Energy Systems, Inc.; and by Division of Materials Science Office of Basic Energy Sciences, U.S. Department of Energy under subcontract DEAC05-84OR21400 with Martin-Marietta Energy Systems,Inc., and by a grant of computer time at NERSC from DOE-BES-DMS.

References

1. J.L. Murray L. H. Bennett and H. Baker, editors, *Binary Alloy Phase Diagrams*, Vol. 2, American Society for Metals (1986)
2. C. E. Dahmani M. C. Cadeville J. M. Sanchez and J. L. Morán-López, *Phys. Rev. Lett.* **55**, 1208 (1985).
3. R. Baudoing Y. Gauthier M. Lundberg and J. Rundgren, *J. Phys. C* **19**, 2825 (1986).
4. A. Mookerjee and R. P. Singh, *J. Phys. C* : **18**, 4261 (1985).
5. D. de Fontaine, *Solid State Physics*, Eds. H. Ehrenreich, F. Seitz and D. Turnbull, Vol. 34, Academic Press, New York, 1979.
6. A. G. Khachaturyan, *Theory of Structural Transformations in Solids*, Wiley, New York, 1983.
7. M. Cyrot and F. Cyrot-Lackmann, *J. Phys. F*, **6**, 2257 (1976).
8. D. G. Pettifor, *Phys. Rev. Lett.*, **42**, 846 (1979).
9. J. Van der Rest F. Gautier and F. Brauers, *J. Phys. F*, **5**, 2283 (1975).
10. V. Heine and J. Samson, *J. Phys. F*, **13**, 2155 (1983).
11. A. Bieber and F. Gautier, *Acta Metall.*, **34**, 2291 (1986).
12. G. Treglia and F. Ducastelle, *J. Phys. F*, **17**, 1935 (1987).
13. F.J. Pinski B. Ginatempo D. D. Johnson J. B. Staunton G. M. Stocks and B. L. Gyorffy, *Phys. Rev. Lett.*, **66**, 766 (1991).
14. Z. W. Lu S.-H. Wei and A. Zunger, *Phys. Rev. Lett.*, **66**, 1753 (1991).

15. Z. W. Lu S.-H. Wei and A. Zunger, 1991, "Reciept of a preprint is gratefully acknowledged"
16. J. W. D. Connolly and A. R. Williams, *Phys. Rev.*, **B27**, 5169 (1983).
17. F. Ducastelle and F. Gautier, *J. Phys. F*, **6**, 2039 (1976).
18. B. L. Gyorffy and G. M. Stocks, *Phys. Rev. Lett.*, **50**, 374 (1983).
19. B. L. Gyorffy D. D. Johnson F. J. Pinski D. M. Nicholson and G. M. Stocks, *Alloy Phase Stability*, Eds. G.M. Stocks and A. Gonis, NATO-ASI, Klewer, Boston, 1989, p. 421.
20. A. Zunger S. H. Wei L.G. Ferreira and J.E. Bernard, *Phys. Rev. Lett.*, **65** 353 (1990).
21. V. Heine, I. J. Robertson and M. C. Payne, *Phil. Trans. R. Soc. Lond.* **A334**, 393 (1991).
22. M. E. Fisher, *J. Math. Phys.*, **5**, 944 (1964).
23. S. C. Moss, *Phys. Rev. Lett.*, **22**, 1108 (1969).
24. W. Hume-Rothery and B. R. Coles, *Atomic Theory for Students of Metallurgy*, Vol. 3, Institute of Metals, 1969, p. 122.
25. D. D. Johnson, unpublished.
26. A. Zunger S. H. Wei A. A. Mbaye and L. G. Ferreira, *Acta Metall.*, **36**, 2239 (1988).
27. S. Marias V. Heine C. Nex and E. Salje, *Phys. Rev.* **B66**, 2480 (1991).
28. J. V. Ashby and B. L. Gyorffy, *Phil. Mag. B.*, **53**, 269 (1986).
29. B. Ginatempo W. A. Shelton E. Bruno, F. J. Pinski and G. M. Stocks, unpublished.

Ultrathin Films of Transition Metals and Compounds: Electronic Structure, Growth and Chemical Order

Daniel Stoeffler and François Gautier

I. P. C. M. S.- GEMME (U. M. R. 46 C. N. R. S.)
Université Louis Pasteur
4 rue Blaise Pascal
67070 Strasbourg
France

Abstract

The growth of deposited and co-deposited epitaxial overlayers of pure transition metals (TM) A and TM ordered compounds ($A_cA'_{1-c}$) is related to their electronic structure and to the phase and surface stabilities of the corresponding bulk materials. We use a simple tight binding model (TBM) and the recursion method to get general trends. We show that a layer by layer or bilayer by bilayer growth of a compound $A_cA'_{1-c}$ is possible even if one of the constituents of the compound has a high surface energy.

I. Introduction

The growth of epitaxial overlayers during the initial stage occurs by one of three well-known mechanisms: the Frank-Van der Merve (FM), the Stranski-Krastanov (SK) and the Volmer-Weber (VM) mode which correspond respectively to a layer by layer deposition, to the nucleation of 3D crystals after a layer by layer deposition of n monolayers on the substrate, and to the formation of 3D crystals onto the substrate. A macroscopic theory of wetting phenomena predicts either FM or VW mode according to the sign of the spreading coefficient:[1]

$$S_{A/B} = \gamma_A + \gamma_{AB} - \gamma_B \tag{1}$$

Structural and Phase Stability of Alloys
Edited by J.L. Morán-López *et al.*, Plenum Press, New York, 1992

γ_B and γ_A are the surface free energies of the substrate and of the film and γ_{AB} is the substrate-film interface free energy. This macroscopic theory, which has been extended to solid films including the elastic strains,[2] cannot be applied in the monolayer (ML) range.

Up to now, neither theoretical microscopic models nor systematic experimental data on interfacial energies are available to understand and *a fortiori* to predict the conditions to get epitaxial ultrathin films and perfect superlattices. However, Bauer and Van der Merve[1,3] introduced simple and relevant phenomenological criteria. Using the condition Eq. (1) they assume that the interfacial energies are much smaller than the surface energies so that the substrate is wetted by A if $\gamma_A < \gamma_B$. However, to epitaxy A on B and B on A, *i. e.* to get superlattices, it is therefore necessary that γ_A and γ_B be of the same order of magnitude ($S_{A/B}$ and $S_{B/A} < 0$). From the experimental data on some systems they suggest that A_mB_n superlattices can be obtained if $|\gamma_A - \gamma_B| \lesssim (\gamma_A + \gamma_B)/2$. Although γ_{AB} is certainly much smaller than the surface energies, the range of validity of this thermodynamic criterion is not clearly established.

In the first part of this paper, we examine the growth mode (FM, SK, VW) of ultrathin epitaxial overlayers of A on a perfect substrate B. We summarize the results we obtained recently[4] from electronic theory. The wetting of ν monolayers is obtained by a balance between the various contributions of the spreading energy, but the phenomenological criterion $\gamma_A < \gamma_B$ is shown to be valid for $\nu = 1$ in most cases. A layer by layer growth is obtained only for $\nu \geq 2$ (SK mode), so that, from these theoretical results, the growth of superlattices is possible only when the Bauer-Van der Merve criterion is satisfied and when a limited interdiffusion is allowed. For illustration, the Co-Ru and Co-Cr systems are discussed.

The second part of the paper is devoted to the growth of coevaporated compounds ($A_cA'_{1-c}$). It is qualitatively related to the surface energies of the bulk compounds $\gamma_{(AA')}$ and to the interfacial energy $\gamma_{(AA')B}$. A preliminary study of $\gamma_{(AA')}$ shows that, for stable compounds, this energy is usually larger than the corresponding average of γ_A and $\gamma_{A'}$ at the same volume and crystalline structure, the lowest surface energy being obtained when the number of A' atoms (with $\gamma_{A'} < \gamma_A$) is the largest. Then, we assume that ν A and A' atoms per interfacial site are simultaneously deposited on the substrate B, the surface diffusion being sufficient to reach the most stable configuration and the interdiffusion being quenched. We present the results for typical situations, *i. e.* for (HfW)/Ta, (IrTa)/W, (IrW)/Re, (IrW)/Ru. We investigate the relevant parameters which determine the progressive growth of 2D compounds and more especially the 2D compound stability as compared to the bulk one. We show that a layer by layer growth of a compound AA' is possible even if one of the constituents (A) of the compound has a high γ_A value.

II.1 Electronic Structure and Growth Model

We consider overlayers of $N_A = N \cdot \nu$ atoms A made of ν monolayers ($\nu = 1, 2, \cdots$) which are assumed to be in epitaxy with a perfect surface of a semi-infinite substrate B. The total energy of this system is:

$$\begin{aligned} E &= N_B \cdot E_B^\circ + N_A \cdot E_A^\circ + N \cdot (\gamma_B + \gamma_\nu^\circ) \\ &= N_B \cdot E_B^\circ + N_A \cdot E_A^\circ + N \cdot (\gamma_A + \gamma_{1,\nu}^\circ) \end{aligned} \tag{2}$$

$N_B E_B^\circ$ and $N_A E_A^\circ$ are the bulk energies of N_B and N_A atoms of the substrate and the adsorbate respectively. N is the number of adsorption sites. We consider here only

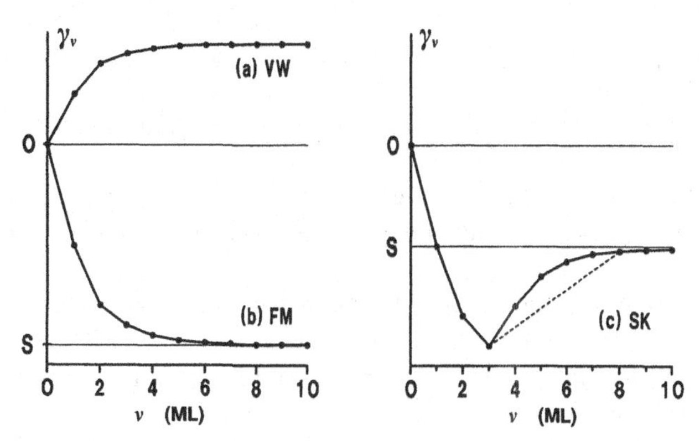

Figure 1. Variation of the spreading energy versus coverage for (a) Volmer-Weber (VW), (b) Franck-Van der Merwe (FM) and (c) Stranski-Krastanov (SK) (with $N = 3$) growth modes.

the thermodynamic limit (N, $N_A \to \infty$ whereas $\theta = N_A/N$ remains finite) so that we do not study the equilibrium shape of two- or three-dimensional clusters. Therefore γ_ν° and $\gamma_{1,\nu}$ are, respectively, the spreading and interface energies (per adsorption site). They become equal to $S^\circ_{A/B}$ and γ°_{AB} in the macroscopic limit (ν, $\theta \to \infty$). Let us now assume, as a first step, that the adsorbate substrate interdiffusion is quenched and that the surface diffusion allows to get the most stable configuration of the deposited atoms. Then, VW growth takes place for $\gamma_\nu^\circ > 0$ whereas a layer by layer (FM) growth occurs for negative and decreasing ν values[1-4] *i. e.* for long ranged attractive substrate-adsorbate interactions. Finally, the SK growth takes place when $\gamma_\nu^\circ < 0$ increases with ν for $\nu > n$ (see Fig. 1). FM growth is certainly an exception if we include the strain energy and if we take into account the fact that the adsorbate-substrate interactions are short ranged in metals and more especially in TM from the efficient secreening by "d" electrons; this is why we found $n \leq 2$ from the electronic theory.[5]

Let us now split the spreading energy γ_ν° into its separate contributions. The bulk A and B crystals have different equilibrium structures (α, β respectively) and atomic volumes (Ω_A, Ω_B respectively). We first transform the crystal A into the structure β at the volume Ω_A (structural contribution $\gamma_\nu^{\text{struct}}$). Then we compress (or expand) it to the volume $\Omega_B(\gamma_\nu^{\text{el}})$, we arrange the atoms on the substrate to get ν A ML without any relaxation (γ_ν^{chem}) and finally the adsorbate and substrate atoms are allowed to relax to their equilibrium position (γ_ν^{rel}). γ_ν° and $\gamma_{1,\nu}^\circ$ are thus split into four different contributions. For example:

$$\gamma_\nu^\circ = \gamma_\nu^{\text{struct}} + \gamma_\nu^{\text{chem}} + \gamma_\nu^{\text{rel}}, \tag{3}$$

with $\gamma_\nu^{\text{el}} = (E_A(\Omega_B) - E_A(\Omega_A))$ where $E_A(\Omega)$ is the energy per atom of the bulk A with the volume Ω. The chemical contribution comes from the adsorbate-substrate interactions at constant atomic volume Ω_B. It can be directly related to the energy changes of each layer induced by the deposition. For example,

$$\gamma_\nu^{\text{chem}} = E_{A/B} - E_A^\beta(\Omega_B) + \sum_{j\geq 1}\left[E_B^j - E_B^{0,j}\right], \tag{4}$$

where $E_{A/B}$ is the energy per atom of the overlayer, E_B^j the energy per atom of the jth underlayer and $E_B^{0,j}$ the energy per atom of the jth substrate layer before deposition. The sign of γ_1^{chem} results from the competition between the energy loss of a bulk atom A when it is deposited on the substrate and the energy gain induced in the substrate by the deposit.

We use the simplest model which enables to obtain the order of magnitude of the various contributions to the spreading energy and the related trends. The energy results from

(i) the band contribution (determining the variation of the cohesive and surface energies as a function of the band filling):[6–7] it is evaluated in the TBM.

(ii) the repulsive contribution (ensuring crystal stability): it is represented by pair potentials of the Born-Mayer type.[7,8]

Each of the contributions to γ_ν° can then be splitted, as the total energy, into its band ($\gamma_{\nu,band}^{chem}$) and repulsive ($\gamma_{\nu,rep}^{chem}$) contributions. Then, the band contribution is determined keeping only d electrons and the bulk d band structure being derived from LMTO calculation.[9] The repulsive energy is deduced from a fit to the experimental lattice parameters and compressibility.[8] We required the local neutrality of each atom up to the fifth layer from the surface to determine the local densities of states (LDOS) and up to the third layer to determine the relaxation.

The comparison between the band contributions to the spreading energy obtained either from a simple (Friedel's) model (SM) assuming rectangular LDOS or from the complete calculation of the LDOS by the recursion method (RM) shows that the sign of this contribution and its order of magnitude are obtained using the SM with sufficient accuracy for our purpose. A comparison between the signs of γ_1°, of the band contributions $\gamma_{1,band}$ and of the calculated values of $\gamma_A - \gamma_B$ for all the transition metals pairs shows that the band contribution determines, in most cases, the sign of γ_ν° and that the phenomenological criterion $\gamma_A < \gamma_B$ for the wetting of B by a monolayer and the experimental data[5] are in agreement with our results of the electronic theory. For example, the wetting condition $\gamma_1^\circ < 0$ is satisfied for the $4d$ transition metals on most of the $5d$ metals, for $3d$ elements on most of the $4d$ and $5d$ elements, and for all the elements on W, if γ_1°, $\gamma_A - \gamma_B$ and $\gamma_A(\Omega_B) - \gamma_B$ have often the same sign, their values can be very different. The interface $\gamma_{1,1}^\circ$ and elastic terms are of the same order of magnitude as the surface energy differences and are thus essential to determine the spreading energy (even if they are competing). The relaxation and structural contributions are much smaller in general (< 0.1 eV/at) and do not contribute significantly to γ_1°.

Up to now, the interdiffusion has been assumed to be quenched. However, the previous atomic configurations can be unstable with respect to

(i) the formation of two dimensional (SD) ordered structures by an exchange of A and B atoms in the first surface layers;

(ii) the formation of three dimensional (3D) compounds deposited on B;

(iii) the disolution of the A atoms in B.

For example, even if $\gamma_1^\circ < 0$, a 3D compound can be observed at the surface of B if $\Delta E_f(AB_n) < \gamma_1^\circ$ where $\Delta E_f(AB_n) = E(AB_n) - E_A^\circ - n \cdot E_B^\circ$ is the formation energy of the compound. For illustration, we examined in Ref. 4 the consequences of the simplest possible assumptions concerning a limited interdiffusion:

(i) For *fcc* (001) substrates, we allow an exchange of A and B atoms on the first two (001) layers only and initiate the Ll_0 structure by occuping the sites in the (100) planes of these layers successively by A and B atoms.

(ii) For *bcc* (100) substrates we allow an exchange of all the deposited A atoms with all the B atoms of the first underlayer to initiate a surface B2 ordered structure.

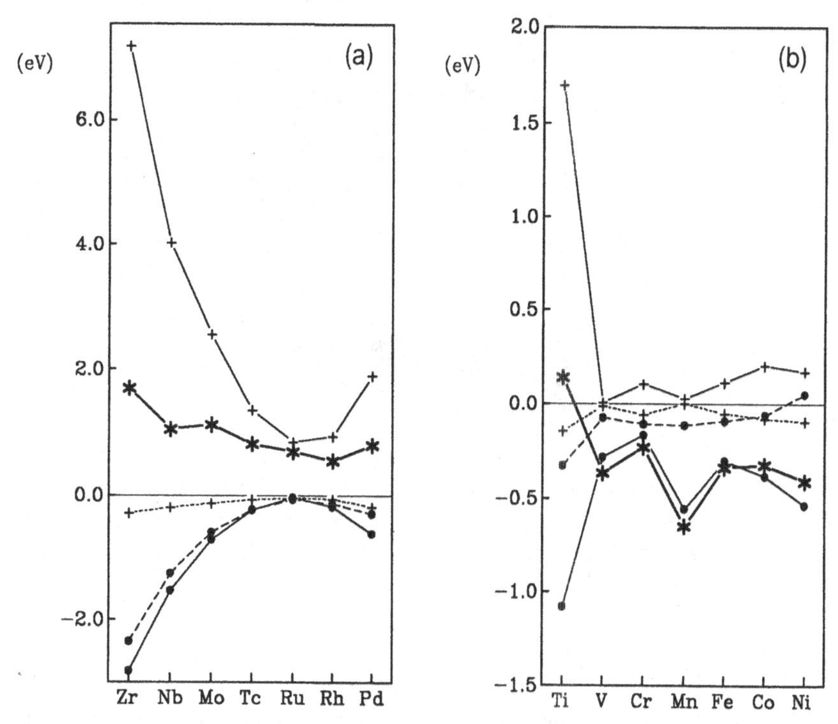

Figure 2. The contributions of the spreading energy γ_1 of (a) the $4d$ elements on Co(111) and (b) the $3d$ elements on Ru(111). The various curves are: $\gamma_A(\Omega_B) - \gamma_B$, continuous line with solid circles; γ_1^{chem}, dashed line with solid circles; γ_1^{el}, dashed line with crosses; γ_1^{rel}, solid line with crosses; and γ_1^0, solid lines with stars.

The general trends for the occurence of stable epitaxial ML are then same as obtained previously assuming a quenched interdiffusion. However,

(i) surface ordered compounds can be stable and "wet" the substrate when a VW growth is predicted ($\gamma_1^\circ > 0$) for quenched interdiffusion (Ir/V, Re/Nb, V/Pd, Cr/Fe).

(ii) The two dimensional ordered structures occur mostly when the bulk ordered compounds are very stable ($|\Delta E_f|$ large) and sometimes when the elements in the bulk are miscible; they are not stable when a miscibility gap is found for the bulk alloys.

II.2 Application to the Growth of Superlattices: Co-Ru and Co-Cr

Recently, *hcp* (0001) CoRu multilayers have been obtained either by sputtering[10] or by MBE.[11] For small Cobalt thicknesses, they exhibit very interesting magnetic properties which result from large Cobalt perpendicular anisotropy and antiferromagnetic interlayer couplings.[11] Let us now relate qualitatively the results of our theoretical analysis of the observed growth of such superlattices (SL). Figure 2 shows the different contributions to γ_1° for ($3d$ TM)/Ru and ($4d$ TM)/Co along the series. For ($3d$ TM)/Ru, γ_1 is mainly given by $\gamma_A(\Omega_B) - \gamma_B$ and is itself very similar to $\gamma_A - \gamma_B$ (macroscopic criterion), the other contributions cancelling each other: a Co ML wets the Ru substrate as suggested by the values of the surface energies. For (4d TM)/Co, the elastic contribu-

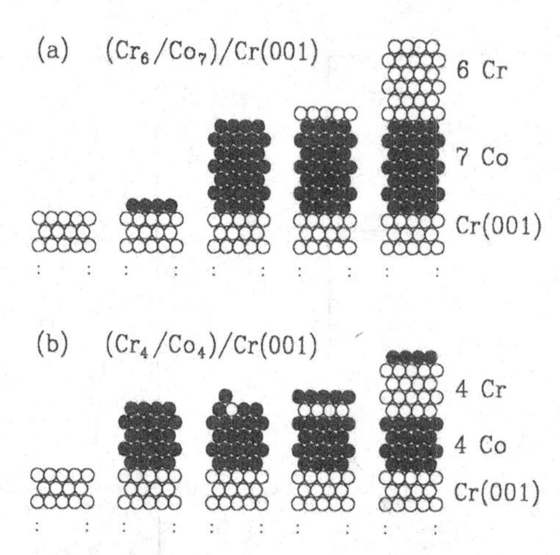

Figure 3. Schematic representation of the suggested grown mode for (a) a Cr_6Co_7 superlattice without interdiffusion (full line) and for (b) a Cr_4Co_4 superlattice. In the last case, each deposited Cr atom is exchanged with a neighbouring Co atom, in such a way that during the process the surface remains covered by a Co ML.

tion γ_1^{el} is the predominant term: a Ru monolayer does not wet a Co substrate because the elastic strain induced by the perfect epitaxy is too large. Strictly speaking, the growth of Co-Ru superlattices with a perfect epitaxy during the growth process is not possible. Effectively, good RHEED patterns are obtained during the Co pseudomorphic deposition on Ru for relative large thicknesses but the patterns reveal[11] an increasing roughness during the Ru deposition on Co. Therefore, our results agree qualitatively with these experimental observations.

A similar situation occurs for *bcc* Co/Cr (100) superlattices. A recent experimental study has shown[12] that

(i) a *bcc Co* phase (with $1 \leq \nu \leq 15$) can be easily grown at room temperature on a Cr (100) surface;

(ii) *bcc* (100) Cr can be epitaxially grown on the *bcc* Co surface only for ultrathin films ($\nu \lesssim 2$), the LEED patterns showing that the Cr film becomes disordered for larger ν values.[12]

Therefore, it is interesting to study theoretically the posibility to get experimentally *bcc* Co/Cr sandwiches and SL. Let us now summarize briefly the results of such a study. The calculation of the spreading energies γ_ν° using either the SM we introduced previously or the RM has been done. Since Co and Cr equilibrium atomic volumes are very close, we have neglected the elastic strains in the overlayers. The comparison between SM and RM band contributions shows that magnetism is a second order contribution which is not essential to get qualitatively the growth mode; the comparison between the band and total contributions for the SM shows that the band contribution is also sufficient to obatain these trends. The results of our simulation of a Co-Cr superlattice growth can be summarized (Figs. 3 and 4) as follows:

(i) A Co ML wets the Cr (001) substrate ($\gamma_1^\circ < 0$) but $\gamma_\nu^\circ \gtrsim \gamma_1^\circ$ ($\nu > 1$) so that, strictly speaking, a SK growth mode occurs. However, the values of $|\gamma_\nu^\circ - \gamma_1^\circ|$ are small (some 10^{-2} eV) and γ_ν converges rapidly ($\gamma_2^\circ \simeq S_{Co/Cr}$) so that the conditions for a predominant lateral growth of flat clusters of monoatomic height are satisfied. A FM like growth mode with limited surface roughness is then predicted.

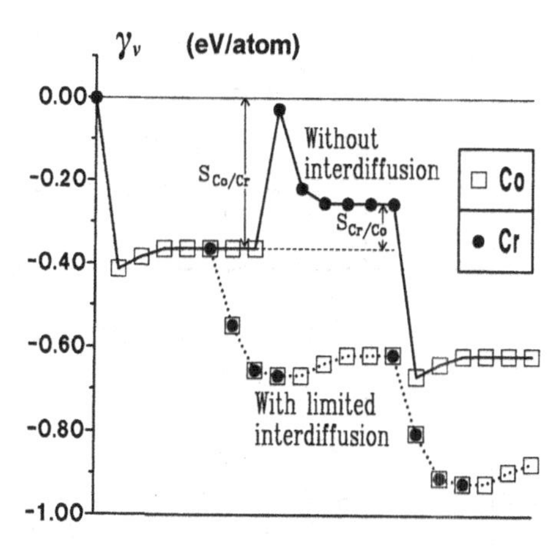

Figure 4. Variation of the spreading energy γ_ν with ν for a (a) a Cr_6Co_7 superlattice without interdiffusion (full line) and for (b) a Cr_4Co_4 superlattice with the suggested limited interdifussion (dotted line).

(ii) A Cr overlayer does not wet neither the Co film deposited on Cr nor the Co substrate ($\gamma_1^\circ > 0$). γ_ν decreases rapidly for increasing values of ν to reach its asymptotic value ($\gamma_4^\circ = 0.108$ eV $\simeq S_{\mathrm{Cr/Co}}$ for Cr/Co) but it is always positive.

(iii) However, if we assume a limited interdiffusion between the deposited Cr and the first Co layers, the most stable state is obtained when a Co ML covers the Cr film during its growth (Fig. 4) and the conditions for a FM mode are satisfied.

The previous analysis shows that a controlled interdiffusion during the growth can be essential to get epitaxial overlayers and superlattices.

III. Epitaxy of a Surface Compound and Coevaporation

III.1 Introduction

Our previous study shows that, in most cases, a ML of an element A with a high surface energy cannot wet the substrate B when $\gamma_{\mathrm{B}} < \gamma_{\mathrm{A}}$; W, for example, whose surface energy is very large, cannot wet *a priori* a TM substrate. In this section, we examine the possibility to get epitaxial films $A_cA'_{1-c}$ by the coevaporation of its elements A and A' even if one of these elements (A for example) has a higher surface energy and does not wet the substrate. As in section II.1, we assume that the surface diffusion allows to get the most stable state, the interdiffusion between the deposited atoms and the substrate being quenched. In order to discuss qualitatively the wetting conditions, we first consider the macroscopic limit ($\nu \to \infty$) (Sec. III.2) and examine briefly the physical quantities which determine such conditions *i. e.*

(i) the formation energy ΔE of the bulk $A_cA'_{1-c}$ compound (or alloy) (Sec. III.3);

(ii) the corresponding surface energy $\gamma_{(AA')}$ and its variation with the nature of the top layer (Sec. III.4);

(iii) the interfacial energy $\gamma_{(\mathrm{AA'})\mathrm{B}}$ between two perfect epitaxial semi-infinite $A_cA'_{1-c}$ and B crystals (Sec. III.5).

Table 1

Growth of TM compounds by coevaporation: calculated surface, interface and formation energies. The spreading energy $S_{(AA')/B}$ given in the last column correspond to the lowest energy state in the macroscopic limit. All quantities are given in eV/atom.

Situation $(AA')/B$ misfit f	Δ N ΔE_f^0 ΔE_f	γ_A $\gamma_{A'}$ γ_B	γ_{AB} $\gamma_{A'B}$	$S_{A/B}$ $S_{A'/B}$	Top Layer and interfacial layer	$\gamma_{AA'}$	$\Delta\gamma_{(AA')}$	$\gamma_{(AA')B}$	$S_{(AA')/B}$	$\gamma_{(AA')/B,1}$ $\gamma_{(AA')/B,2}$ $S_{(AA')/B}$
(HfW)/Ta	2	0.31	−0.07	−1.22	(Hf)	1.33	−0.04	−0.62	−0.75	−0.39
BCC (001)	−0.25	2.03	−0.44	0.13	(W)	1.74	0.37	0.80	1.08	−0.79
f = 0.4%	−0.25	1.46			$(HfW)_{(\sqrt{2}\times\sqrt{2})}$	1.43	0.06	−0.03	−0.06	−0.75
(WIr)/Re	3	1.09	−0.11	−0.21	(W)	1.31	0.34	−0.13	−0.01	−0.24
FCC (001)	−0.44	0.88	−0.07	−0.38	(Ir)	0.98	0.01	0.43	0.22	−0.28
f = 0.3%	−0.44	1.19			(WIr)	1.02	0.05	0.02	−0.15	−0.34
(WIr)/Ru	3	0.83	−0.41	−0.41	(W)	1.18	0.21	−0.31	0.04	−0.16
FCC (001)	−0.44	0.81	−0.01	−0.03	(Ir)	0.82	−0.15	0.43	0.42	−0.33
f = 3.0%	−0.32	0.83			(WIr)	0.87	−0.10	−0.03	0.01	−0.32
(TaIr)/W	4	0.58	−0.18	−0.84	(Ta)	1.31	0.56	0.38	0.45	−0.26
FCC (001)	−1.01	0.89	−0.29	−0.64	(Ir)	1.20	0.45	0.25	0.21	−0.16
f = 0.2%	−1.01	1.24			(TaIr)	1.00	0.26	0.03	−0.21	−0.21

Then we examine the conditions to get a layer by layer growth of ultrathin films of compounds. We summarize the results we obtained for some representative situations using the TBM and the recursion method for determining the LDOS (Sec. III.6) and we show that coevaporation can be very useful to obtain epitaxial compounds.

III.2 Macroscopic limit

Let us first choose as reference the state (labelled 0) in which A and A', when deposited on the substrate, form separated isolated clusters. In the macroscopic limit, we have to consider the stability of two other states with respect to the reference state:

(1) A and A' form 3D clusters of disordered alloys or compounds $A_cA'_{1-c}$ if these clusters are stable for the considered concentration: in such a case, the energy gain per atom (ΔE°) with respect to the reference state is equal to the mixing energy of the alloy (ΔE°_m) or to the formation energy (ΔE°_f) of the considered compound;

(2) A and A' form an epitaxied disordered alloy or compound: the corresponding energy gain, $\Delta E_\nu = (E_\nu - E_o)/N$ per interfacial site, can be expressed in terms of ΔE° and of the spreading energy $S_{(AA')/B}$:

$$\Delta E_\nu = \nu \cdot \Delta E^\circ + S^\circ_{(AA')/B}, \tag{5a}$$

$$\Delta E^\circ = E_{(AA')} - c \cdot E^\circ_A - (1-c) \cdot E^\circ_{A'}, \tag{5b}$$

$$S_{(AA')/B} = \gamma^\circ_{(AA')} - \gamma_B + \gamma_{(AA')B}. \tag{5c}$$

In practice, the equilibrium state of an interface between two semi-infinite crystals is characterized by a periodic array of misfit dislocations whose energy must be calculated separately. In the present paper, we consider only an intermediate situation for which the films (AA') have a finite height $h = \nu \cdot d$ (d is the interlayer distance) and are in perfect epitaxy with the substrate. If h is sufficiently small ($h < h_c$) such that misfit dislocations are not stable and sufficiently large that the electronic structure of the film is bulk-like except near the interfaces (*i. e.* typically $h \simeq 5$ to $10\ d$ for small misfits), the interface binding energy is limited to the interfacial planes, the film deformation is homogeneous and the substrate remains underformed. In such a case ΔE_ν is expressed as previously in terms of volume (ΔE) and of spreading (S) energies:

$$\Delta E_\nu = \nu \cdot \Delta E + S_{(AA')/B}, \tag{6a}$$

$$\Delta E = \Delta E^\circ + \Delta E^{el}, \tag{6b}$$

$$S_{(AA')/B} = \gamma_{(AA')} - \gamma_B + \gamma_{(AA')B}, \tag{6c}$$

ΔE^{el} is the elastic energy which is stored in the strained overlayers; it can be determined macroscopically from the bulk elastic constants of $A_cA'_{1-c}$ assuming that the stress component perpendicular to the film is zero; here, it is determined directly from the TBM parameters. Finally, the surface and interface energies which define S in Eq. (6c) are relative to the strained overlayer.

The wetting of the substrate by the alloy $A_cA'_{1-c}$ is determined by a competition between the interface and surface energies of the strained film; its stability ($\Delta E_\nu < 0$) results from the interplay between the two terms of Eq. (6a). Such epitaxial overlayers can be obtained *a priori* when:

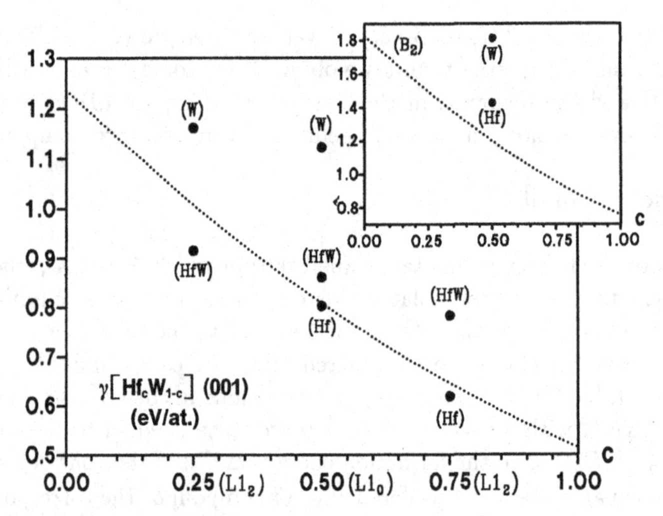

Figure 5. The (001) surface energies of Hf_cW_{1-c} $L1_0$, $L1_2$ and B_2 (in the insert) phases and for pure (Hf), (W) or mixed (HfW) top layers. The dashed line corresponds to the average surface energy $\bar{\gamma}_{HfW}$ calculated at the average volume $\bar{\Omega} = c\Omega_A + (1-c)\Omega_{A'}$.

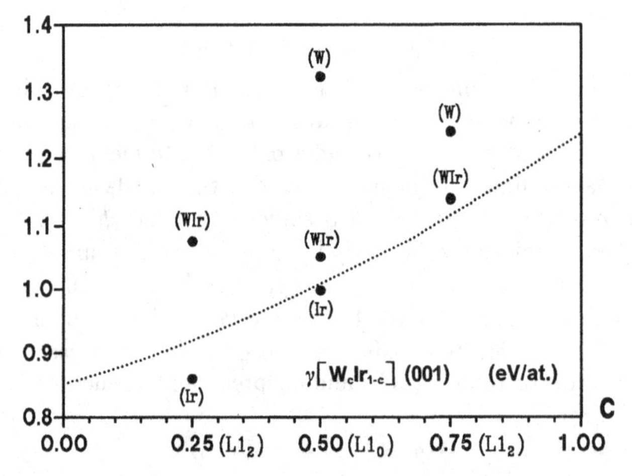

Figure 6. The (001) surface energies of W_cIr_{1-c} $L1_0$, $L1_2$ and B_2 phases and for pure (W), (Ir) or mixed (WIr) top layers. The dashed line corresponds to the average surface energy $\bar{\gamma}_{WIr}$ calculated at the average volume $\bar{\Omega} = c\Omega_A + (1-c)\Omega_{A'}$.

(i) the mixing (formation) energy $|\Delta E| > 0$ is large;
(ii) the misfit f between the substrate and $A_cA'_{1-c}$ in its equilibrium state is small;
(iii) the top and interface layers are such that both the surface and interface energies are small.

Let us consider now separately these quantities and comment our results from some representative situations. The substrate B and the deposited atoms are chosen among

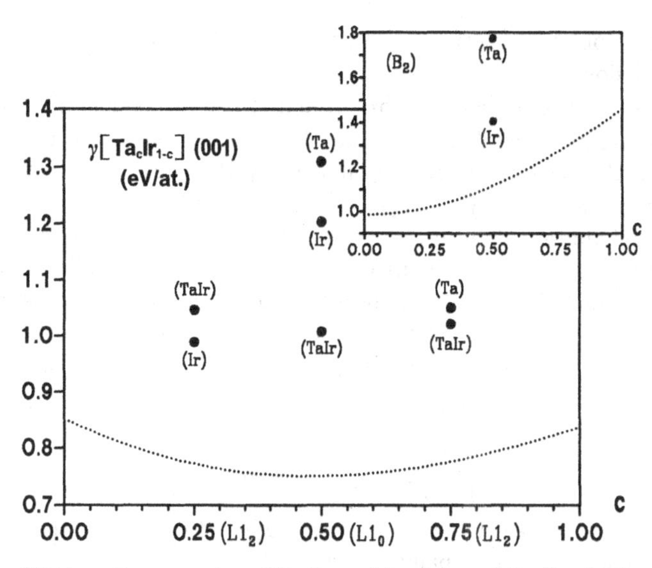

Figure 7. The (001) surface energies of Ta_cIr_{1-c} $L1_o$, $L1_2$ and B_2 (in the insert) phases and for pure (Ta), (Ir) or mixed (TaIr) top layers. The dashed line corresponds to the average surface energy $\bar{\gamma}_{TaIr}$ calculated at the average volume $\bar{\Omega} = c\Omega_A + (1-c)\Omega_{A'}$.

the elements of the third series of TM to avoid magnetic and correlation effects. Moreover, the parametrization of the band and repulsive contributions to the total energy of the pure metals is well established and allows to get a consistent representation of the surface relaxation and reconstruction.[13] The elements A and A' were chosen in order to get simple stable ordered structures ($L1_0$ or B2) as deduced from the generalized structural maps.[14] Finally, in three of these examples we chose to deposit W on a substrate in order to test the possibility to get by coevaporation epitaxial overlayers including an element with a high surface energy. The calculated quantities are given in Table I.

III.3 Mixing and formation energies

The mixing and formation energies have been extensively studied either by TBM or by *ab-initio* calculations.[14,15] The relevant parameters which determine qualitatively the range of stability of the TM compounds are the average number of electrons per atom and the difference of electronegativity of the alloy constituents. Generalized structural maps have been built from the electronic structure (TBGPM)[16] to determine the range of stability of the bulk compounds (see Ref. 14 for a general discussion).

III.4 Surface energies for ordered compounds and disordered alloys

Up to now, to the authors' knowledge, there is no systematic study of alloy surface energies. This is why we have initiated such a study in the TBM. Figures 5, 6 and 7 show the surface energies of several ordered compounds ($L1_0$, $L1_2$, B2) for simple crystallographic surfaces and for all possible top layers. In these calculations we assume that the crystallographic and chemical order remain perfect up to the surface (neglect of surface segregation and of surface relaxation and reconstruction). The results can be summarized as follows:

(1) the surface energy comes as usual from the first two layers near the surface;

(2) the surface energy depends strongly on the composition of the surface: it is the lowest when this surface is occupied mostly by the element A' with the smallest surface energy ($\gamma_{A'} < \gamma_A$) (see Figs. 5, 6, and 7 and Table I);

(3) the band contribution determines mostly the relative order of magnitude of the surface energies even if the repulsive term can be important in some cases. As a first approximation, $\gamma_{(AA')}$ can be obtained from the average surface energy: $\bar{\gamma}_{(AA')} = c.\gamma_A + (1-c).\gamma_{A'}$. Our calculations show that the deviations from the average value is the largest for the largest formation energy (Figs. 5, 6, and 7 and Table I):

$$\gamma_{(AA')} = \bar{\gamma}_{(AA')} + \Delta\gamma_{(AA')}, \tag{7a}$$

$$\Delta\gamma_{(AA')} = -\beta \times \Delta E^{\circ} \ (0 < \beta < 1), \tag{7b}$$

such a relation can be easily deduced for a disordered alloy using the Friedel's model for the cohesion and surface band energies:

$$\Delta\gamma^{\text{band}}_{(AA')} \simeq -\sqrt{1 - \frac{Z_b}{Z}} \cdot \Delta E^{\text{band}}_m (A_c A'_{1-c}). \tag{8}$$

In this formula Z and Z_b are respectively the number of nearest neighbours and of surface broken bonds.

III.5 The interfacial energies

The interfacial energies have been qualitatively estimated[5] for two semi-infinite crystals using a simple TBM; they are much smaller than the surface energies and their values are in agreement with estimates based on Miedema's model.[17] This is why, in most cases, the wetting condition is in agreement with the thermodynamic criterion ($\gamma_A < \gamma_B$). However, γ_{AB} can be of the same order of magnitude as $\gamma_A - \gamma_B$. Moreover, for the interfaces between an alloy $A_cA'_{1-c}$ and a pure metal B, the interfacial energy $|\gamma_{(AA')B}|$ can be large and has been shown to depend strongly on the nature of the interfacial planes. Assuming as previously that there is neither interfacial segregation nor relaxation, the examples we investigated (see Table I) show that when $\Omega_A < \Omega_B < \Omega_{A'}$ the interfacial energy is the lowest for mixed interfacial planes with both "larger" and "smaller" atoms than the substrate. On the contrary, the interfacial energy can be large and even change the sign of $S_{(AA')/B}$ when the interfacial plane is pure.

III.6 Growth and stability of ultrathin films of TM compounds

The growth mode can be derived as in section II.1 from a study of the variation of the spreading energy $\gamma_{(AA')/B,\nu}$ with ν defined as in Eq. (2) by:

$$E = N_B \cdot E^{\circ}_B + N \cdot \nu \cdot E^{\circ}_{(AA')} + N \cdot (\gamma_B + \gamma^{\circ}_{(AA')/B,\nu}), \tag{9}$$

$\gamma^{\circ}_{(AA')/B,\nu}$ becomes equal to the spreading energy $S^{\circ}_{(AA')/B}$ in the macroscopic limit. Therefore, the conditions which determine the growth mode (FM, SK or VW) can be deduced, as previously, if the deposit is a 2D or a 3D compound with a well defined structure, composition and order; for example, a layer by layer growth occurs if $\gamma^{\circ}_{(AA')/B,\nu} < 0$ decreases for increasing ν values. However, a complete study must

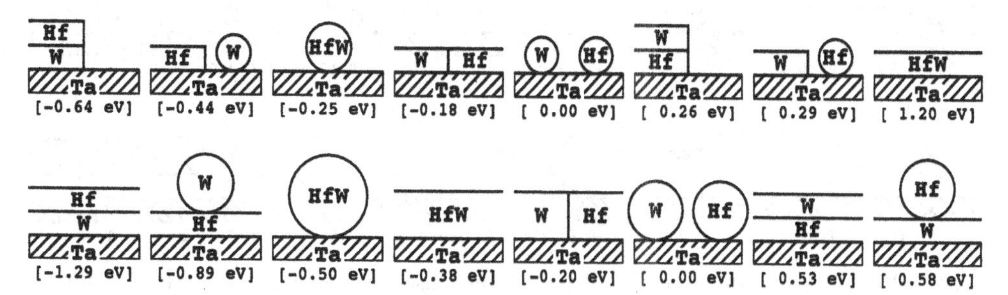

Figure 8. Relative stability of the different states considered for $(Hf_{1/2}W_{1/2})/Ta(bcc\,(001))$. These states are classified from the most stable to the less stable state for $\nu = 1$ (upper set) and $\nu = 2$ (lower set).

Table 2

The seven states considered in this study with their corresponding energy in the macroscopic limit.

State	Schematic representation	Energy
0	A′ A / B	0 (reference)
1	AA′ / B	ΔE_m^0 or ΔE_f^0
2	AA′ / B	$\Delta E_m + \gamma_{(AA')} - \gamma_B + \gamma_{(AA')B} = \Delta E_m + S_{(AA')/B}$
3	A / A′ / B	$\gamma_A - \gamma_B + \gamma_{A'B} + \gamma_{AA'}$ or A↔A′
4	A / A′ / B	$\gamma_A - \gamma_B + \gamma_{AB}$ or A↔A′
5	A A′ / B	$\frac{\gamma_A - \gamma_B + \gamma_{AB}}{2}$ or A↔A′,
6	A′ A / B	$\frac{\gamma_A + \gamma_{AB}}{2} + \frac{\gamma_{A'} + \gamma_{AA'}}{2} - \gamma_B.$

take into account the existence of all the possible film heterogeneities (height, composition, chemical order, ...). In this preliminary study, we consider only an epitaxial film $A_cA'_{1-c}$ having the same crystallographic structure as the substrate and a fixed composition $c = 0.5$; we examine its stability with respect to films of constant height and 3D clusters whose composition is $c = 0$, 1 or 0.5. The epitaxial film (state 2 in Table II) can then be unstable with respect to the occurence of (1) 3D clusters of A and A'; (2) 3D clusters with the same structure and composition as the deposit; (3) epitaxial bilayers on B(A/A'/B or A'/A/B); (4) 3D clusters of A(A'), on an epitaxial A'(A) overlayer; (5) epitaxial A(A') films and A'(A) 3D clusters on the substrate; and (6) A

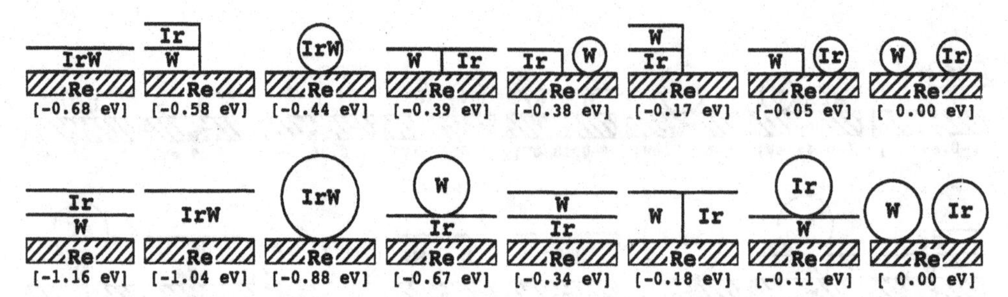

Figure 9. Relative stability of the different states considered for $(Ir_{1/2}W_{1/2})$/Re(*fcc* (001)). These states are classified from the most stable to the less stable state for $\nu = 1$ (upper set) and $\nu = 2$ (lower set).

and A' films epitaxied separately on the substrate. For the intermediate film heights we considered in the section III.2, the energies of these states depend on ΔE and on the surface (γ_{A}, $\gamma_{A'}$, γ_{B}, $\gamma_{(AA')}$) and interfacial energies (γ_{AB}, $\gamma_{A'B}$, $\gamma_{(AA')B}$) (see Table II) so that qualitatively the $A_cA'_{1-c}$ film is stable, as discussed previously, for large ΔE and for surface and interface energies as amall as possible. Let us finally recall that we consider here only epitaxial films in perfect registry with the substrate: therefore, we split the elastic term ΔE^{el} from γ° and discuss as previously [see Eq. (6)] the growth mode and film stability in terms of $\gamma_{(AA')/B,\nu}$ defined by:

$$\Delta E_{\nu} = \nu \cdot \Delta E + \gamma_{(AA')/B,\nu} \tag{10}$$

Let us now summarize the results we obtained for some representative situations (see Table I and Fig. 10).

III.6a. $(Hf_{1/2}W_{1/2})$/Ta(*bcc*, (001)): FM like growth

The substrate is wet by Hf for small ν values whereas it is not wetted at all by W whose surface energy is much too large (see the $S_{A/B}$ values in Table I and Ref. 4). The relative stability of the states we consider when Hf and W are coevaporated on Ta is given in Fig. 8 for $\nu = 1$ and 2 and the corresponding spreading energies are given in Table 1. These results show that for $\nu = 1, 2$ is possible to stabilize W by coevaporation into a 2D B2-like ordered compound which is grown approximately layer by layer. For such ultrathin films W is located between the substrate (Ta) and a ML of Hf whose surface energy is relatively low. However, the interfacial energy $\gamma_{(HfW)Ta)}$ is large when the interfacial layer is a W layer so that for large ν values the lowest energy state must be a B2 film with top and interface Hf layers. This suggests that there is a transition between both states for increasing ν values, the growth remaining approximately layer by layer since, as in the CoCr case (Sec. II.2), the values of $|\gamma_{(AA')/B,\nu} - S_{(AA')/B}|$ are small (see Fig. 10).

III.6b. $(Ir_{1/2}W_{1/2})$/Re(*fcc*, (001)): FM growth

Here, the substrate surface energy γ_{Re} is large ($\gamma_{Re} \simeq \gamma_{W}$) and we find that it is wetted both by Ir and W. A study of the stability of the states obtained by coevaporation

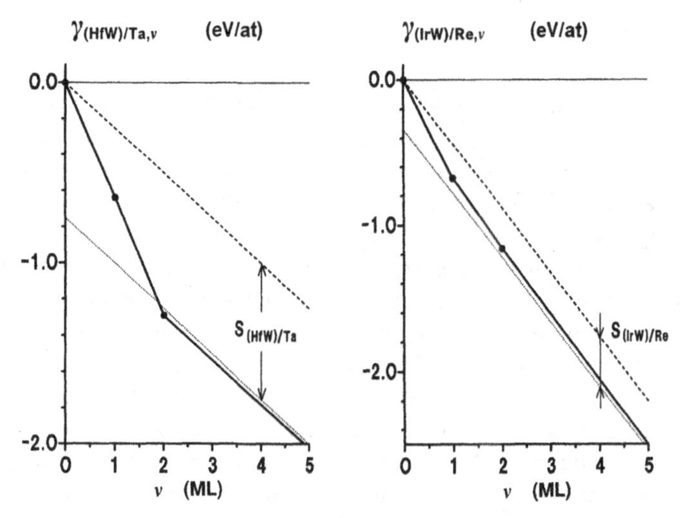

Figure 10. Speading energy of the most stable states $\gamma_{(AA')/B,\nu}$ versus ν for (a) (HfW)/Ta and for (b) (IrW)/Re; the dashed line corresponds to $\nu\Delta E_f$ and the dotted line corresponds to $\nu\Delta E_f + S_{AA')/B}$.

(Fig. 9) shows that (i) for $\nu = 1$, the most stable state is a mixed IrW plane from which the stable bulk $L1_0$ can be grown; (ii) for $\nu = 2$, the state obtained by the growth of a second mixed IrW plane is less stable than a film with pure W and Ir plane: this exchange of stability results qualitatively from the fact that, for thick films, the surface (interface) energy is (respectively) the smallest for Ir(W) top (interfacial) layers and is similar to the one mentioned for (HfW)/Ta. Finally, the value of the spreading coefficient for such top and interfacial layers ($S = -0.34$ eV) is consistent with a layer by layer growth of a $L1_0$ IrW compound (Fig. 10).

III.6c. $(Ir_{1/2}W_{1/2})/Ru$(*fcc*, (001)): bilayer by bilayer growth

In such a case the misfit between the $L1_0$ compound and the substrate is relatively large ($f \simeq 3\%$) so that elastic effects are important ($\Delta E^{el} \simeq 0.12$ eV/at) and can prevent the stability of epitaxial films. However, our results for $\nu \leq 2$ show that the substrate is progressively covered by a Ir/W bilayer. Moreover, the values of the spreading energies $\gamma_{(IrW)/Re,\nu}$ suggest that an epitaxial $L1_0$ compound can be grown bilayer by bilayer to minimize both the surface and interface energies.

III.6d. $(Ta_{1/2}Ir_{1/2})/W$(*fcc*, (001)): SK growth

This last case is only given here for illustration since it corresponds to the growth on a fictitious *fcc* W(001) substrate. The most stable homogenous states for $\nu \leq 2$ correspond to the layer by layer growth of mixed TaIr planes which initiate the stable bulk compound. However, $\gamma_2 > \gamma_1$ so that the growth mode is found to be of the SK type with $n = 1$.

IV. Conclusion

In this paper, we have shown that (1) the growth mode can be derived from the variations of the spreading energy γ_ν with the layer thickness ν; (2) the spreading energy γ_ν at $T = 0$ K can be derived from the electronic theory and general trends can be obtained using a simple TBM. For pure A overlayers, the layer by layer growth is found to be energetically favorable only for $\nu = 1$ or 2 ML (SK mode) and the macroscopic criterion ($\gamma_A < \gamma_B$) is often qualitatively sufficient to "predict" the wetting of the substrate. However, we pointed out the possible importance of limited interdiffusion either to get a surface ordered compound when a Volmer Weber growth is predicted for quenched interdiffusion, or to allow the growth of superlattices; this last possibility has been illustrated by the simulation of the growth of CoCr superlattices.

We presented in the second part of this paper a first approach of the growth modes of transition metal compounds $A_cA'_{1-c}$ when both A and A' constituents are coevaporated on the substrate B and when surface diffusion allows to get the most stable atomic configuration. It has been shown that ordered compounds including elements with large surface energies can be grown layer by layer; the most stable state corresponding to a configuration minimizing the surface energies $\gamma_{(AA')}$ for a given composition and long range order. The interfacial planes can be either pure or mixed and the growth can be realized, if necessary, bilayer by bilayer as suggested by the (IrW)/Re case. Therefore, coevaporation can be very useful to get directly (without thermal annealing) compounds with well defined interfaces and interesting transport or magnetic properties. Such a growth process has been used successfully to obtain transition metal or rare earth silicates on silicon. It can be also used to get directly magnetic compounds (such as CoPt) with high magnetic anisotropies and interesting magneto-optical properties.

This first study of the wetting of a substrate by alloys and of the phase stability of deposited alloys is very partial: for example, the stability of the considered films with respect to concentration and height fluctuations has not been fully take into account. Finally, it is important to point out that we used only phase stability arguments. Moreover, all the calculations of this paper being done at zero Kelvin, the results we obtain must be considered as a guide for determining metastable phases blocked near the surface by kinetic conditions. Finite temperature calculations must be done to include the dynamics and to describe precisely the growth process in relation with electronic theory.

References

1. U. Gradman and M. Przybylski in *Thin Film Growth Techniques for Low Dimensional Systems*, edited by R. F. C. Farrow, S. S. P. Parkin, P. J. Dobson, J. H. Neave, and A. S. Arrott, Plenum Press, New York, 1987. p. 261; E. Bauer, *Z. Kristallogr.* **110**, 372 (1958); R. Kern, *Bull. Mineral.* **101**, 202 (1978); S. Dietrich in *Phase Transitions and Critical Phenomena*, edited by S. Domb and J. L. Lebowitz, Vol. 12, (1988) p. 2.
2. R. Bruisma, A. Zangwill, *Europhys. Lett.* **4**, 729 (1987).
3. E. Bauer, J.H. Van der Merve, *Phys. Rev.* **B33**, 3657 (1986).
4. F. Gautier and D. Stoeffler, *Surface Science* **249**, 265 (1991); F. Gautier in *Metallic Multilayers*, edited by A. Chamberod and J. Hillairet, Material Science Forum **59, 60** Transtech Publications, 1990. p. 360–437.
5. F. Gautier and A.M. Llois. *Surf. Sci.* **245**, 191 (1991).
6. J. Friedel in *The Physics of Metals*, edited by J. M. Ziman, Cambridge University Press, 1969. p. 494

7. D. G. Pettifor, *Sol. State Phys.* **40**, 43 (1987).
8. J. S. Luo and B. Legrand, *Phys. Rev.* **B38**, 1728 (1988).
9. O. K. Andersen, O. Jepsen, and D. Gloetzel in *Highlights of Condensed Matter Theory*, edited by F. Bassani, F. Fumi, and M. P. Tosi, North-Holland, Amsterdam, 1985.
10. S. S. P. Parkin, N. More, and K. P. Roceh, *Phys. Rev. Lett.* **64**, 2304 (1990).
11. D. Muller, K. Ounadjela, P. Vennegues, V. Pierron-Bohnes, A. Arbaoui, J. P. Jay, A. Dinia, P. Etienne, and P. Panissod, to be published in the proceedings of ICM'91 (J. M. M. M.); K. Ounadjela, D. Muller, A. Dinia, A. Arbaoui, and P. Panissod, submitted to *Phys. Rev. Lett.*
12. F. Scheurer, B Carriere, J. P. Deville, and E. Beaurepaire, *Surf. Sci.* **245**, L175 (1991); and private communication.
13. B. Legrand, G. Treglia, M. C. Desjonqueres, and D. Spanjaard, *Phys. Rev.* **B40**, 6440 (1989).
14. A. Bieber and F. Gautier, *Acta Metall.***34**, 2291 (1986).
15. See for example the papers in *Alloy Phase Stability* edited by G. M. Stocks and A. Gonis, Kluwer Academic Publishers, Dordrecht, 1989.
16. F. Gautier, F. Ducastelle, and J. Giner, *Phil. Mag.* **31**, 1373 (1975); F. Ducastelle and F. Gautier, *J. Phys. F: Met. Phys.* **6**, 2039 (1976).
17. J. Gemerka and A.R. Miedema, *Surf. Sci.* **124**, 351 (1983).

Alloy Surface Behavior: Experimental Methods and Results

T. M. Buck

Laboratory for Research on the Structure of Matter
University of Pennsylvania
Philadelphia, PA
U.S.A.

I. Introduction

While the title of this conference does not mention alloy surfaces, and most of the contributions will be concerned with bulk properties of alloys, the surface behavior of alloys has received considerable attention, both theoretical and experimental. This has come about in the past 20–25 years, as new surface analysis techniques and ultra-high vacuum (UHV) technology have been used to investigate surface phenomena important in such fields as corrosion,[1] catalysis,[2] microelectronics,[3,4] and self-renewing coatings.[5] The use of ordered alloy systems for higher temperature, high ductility applications, such as turbine blades in jet engines has generated interest in the ordered surfaces of NiAl.[6]

Alloy surface phenomena which have been revealed and studied by surface analytical techniques include surface segregation, order-disorder transformation, miscibility gaps, structural rearrangements in which particular atomic species do not occupy the same lattice sites at the surface which they do in the bulk, and preferential adsorption of gases on particular components of alloys. Examples of these surface phenomena will be presented after brief introductions to the techniques which have been developed to investigate them.

II. Surface Analysis Techniques

II.1 Auger Electron Spectroscopy (AES)

AES is the most widely used technique for composition analysis of solid surfaces and has been used extensively in studies of surface segregation.[7–10] Elemental detection

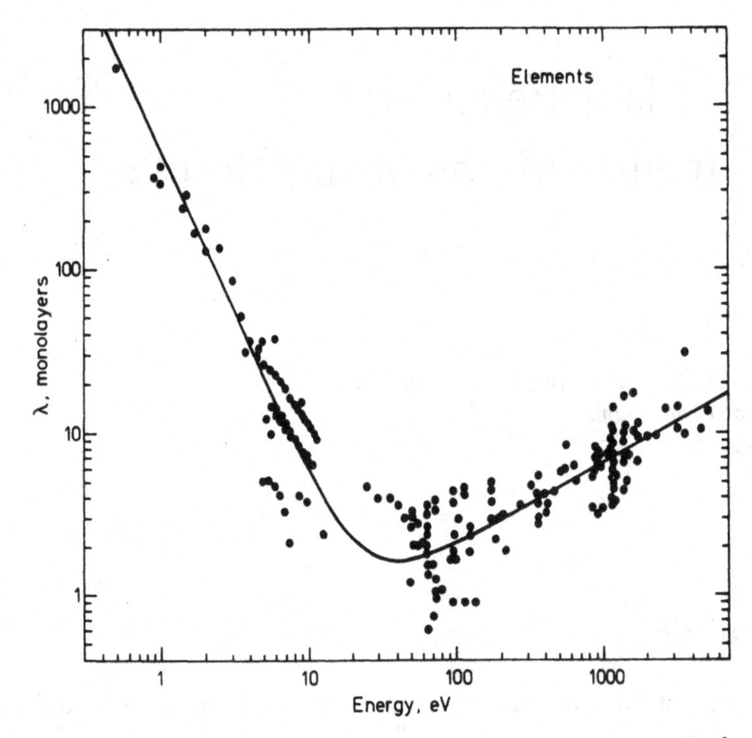

Figure 1. The dependence of λ (escape depth) on electron energy.[8]

is based on the characteristic energy distribution of electrons emitted by an atomic species with its unique energy level structure that has been excited by 1–10 keV incident electron beam creating a hole in an inner atomic shell. AES has a high degree of surface sensitivity arising from the fact that Auger electrons of typical energies, 40–2000 eV have short escape depths, λ (see Fig. 1). Energies of 40–200 eV provide the shortest λ's of around 5–6 Å. AES is convenient, widely available, has good elemental specificity and lateral resolution.

II.2 Low Energy Ion Scattering (LEIS)

This technique, also known as ISS, has single layer surface selectivity and can supply structure as well as composition information about the first few atom layers. There are at least three experimental variations of LEIS: the earliest,[11] using noble gas ions, e. g. He^+ or Ne^+, and an electrostatic analyzer (ESA) for energy analysis of the scattered ions[12–14] another using time-of-flight energy analysis and noble gas ions[15,16] and a third using alkali ions with an ESA[17–21] (ALICISS). The TOF method collects both the scattered neutrals and the ions, thereby avoiding uncertainties due to neutralization of ions in scattering. Differences in neutralization by different target atoms do not interfere with quantitative analysis as may happen in electrostatic analysis of noble gas ions. The same is true of ALICISS since alkali ions undergo little or no neutralization in scattering.

The basis concepts of LEIS are quite simple. As illustrated in Fig. 2, a monoenergetic, well-collimated ion beam with energy of 0.5–10 keV strikes the target surface,

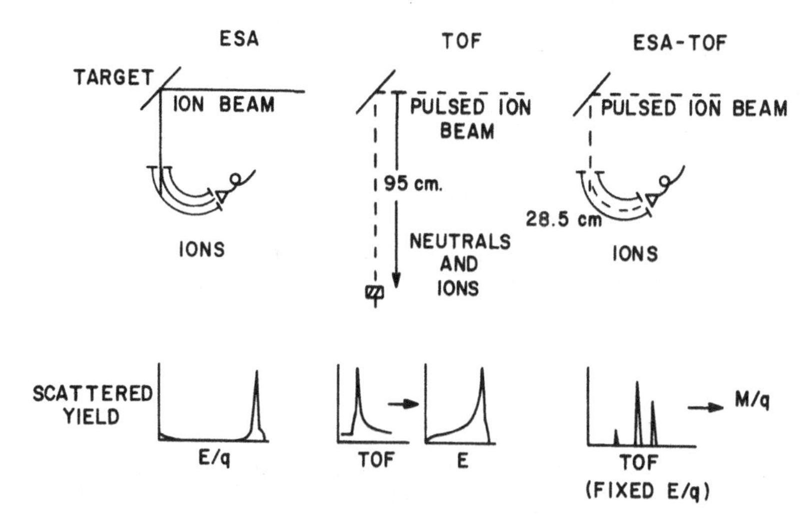

Figure 2. Schematic diagrams of electrostatic analyser (ESA) and time-of-flight (TOF) techniques for low-energy ion scattering. Typical spectra below. The ESA-TOF method alloys m/q identification of secondary ions removed by the beam.

and the energy or time-of-flight distribution of ions, or ions and neutrals, scattered off at a particular scattering angle θ_L, is obtained. The scattering events are binary collisions between the ions and individual atoms, and are generally elastic, ions or neutrals scatter off at angles and energies governed by conservations of kinetic energy and momentum. Single scattering (SS) peaks appear in a spectrum at energies given by

$$\frac{E_1}{E_0} = \frac{M_1^2}{M_1 + M_2}\left[\cos\theta_L + \left(\frac{M_2^2}{M_1^2} - \sin^2\theta_L\right)^{\frac{1}{2}}\right]^2. \tag{1}$$

For $\theta_L = 90°$ this reduces to

$$\frac{E_1}{E_0} = \frac{M_2 - M_1}{M_2 + M_1}, \tag{2}$$

where E_1 and E_0 are the scattered and incident energies, respectively; M_1 and M_2 are masses of the incident particle and target atom and θ_L is the laboratory scattering angle. The SS peaks may ride on a background from multiple scattering and deeper layer scattering, especially in LEIS (TOF) or ALICISS where those events would not be hidden by neutralization. TOF spectra, extending over ranges of several microseconds are approximately mirror images of the corresponding energy spectra, and may be transformed by means of K. E.= $\frac{1}{2}mv^2$.

In LEIS (TOF) which was used in most of the experiments to be discussed, and considering a single-crystal binary alloy A_xB_{1-x} with unknown surface composition, the first layer N_A/N_B ratio is given by

$$\frac{N_1^A}{N_1^B} = \frac{1}{d}\frac{Y_1^A}{Y_1^B}\frac{\sigma_B}{\sigma_A}, \tag{3}$$

in which $N_1^{A(B)}$ is the number of A or B exposed to the beam; $Y_1^{A(B)}$ the measured scattering yield in the A or B peak; $\sigma_{A(B)}$ the differential scattering cross-section de-

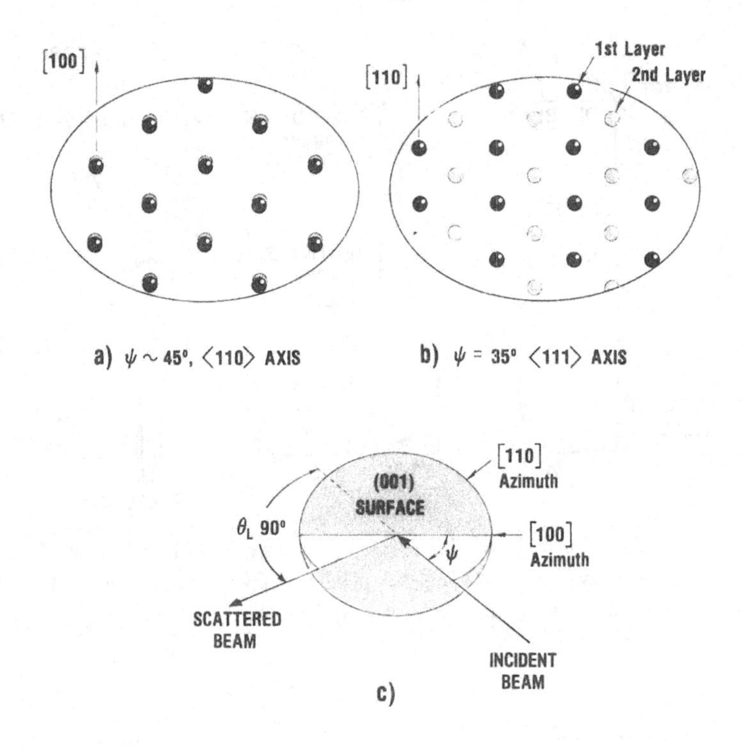

Figure 3. Models of *fcc*(001) surface atoms exposed to ion beam in two directions: (a) along [100] azimuth at ψ=45° incidence angle, and (b) along [110] azimuth at ψ=35°. Angles are defined in (c).

rived from the Moliere approximation to the Thomas-Fermi potential,[22] and d corrects for the fall-off in detector sensitivity with energy between the A and B peaks. If no foreign atoms or vacancies are present the concentration of A is then

$$X_1^{\mathrm{A}} = \frac{N_1^{\mathrm{A}}/N_1^{\mathrm{B}}}{1 + N_1^{\mathrm{A}}/N_1^{\mathrm{B}}}. \tag{4}$$

To ensure that first layer atoms only are exposed to the ion beam, one uses an orientation in which first layer atoms shadow deeper layers. For *fcc* (100) surface (Fig. 3) this would be true with the scattering plane, defined by the ion beam and detector, parallel to [100] and beam incident at $\Psi = 45°$, denoted by [100]45 below.

Second layer composition is determined, in the *fcc*(100) case, by turning the crystal to [100] with 35° incidence to expose the second layer as well as the first, while deeper layers are shadowed. The scattering yields form the second layer, Y_2^{A} and Y_2^{B} are then obtained, essentially, by substracting Y_1^{A} and Y_1^{B} from the measured yields of both layers, Y_{1+2}^{A} and Y_{1+2}^{B} while correcting, if necessary for "focussing" effects,[23–25] detector energy dependence and different incidence angles for the two measurements at [100]45 and [110]35.

The "focussing effect" mentioned above refers to ion flux enhancement at the edge of a shadow-cone [envelope of trajectories of ions deflected by the screened Coulomb potential of the shadowing atom (Fig. 4)]. Shadow cones are very important in ion

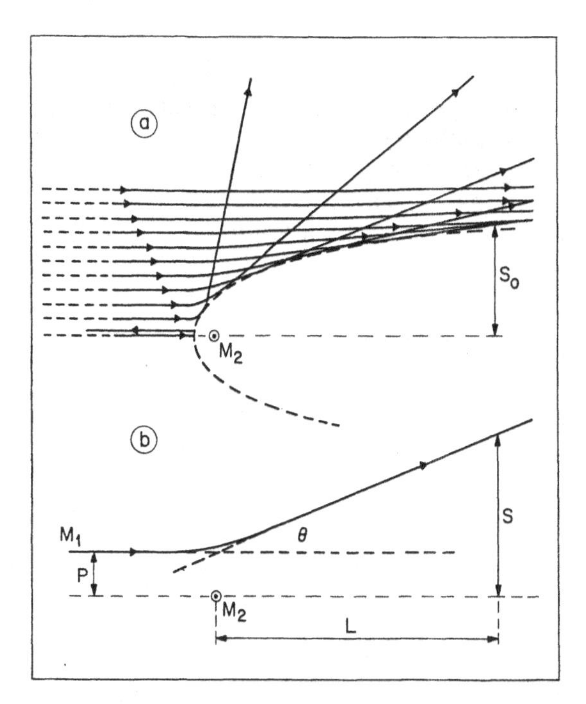

Figure 4. Possible trajectories of projectile M_1 scattered by a target atom M_2. In (a) the shadow cone behind M_2 is shown. In (b), S gives the position of the scattered particle for an impact parameter P and scattering angle θ, at a distance L behind the scattering center. The example shown is calculated for 5-keV Ne^+ ions on a Cu atom with impact parameters varying between 0.1 and 1 Å.[88]

scattering, in all three energy ranges: LEIS (0.5–10 KEv), MEIS (50–300 keV), and HEIS (1–3 MeV) also called Rutherford back scattering (RBS). Shadow cones in LEIS are larger, with radii in the neighborhood of 1 Å, and they contribute to the outstanding surface sensitivity. Shadow cone calculations are used in analysis of sub-surface layers[25–27] and in the surface structure analysis method called ICISS[28] in which polar or azimuthal rotations of a sample cause alternate supression and enhancement of scattering intensity from surface atoms as they are hidden by and then emerge from shadow cones of neighboring atoms. Examples of both uses of shadow cones will be shown.

II.3 Atom-Probe Field-Ion Microscope

This technique appears to be one of the three most popular techniques for composition analysis of segregated alloy surfaces, along with with AES and LEIS. It has single atom-layer depth resolution over as many as 13 layers.[29] Layers of atoms are peeled off a pointed tip, in UHV, by a high electric field and mass-analyzed by time-of-flight. Accumulation of adequate statistics was a problem earlier, but improvements in detection efficiency have alleviated that problem.[30,31]

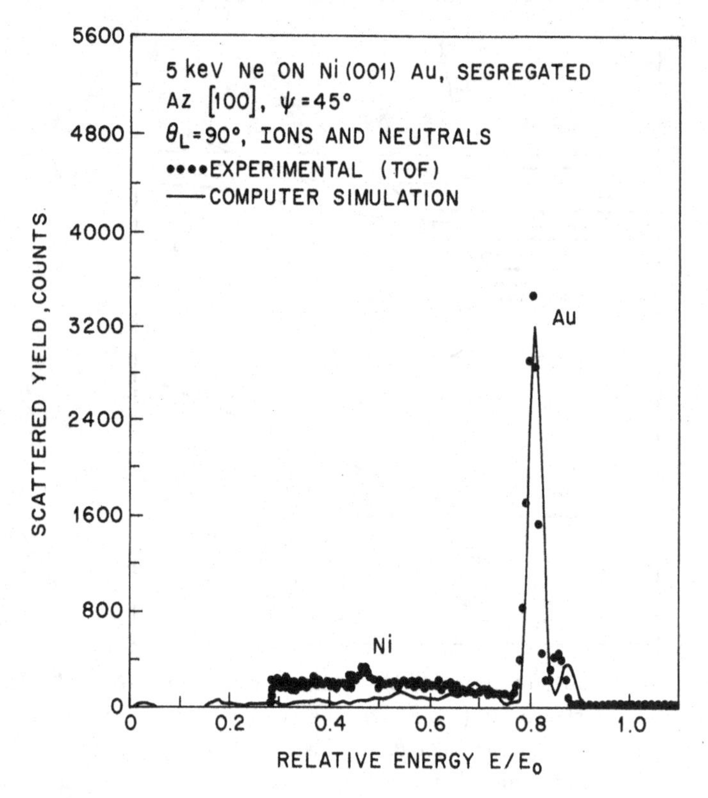

Figure 5a. Experimental and simulated energy spectra of Ne scattered from Ni(001)Au (segregated) surface. Beam direction [100]45. E_o=5keV.[49]

II.4 Low Energy Electron Diffraction (LEED)

LEED patterns have been, since the early 1960's, the basic standard for characterizing clean crystal surfaces, and also adsorbed gas layers thereon.[32–34] An ordered, unreconstructed surface yields diffracted beams at angles determined by constructive interference between electron waves scattered from the evenly spaced surface atoms. These beams produce a (1×1) pattern on the LEED screen for an unreconstructed surface. A (1×1) surface would also be seen for a disordered but unreconstructed alloy surface. An ordered alloy surface produces extra spots, fractional-order (FO) spots, due to the additional periodicity introduced, e. g., by alterning A and B atoms on the lattice sites. The degrees of long-range order is indicated by the parameter $M_1 \equiv \langle p \rangle$ where $\langle\ \rangle$ denotes a value averaged over sites, and for each site p can be $+1$ or -1: $p = +1$ if an A(B) atom is on an A(B) site, otherwise $P = -1$. In kinematical LEED theory the intensities of FO beams are proportional to M_1^2. So the variation of the long-range order parameter with temperature T may be determined by measuring FO beam intensities.

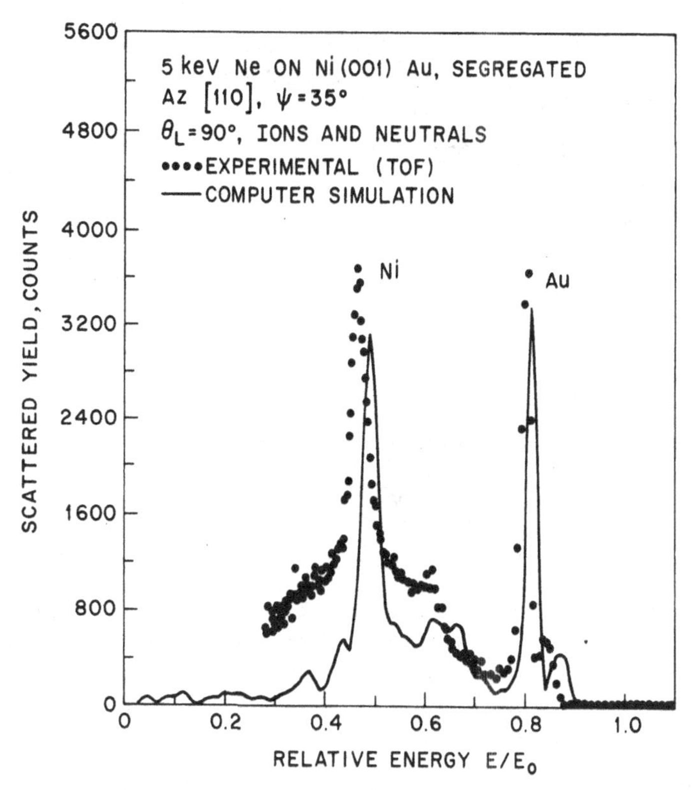

Figure 5b. Experimental and simulated energy spectra of Ne scattered from Ni(001)Au (segregated) surface. Beam direction [110]35. E_o=5keV.[49]

III. Alloy Surface Phenomena

III.1 Surface Segregation

Segregation at Cu-Ni surfaces was one of the earliest cases to receive considerable experimental attention owing to observations in catalysis research.[2,35,36] AES [37,38] and LEIS[39,40] were in good agreement that Cu was enriched at the surface, as predicted by the regular solution models.[41] An early AES study[42] and a recent investigation by the atom-probe field-ion micorscope,[31] found a reversal in Cu-Ni segregation; the surface was enriched in the minority component of the bulk in very dilute alloys. Two more recent studies by LEIS,[43] however, found no evidence of the reversal. Reversal of the segregating element in co-deposited Cu-Mn alloy films has been observed recently using LEIS.[44] Theoretical predictions had differed, one predicting that solute Cu segregates in Mn-rich alloys,[45] and the other[46] that Mn segregates in Cu-rich alloys, but no Mn segregates at the other end. Katayami *et al.*[44] note that lattice strain effects can account for the segregation of the solute in dilute Cu-Mn alloys but that the observed Mn segregation at bulk composition of up to 80% could not be explained by lattice strain.

Au-Ni alloys have been studied extensively.[47–53] Strong segregation of Au is expected on the basis of atom size mismatch and surface energy plus the positive en-

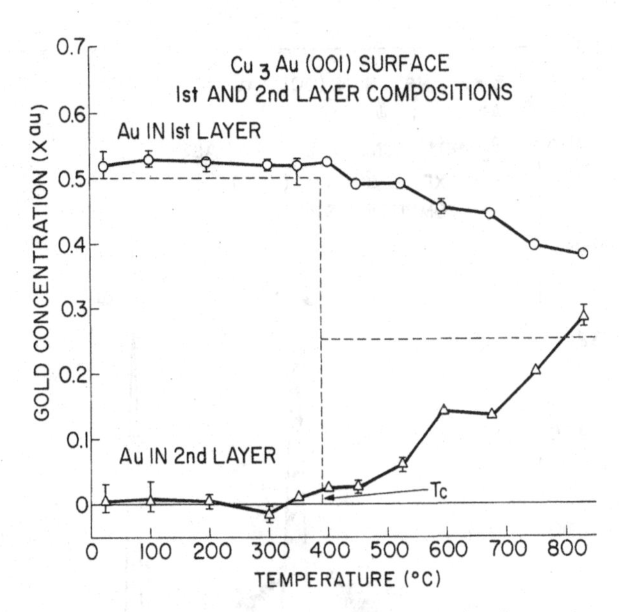

Figure 6. Composition of first and second layers of $Cu_3Au(001)$ as a function of annealing temperature. Error bars and symbols without bars indicate the spread in three or more measurements on two or three spots. Dashed lines depict hypothetical abrupt transition.[25]

thalpy of mixing. Fig. 5 illustrates the point. The (100) surface of a Ni-0.8 at% alloy was studied, after heating in UHV at 600° C. Fig. 5a shows the first layer spectrum taken with the [100]45 orientation.[49] There is a large gold peak and a very small Ni peak, corresponding to about 11% Ni in the first layer. Fig. 5b is the spectrum for the [100]35 orientation which exposed the second layer in addition to the first. The Au peak has not changed significantly but the Ni peak has, dramatically. The second layer is essentially pure Ni. The pointed shoulder to the right of the main Ni peak represents multiple scattering in the surface semi-channel, according to the computer simulation analysis. Although much of the ion scattering data was consistent with Au atoms being in substitutional sites of an unreconstructed *fcc* (100) Ni surface, LEED patterns were indexed as approximately (2×6) rather than (1×1) so there must have been some deviation, perhaps slight, from the ideal *fcc* (100) structure, which is not surprising in view of the difference in atomic radii. Segregation of Au to the (110) surface of a Ni-1% Au crystal also showed evidence that at high concentration the Au overlayer could not take up substitutional sites. In this case[50–52] LEED and LEIS, together with RBS which determined a 1 monolayer Au coverage on the Ni (100) surface, gave evidence of (7×7) and *c*(2×4) unit meshes associated with a hexagonal arrangement of six Au atoms having a seventh Au atom at its center, the center atom being higher than its nearest neighbors in the [100] directions which were in turn higher than their nearest neighbors. In this work the LEIS (TOF) results consisted of azimuthal scans at low incidence angles $\Psi = 10$ or 13.5° and large scattering angle $\Theta_L = 135°$. This method can be thought of as demonstrating 2-D channeling. It falls under the general heading of ICISS.[28] A study by STM (Scanning Tunneling Microscope) showed the first direct image of a segregated layer and partially explained the LEED and LEIS results, but also revealed scattered "bridges" consisting of double rows of Au atoms

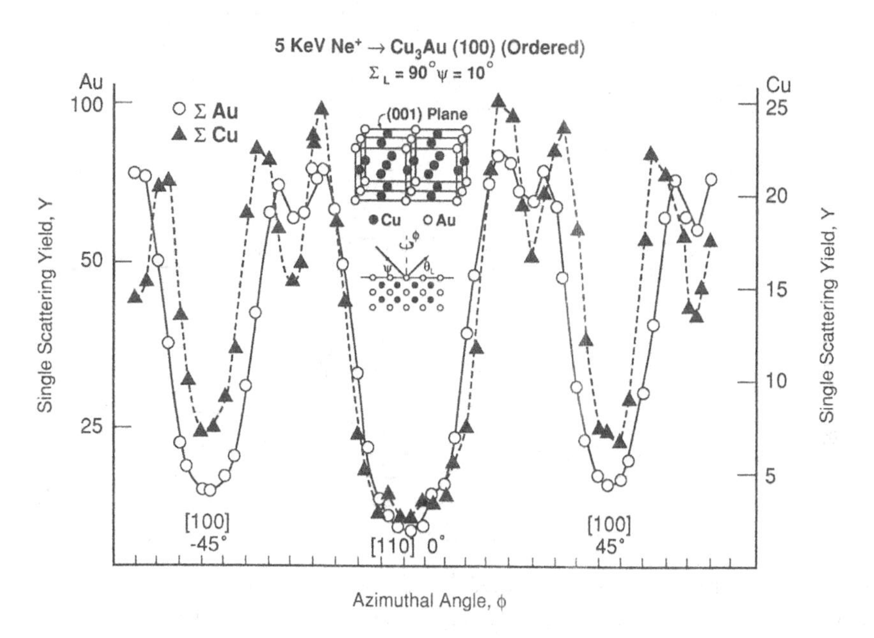

Figure 7. Azimuthal dependence of Ne single scattering intensities from Cu and Au atoms in an ordered $Cu_3Au(100)$ surface, at low incidence angle ($\psi = 10°$). Different scales for Cu and Au are due to different scattering cross-sections.[70]

in the [110] direction which had not been detected by LEED or LEIS. These began to appear when the coverage by Au exceed ~ 0.7 ML [1 ML=1 monolayer= the number of surface Ni atoms per unit area = $1.141\times10^{15}/cm^2$ on Ni (110)].

Multilayer segregation in which the segregation element accumulates in more than one layer near the surface has been observed for Cu-Ni alloys by the Atom-Probe Field-Ion technique[29] and in Fe-Sn alloys by AES[54–58] and LEIS (TOF).[26,27] Surface coverages of two monolayers on polycrystalline Fe-0.5% Sn were reported in the earlier work,[54–56] and $1.4\pm0.1\times10^{15}/cm^2$ Sn atoms on the (100) surface of Fe-1.9% Sn in later work[57–58] by AES, LEED, EELS, and UPS. A coverage of 1.7×10^{15} Sn atoms/cm^2 was determined by Rutherford back-scattering on the (100) surface of an Fe 1.7%at Sn crystal.[27] This corresponds to 1.4 ML. The LEIS measurements indicated that at saturation the top full atom layer was virtually 100% Sn in Fe lattice sites; the second layer was virtually 100% Fe and the location of the other 0.4 monolayer was not determined. Later studies[59–61] of the (100) and (111) surfaces of an Fe-1.3 at% Sn alloy have found Sn enriched in the 1st, 2nd and 3rd layers, and the segregation proceeded in time by nearly filling the outer layer after which progressively smaller concentrations appeared in the second and third layers as equilibrium was reached. This behavior correlated with a shift in the Sn $^4D_{5/2}$ binding energy observed by UPS in other work.[62]

Given the differences in experimental results reported, it appears that the three types of behavior described above —reversal of the segregating component, surface structure change in segregation, and multi-layer segregation— are all worthy of further experimental and theoretical investigation.

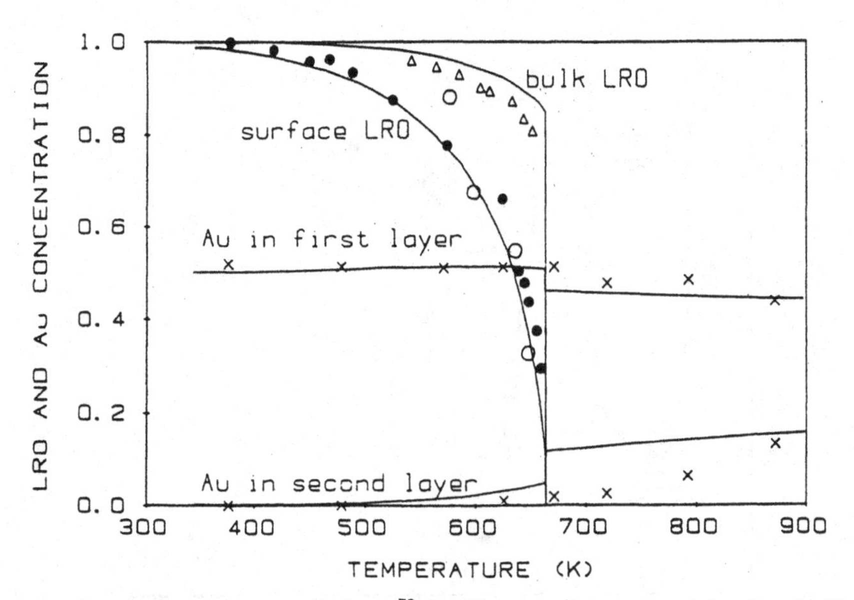

Figure 8. Comparison between calculated[73] (full lines) and experimental values in Cu_3Au: triangles, the long-range order parameter in the bulk[87]; open and closed circles, the long-range order parameter at the (001) surface, Ref. 65 and 67, respectively; crosses, the Au concentration in the first two layers.[25]

III.2 Order-Disorder Transitions at Surfaces

Surfaces of ordering alloys present interesting questions, e.g. (a) Does the ordered arrangements of atoms in the bulk, which is often known from x-ray studies, extend to the surface?; (b) In an AB alloy with alternating A and B layers which is the top layer at the surface?; (c) Does the surface order-disorder transition occur at the same temperature and with the same order, discontinuous (first-order) or continuous (higher-order)?; (d) Does surface segregation of either component enter the picture? Such questions had been addressed in a few theoretical[65,66] and experimental[65,66] studies in the 1970's and in more which have followed since.

Experimental results on Cu_3Au (100) are shown in Figs. 6–9. Equilibrium concentrations of Au, x^{Au}, in the first and second atom layers are shown as functions of temperature in Fig. 6.[25] The first layer was essentially 50% Au and the second layer pure Cu from 25° up to and somewhat beyond 400° C, *i.e.*, well past the bulk T_C at $\sim$ 390° C. There was little or no intermixing between the first two layers, presumably because of the Au-Cu bonding preference, although 1/2 order LEED spots disappear close to the bulk T_C at 390° C.[65–67] The disorder is apparently confined to the first layer. The Au concentration is apparently held up beyond T_C by the Au segregation tendency,[46,68,69] which may also lead to the fact that the 50–50 layer was invariably on top after sputter cleaning and annealing. Evidence of the ordered arrangement in the first three layers, for $T < T_C$ was found in 3D azimuthal scans in which the experimental single-scattering intensity from Cu and Au, at $E_\circ = 5$ keV and 9.5 keV agreed well with shadow cone analysis based on the ideal ordered arrangement. Additional evidence of the ordering in the first layer is shown in Fig. 7, an example of

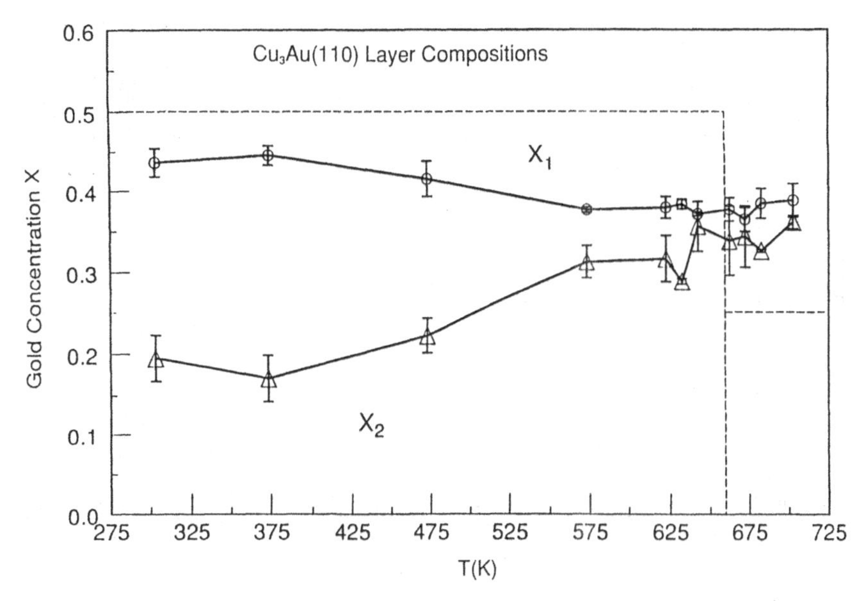

Figure 9. Au atom fractions $X_1(X_2)$ in the first (second) layers determined by LEIS using a 5 kev Ne^+ beam and 90° scattering angle. Error bars indicate the variance of four consecutive 5 min determinations at contiguous beam locations on the surface. Broken lines indicate atom fractions for an ideal ordered crystal with Au-rich termination ($T < T_C$) and an ideal disordered crystal ($T > T_C$).[76]

the "2D channeling". Single scattering yields from Au and Cu are plotted against azimuthal orientation[70]. Incidence angle Ψ is low, 10° measured from the surface so that atoms in the major rows of atoms at [100] and [110] shadow their nieghbors in line with the beam and create deep minima in the scattering intensity at those azimuths. At [100] azimuths Cu dips are narrower than Au because Cu atoms shadow Cu atoms, Au shadows Au, and Cu shadow-cone radii are smaller than Au. At [110] shadowing is reversed, Au on Cu and Cu on Au, so the Cu dip is broader than the Au. Estimates of the degree of short-range order might be provided by computer simulation of the scattering such as Derks *et al.*[71] have used for pure metal surfaces using the Marlowe[72] program. Analysis of the 1/2 order LEED beams favored a continuous transition rather than a first-order transition as found for the bulk transition. LEED data and layer compositions, x_1^{Au} and x_2^{Au}, are plotted against temperature in Fig. 8 together with cluster theory predictions by Sanchez and Morán-López.[73] The agreement is good, although the LEIS results were not precise enough to check the predicted small steps in x^{Au} at T_C. More recently Monte Carlo simulations in conjunction with the embedded atom method[74] gave good agreement with the first layer experimental results, and the tetrahedron approximation of the cluster variation method has been used[75] to calculate phase diagrams of the *fcc* (100) surface ordering alloys of different ratios of the ordering energy to surface segregation energy.

Distinctly different behavior has been found on the (110) surface of Cu_3Au.[76] The equilibrium first and second layer Au concentrations are 0.45 and 0.20, respectively, at room temperature (Fig. 9) and converge toward 0.35 near 660 K, in contrast to the constant 0.52 and 0.0 values of x_1^{Au} and x_2^{Au} seen on the (100) surface up to 663 K.

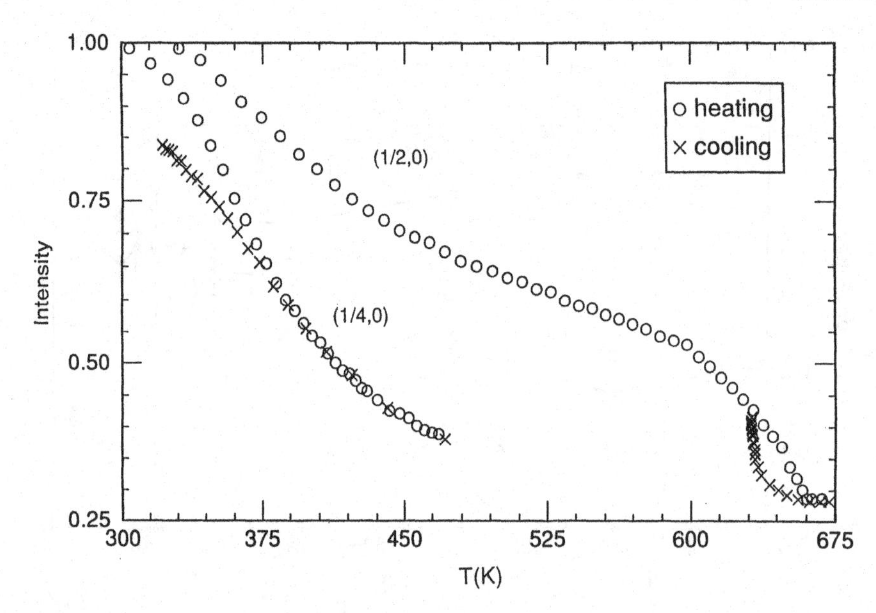

Figure 10. Observed variation of LEED intensity versus crystal temperature T for the beams indicated. The indexing is such that [10] and [01] surface vectors correspond respectively to $[\bar{1}10]$ and [001]. Electron energy 23 eV, diffraction angle relative to the incident beam 45°, heating and cooling rates 0.2 K/s near 660 K.[76]

LEED patterns at different temperatures (Fig. 10) indicate a (4×1)→(2×1) transition at ~ 425 K. (2×1) is expected for the ideal ordered (110) surface. The (2×1)→(1×1) transition exhibits hysteresis and close inspection shows that it occurs six degrees below the bulk critical temperature, T_C.

Fig. 11 shows a plot of (1/2,0) beam intensity as T was increased slowly, with pauses at 649 K, 651 K, and 654 K, long enough for intensity to descend to a steady value. At each of the first two pauses there is a step-like drop of intensity. Each step corresponds to a slow approach of the surface to its disordered state at that temperature. At the third pause, 654 k, there is no intensity step, showing that at about 654 K the surface had already reached the disordered state. 654 K is an upper limit to T_1, the surface transition temperature. The inset shows (1/2,0) beam intensity against log time during one of the pauses and during ordering at the same temperature. Disordering is an order of magnitude faster than ordering. This and the finding $T_1 < T_C$ are consistent only with a discontinuous transition. The (2×1)→(1×1) behavior can be understood by analogy with the two-phase region in an off-stoichiometry discontinuous transition in the bulk crystal (Fig. 12). The average enrichment of the first two layers, to 0.33, above the ideal stoichiometric Au atom fraction, 0.25, causes the order-disorder transiton to occur below the maximum bulk disordering temperature, T_C. Questions remain about the Cu_3Au (110) surface. No explanation has been found for the (4×1) LEED pattern at low temperatures. Low incidence angle azimuthal scans, similar to those of Fig. 7 on the (100) surface, were made on the (110) surface in the (4×1) state. Comparison with other results on known *fcc* (100) surfaces[70,71] seems to confirm that it is not reconstructed. If it were reconstructed in the (4×1) state that could invalidate our layer analysis based on the geometry of the (2×1) arrangement, below 450 K.

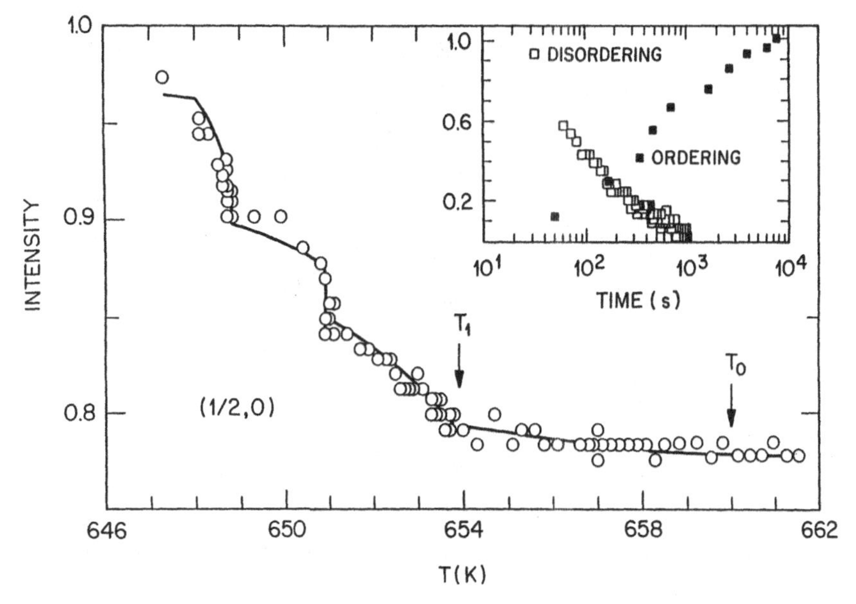

Figure 11. Dependence of (1/2,0) beam intensity on the crystal temperature T. The intensity without background subatraction is indicated with arbitrary units. The crystal was annealed at 630 K, heated rapidly (rate¿0.01 K/s) to 648 K and then heated slowly (rate¡0.01 K/s) with 10×10^2 s pauses at 649, 651 and 654 K. Diluted data points and a smooth computed line for the original data points are shown. Arrows indicate estimated transition temperatures for the bulk (T_C) and surface (T_1). The inset shows the dependence of the (1/2,0) beam intensity on time during ordering and disordering at 651 K. The intensity scale is such that the total change of intensity is unity. The time scale shows elapsed time since the start of ordering/disordering plus 30 s. Before the ordering run, the crystal was annealed for 10×10^2 s at 657 K, then cooled rapidly (rate¿0.1 K/s) 651 K.[76]

III.3 Preferential Adsorption of CO on NiAl(110)

Studies of NiAl surfaces have been pursued in several laboratories lately. The clean NiAl (110) surface is ordered, with alternating rows of Ni and Al atoms, and the surface is rippled, with Al atoms ~0.22 Å higher than Ni.[78–80] Fig. 13 shows a top view. Studies of the NiAl (110)/CO surface had led to postulate that only one CO adsorption site was occupied and that it was most likely a terminal (on-top) site,[81,82] but the adsorbing atom, Ni or Al, was not identified, He ion scattering in the TOF mode was therefore applied to the problem in hopes that shadowing effects would reveal the adsorption site.[83] Fig. 14 shows TOF spectra of the surface taken with beam parallel to the [001] and [$1\bar{1}0$] azimuths, as function of CO exposure. Along the [001] azimuth the Ni peak was strongly reduced but the Al peak was not (Fig. 15 shows the peak areas *vs.* CO exposure). The [001] behavior shows that CO is adsorbing along Ni rows but does not distinguish between an on-top or [001] Ni bridge site. Along [$1\bar{1}0$] the Ni peak is again strongly reduced but the Al peak is also reduced for heavy coverages. The greater attenuation of Ni compared to Al in both directions especially at low coverage suggests that CO is in an on-top Ni site, not a Ni-Ni bridge. In the

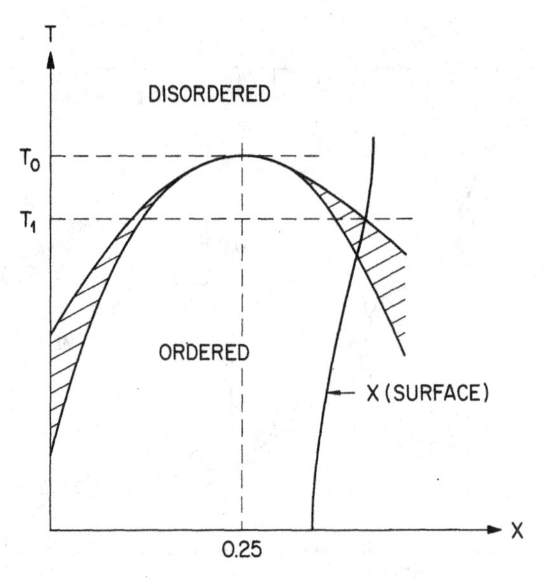

Figure 12. Schematic interpretation of the surface phase transition.[76]

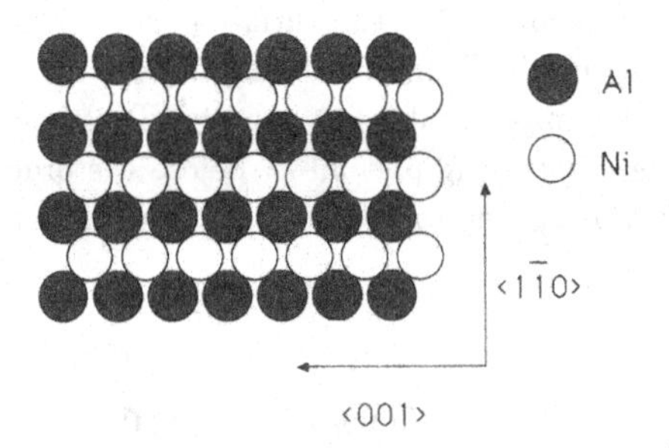

Figure 13. Top view of atomic arrangement in the first layer of NiAl(110). Hollow circles: Ni. Solid circles: Al.

latter case Al should be very strongly shadowed along one of the azimuths. This was confirmed by an experiment in which 5 keV Ne^+ was directed nearly normal to the surface with beam and detector in the plane of surface normal and [110] azimuth. This experiment showed that Ni peak area attenuation was proportional to CO coverage, confirming the on-top Ni site for the CO adsorption. The Al peak attenuation along [110] was attributed to molecular vibrations of CO closing up the narrower space between [110] rows of Ni. The CO adsorption was also found to attenuate significantly the outward relaxation of Al atoms in the clean surface.

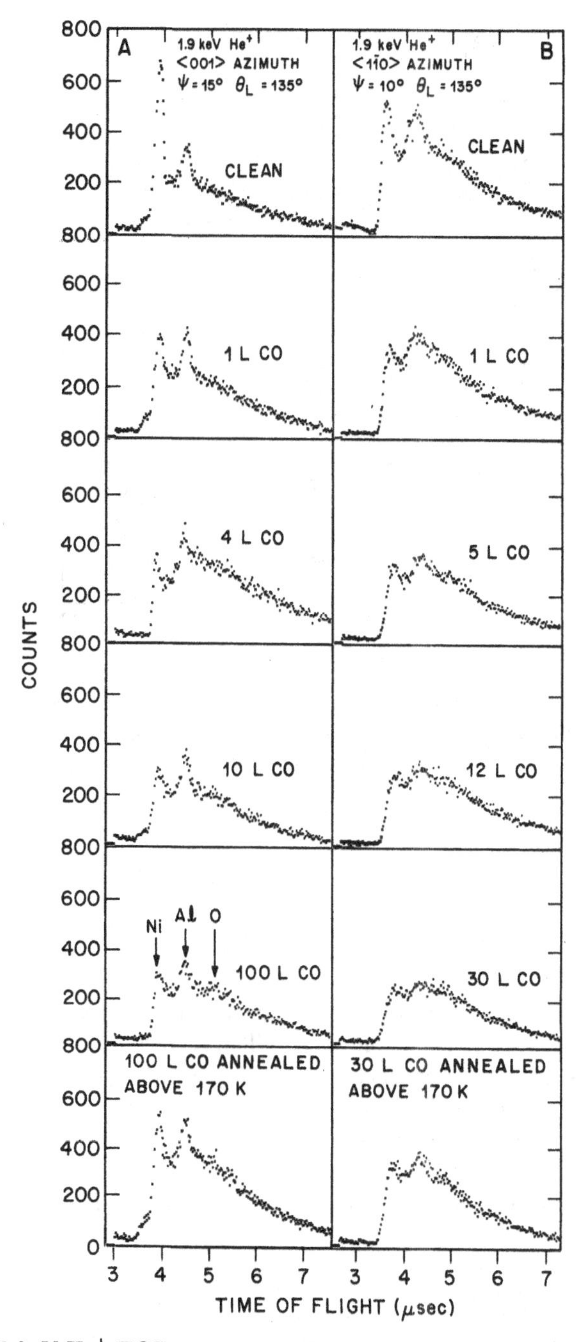

Figure 14. (A) 1.9 keV He^+ TOF spectra as a function of CO exposure with the ion beam incident along $\langle 001 \rangle$ azimuth, $\psi = 15^o$, $\theta_L = 135^o$, and (B) the $\langle 1\bar{1}0 \rangle$ azimuth, $\psi = 10^o$, $\theta_L = 135^o$. Sample temperature 100 K.[83]

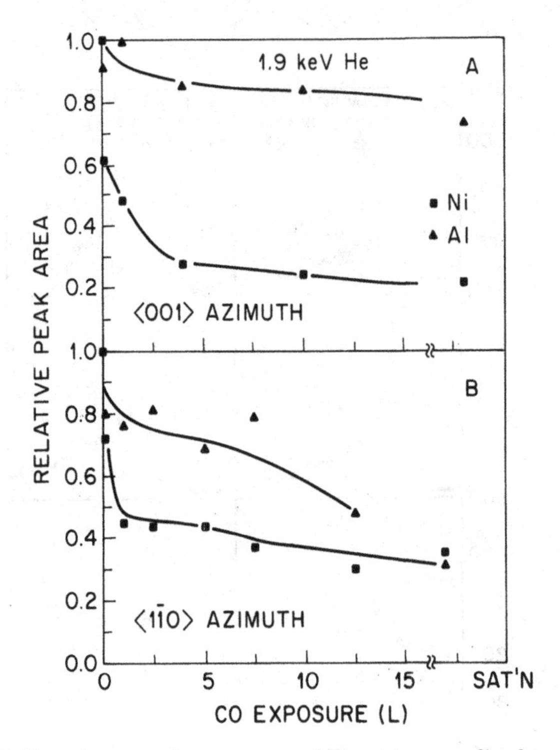

Figure 15. Ni and Al peaks areas from spectra of Fig. 14, normalized to clean surface peaks areas for 1.9 keV He^+ beam incident along (A) the ⟨001⟩ azimuth, $\psi = 15^\circ$, $\theta_L = 135^\circ$, and (B) the ⟨1$\bar{1}$0⟩ azimuth, $\psi = 10^\circ$, $\theta_L = 135^\circ$. Sample temperature 100 K.[83]

III.4 Surface Miscibility Gaps

Surface phase transitions associated with surface miscibility gaps have been observed in interstitial metal-carbon solid solutions.[84] The surface phase boundaries have been derived within regular solution theory.[85] Quite recently Liu and Wynblatt,[86] using AES, found a surface miscibility gap in a Cu-0.3 at% Ag alloy. There were abrupt changes from a Cu-rich to a Ag-rich surface (or *vice versa*) as temperature was changed in the vecinity or 500° C. Using Auger line scans with a high-resolution probe they observed the formation and growth of silver-rich patches in a copper-rich surface, and *vice versa*.

IV. Conclusions

Familiar surface analytical techniques, notably Auger electron spectroscopy, low-energy ion scattering, LEED and field-ion atom-probe continue to reveal interesting and important information about alloy surfaces. Similarities and differences are found between surface and bulk behavior. A new technique, scanning tunneling microscopy (STM), which can give a direct image of a segregated layer is very promising.

References

1. G. W. Graham, *Surface Sci.* **137**, L79 (1984).
2. J. H. Sinfelt, J. L. Carter, D. J. C. Yates, *J. Catalysis.* **24**, 283 (1972).
3. A. E. T. Kuiper, G. C. J. van der Ligt, W. M. van de Wijgert, M. F. C. Willemsen, and F. H. P. M. Habraken, *J. Vac. Sci. Technol. B* **3**, 830 (1985).
4. E. J. van Loenen, L. W. M. Frenken, and J. F. van der Veen, *Appl. Phys. Lett.* **45**, 41 (1984).
5. ASM 1985 International Conference on Surface Modifications and Coatings. Toronto, Canada. (Oct. 14, 1985)
6. C. L. White, R. H. Padgett, C. T. Liu and S. M. Yalisove, *Scripta Metall.* **18**, 1417 (1984).
7. A. Joshi in *Interfacial Segregation*, W. C. Johnson and J. M. Blakely, Editors, American Society for Metals. **39**, (1979).
8. *Practical Surface Analysis*, D. Briggs, and M. P. Seah, Editors, John Wiley and Sons, Ltd. (1983).
9. M. A. Hoffmann, S. W. Bronner and P. Wynblatt, *J. Vac. Sci. Technol. A* **6**, 2253 (1988).
10. *Surface Segregation Phenomena*, P. A. Dowben and A. Miller, Editors, CRC Press (1990).
11. D. P. Smith, *J. Appl. Phys.* **38**, 340 (1967).
12. H. H. Brongersma and P. M. Mul, *Chem Phys. Lett.* **14**, 380 (1972).
13. D. J. Ball, T. M. Buck, D. MacNair and G. H. Wheatley, *Surface Sci.* **30**, 69 (1972).
14. W. Heiland and E. Taglauer, *Surface Sci.* **33**, 27 (1972).
15. Y. S. Chen, D. A. H. Robinson, G. L. Miller, G. H. Wheatley, and T. M. Buck, *Surface Sci.* **63**, 133 (1977).
16. S. B. Luitjens, A. J. Algra, E. P. Th. Suurmeijer, A. L. Boers, *Appl. Phys.* **21**, 205 (1980).
17. I. Terzic, D. Ciric, B. Perovic, *Surface Sci.* **85**, 149 (1979).
18. A. J. Algra, E. van Loenen, E. P. Th. Suurmeijer and A. L. Boers, *Rad. Effects.* **60**, 173 (1982).
19. E. Taglauer, W. Englert, W. Heiland, and D. P. Jackson, *Phys. Rev. Lett.* **45**, 740 (1980).
20. S. H. Overbury, W. Heiland, D.M. Zehner, S. Datz, and R. S. Thoe, *Surface Sci.* **109**, 239 (1981).
21. H. Niehus and G. Comsa, *Surface Sci.* **18**, 140 (1984).
22. M. T. Robinson and I. M. Torrens, *Phys. Rev.* **B9**, 5008 (1974).
23. T. M. Buck, G. H. Wheatley, and L. K. Verheij, *Surface Sci.* **90**, 635 (1979).
24. T. M. Buck, G. H. Wheatley, D. P. Jackson, *Nucl. Inst. Meth. Phys. Res.* **218**, 257 (1983).
25. T. M. Buck, G. H. Wheatley, and L. Marchut, *Phys. Rev. Lett.* **43**, 51 (1983).
26. L. Marchut, T. M. Buck, G. H. Wheatley, and C. J. McMahon, Jr., *Surface Sci.* **141**, 549 (1984).
27. L. Marchut, T. M. Buck, C. J. McMahon, Jr., G. H. Wheatley, and W. M. Augustyniak, *Surface Sci.* **180**, 252 (1987).
28. M. Aono, C. Oshima, S. Zaima, S. Otani and Y. Ishizawa, *Japan J. Appl. Phys.* **20**, L829 (1981).
29. Y. S. Ng, T. T. Tsong, and S. B. McLane, Jr., *Phys. Rev. Lett.* **42**, 588 (1979).

30. T. Sakurai, T. Hashizume, and A. Jimbo, *Appl. Phys. Lett.* **38**, 44 (1984).
31. T. Sakurai, T. Hashizume, A. Jimbo, A. Sakai, and S. Hyodo, *Phys. Rev. Lett.* **55**, 514 (1985).
32. J. J. Lander, in *Progress in Solid State Chemistry*, H. Reiss, Editor, Pergamon Press, New York. 26–116 (1965).
33. L. J. Clarke, *Surface Crystallography*, John Wiley and Sons, New York (1985).
34. P. J. Estrup and E. G. McRae, in *Modern Methods of Surface Analysis*, P. Mark and J. D. Levine, Editors, North-Holland Publ. Co. 1-52 (1971). Reprinted from *Surface Sci.* **25** (1971).
35. P. van der Plank and W. M. H. Sachtler, *J. Catalysis.* **12**, 35 (1968).
36. V. Ponec and W. M. H. Sachtler, *J. Catalysis.* **24**, 250 (1972).
37. C. R. Helms, *J. Catalysis.* **36**, 114 (1975).
38. J. J. Burton, C. R. Helms, and R. S. Polizzotti, *J. Vac. Sci. Technol.* **13**, 204 (1976).
39. H. H. Brongersma and T. M. Buck, *Surface Sci.* **53**, 649 (1975).
40. H. H. Brongersma, M. J. Sparnaay, and T. M. Buck, *Surface Sci.* **71**, 657 (1978).
41. F. L. Williams and D. Nason, *Surface Sci.* **45**, 377 (1974).
42. Y. Takasu and H. Shimizu, *J. Catalysis.* **29**, 479 (1973).
43. L. E. Rehn, H. A. Hoff, and N. Q. Lam, *Phys. Rev. Lett.* **57**, 780 (1986) and H. H. Brongersma, P. A. J. Ackermans and A. D. van Langefeld, *Phys. Rev.* **B34**, 5974 (1986).
44. I. Katayama, K. Oura, F. Shoji and Teruo Hanawa, *Phys. Rev.* **B38**, 2188 (1988).
45. S. Mukherjee and J. L. Morán-López, *Surf. Sci.* **189/190**, 1135 (1987).
46. A. R. Miedema, *Z. Metallkd.* **69**, 455 (1978) and J. K. Chelikowsky, *Surface Sci.* **139**, L197 (1984).
47. P. Wynblatt and R. C. Ku, *Surface Sci.* **65**, 511 (1977).
48. P. Biloen, R. Bouwman, R. A. van Santen, and H. H. Brongersma, *Appl. Surf. Sci.* **2**, 532 (1979).
49. T. M. Buck, I. Stensgaard, G. H. Wheatley, and L. Marchut, *Nucl. Inst. and Meth.* **170**, 519 (1980).
50. Boerma, T. M. Buck, E. G. McRae, R. A. Malic, and G. H. Wheatley, unpublished.
51. T. M. Buck and E. G. McRae, *Proceedings of ASM International Conference on Surface Modifications and Coatings*, 1985.
52. E. G. McRae and R. A. Malic, *Surface Sci.* **53**, 177 (1986).
53. Y. Kuk, P. J. Silverman and T. M. Buck, *Phys. Rev.* **B36**, 3104 (1987).
54. M. P. Seah and E. D. Hondros, *Proc. Roy. Soc.* **191**, A335 (1973) London.
55. C. Lea and M. P. Seah, *Surface Sci.* **53**, 272 (1975).
56. M. P. Seah and C. Lea, *Phil. Mag.* **31**, 627 (1975).
57. H. Viefhaus and M. Rusenberg, *Surface Sci.* **159**, 1 (1985).
58. H. Viefhaus, G. Tauber and H. J. Grabke, *Scripta Met.* **9**, 1181 (1985).
59. R. Hsiao, Ph. D. Thesis, University of Pennsylvania, 1986.
60. R. Hsiao, C. J. McMahon, Jr., E. W. Plummer and T. M. Buck, *J. Vac. Sci. Technol. A*5, 887 (1987).
61. R. Hsiao, T. M. Buck and C. J. McMahon, Jr., to be published.
62. R. Hsiao, C. J. McMahon, Jr. and E. W. Plummer, to be published.
63. J. L. Morán-López and K. H. Bennemann, *Phys. Rev.* **B15**, 4769 (1977).
64. J. L. Morán-López and L. M. Falicov, *Phys. Rev.* **B18**, 2542 (1978).
65. V. S. Sundaram, R. S. Alden, and W. D. Robertson, *Surface Sci.* **46**, 653 (1974).
66. H. C. Potter and J. M. Blakely, *J. Vac. Sci. Technol.* **12**, 635 (1975).

67. E. G. McRae and R. A. Malic, *Surface Sci.* **148**, 551 (1984).
68. J. M. McDavid and S. C. Fain, Jr., *Surface Sci.* **52**, 161 (1975).
69. M. J. Sparnaay and G. E. Thomas, *Surface Sci.* **135**, 184 (1983).
70. T. M. Buck, W. F. Flood, J. Baraga, unpublished.
71. D. Derks, W. Hetterich, E. van de Riet, H. Niehus, and W. Heiland, *Nucl. Instrum. Meth.* **B48**, 315 (1990).
72. M. T. Robinson and I. M. Torrens, *Phys. Rev.* **B9**, 5008 (1974).
73. J. M. Sanchez and J. L. Morán-López, *Surface Sci.* **157**, L297 (1985).
74. M. A. Hoffmann and P. Wynblatt, *Surface Sci.* **236**, 369 (1990).
75. Y. Teraoka, *Surface Sci.* **232**, 193 (1990).
76. E. G. McRae, T. M. Buck, R. A. Malic, W. E. Wallace and J. M. Sanchez, *Surface Sci. Lett.* **238**, L481 (1990).
77. W. E. Wallace and T. M. Buck, unpublished.
78. H. L. Davis, and J. R. Noonan, *Phys. Rev. Lett.* **54**, 566 (1985).
79. S. M. Yalisove and W. R. Graham, *Surface Sci.* **183**, 556 (1987).
80. D. R. Mullins and S. H. Overbury, *Surface Sci.* **141**, 199 (1988).
81. J. M. Mundenauer, R. H. Gaylord, S.C. Lui, E. W. Plummer, D. M. Zehner, W. K. Ford and L. G. Sneddon, *MRS Proc.* **54**, 83 (1987).
82. J. M. Mundenauer, Ph.D. Thesis, University of Pennsylvania (1988).
83. C. H. Patterson and T. M. Buck, *Surface Sci.* **218**, 431 (1989).
84. J. C. Shelton, H. R. Patil and J. M. Blakely, *Surface Sci.* **43**, 493 (1974).
85. C. R. Helms, *Surface Sci.* **69**, 689 (1977).
86. Y. Liu and P. Wynblatt, *Surface Sci.* **241**, L21 (1991).
87. R. Feder, M. Mooney, and A. S. Nowick, *Acta Metall.* **6**, 266 (1958).
88. A. G. J. de Wit, R. P. N. Bronckers, and J. M. Fluit, *Surface Sci.* **82**, 177 (1979).

Structural Phase Transformations in Alloys: An Electron Microscopy Study

G. Van Tendeloo[1], D. Schryvers[1], L. E. Tanner[2], D. Broddin[1]*, C. Ricolleau[3], and A. Loiseau[3]

[1] *University of Antwerp (RUCA) Groenenborgerlaan, 171 B2020 Antwerp, Belgium*

[2] *Lawrence Livermore National Laboratory Livermore, CA 94550, USA*

[3] *ONERA, BP 72 F-92322 Chatillon/Bagneux, France*

Abstract

Electron microscopy, including atomic scale high resolution observations, combined with electron diffraction are highly qualified to provide structural data on different aspects of phase transformations in alloys. Two different topics are described: a) the structural changes occurring at interfaces, particularly antiphase boundaries, close to the order disorder temperature b) the pre-transition modulations associated with the displacive austenite-martensite transformation.

I. Introduction

Phase transitions in alloys can be subdivided in many ways and into many categories; for this overview we will consider:

a) non-displacive, diffusion controled transformations; we will investigate experimentally by means of high resolution electron microscopy the fine structure of ordering

* Present Address: AGFA-GEVAERT, B 2640 Mortsel, Belgium

defects such as antiphase boundaries and their behaviour at temperatures close to the order-disorder temperature. The results in Cu-Pd and Co-Pt can be compared with Monte Carlo simulations statistical mechanics calculations.

b) displacive, diffusionless transformations; we will treat as an example the austenite-martensite transformation in a nickel rich Ni-Al alloy. Special attention will be focused on the microstructure or "tweed" structure of the premartensitic state. This modulated structure is known to be associated with unusual variations of physical properties of the parent phase.

II. Interface Wetting in Ordered Alloys

We will first consider a very simple compound, Pt-Co , at a composition $Pt_{70}Co_{30}$. Below 755° C this *fcc* solid solution becomes ordered with the $L1_2$ structure. Details on the phase diagram are to be found in Ref. 1 Within the ordered phase, antiphase boundaries with a displacement vector of the type $\mathbf{R} = 1/2\langle 110\rangle$ appear as a consequence of the decrease in translation symmetry. When heat treated at temperatures well below Tc (e.g. 50°) these antiphase boundaries show a normal behaviour under the electron microscope. Under two beam conditions with a reflection $\mathbf{g}$ (and $\mathbf{0}$) and for $\mathbf{g} \cdot \mathbf{R} = n$ the boundary is out of contrast. When the antiphase boundary is imaged under high resolution conditions as in Fig. 1a this displacement vector can immediately be identified. Moreover it becomes clear that the structure is perfect up to the boundary plane and that the width of the interface is restricted to a single atomic plane; this is the normal behaviour for an ordering defect. Note that in this image the bright dots correspond to the minority atom configuration, i.e. they figure the projected Co-atom configuration when projected along one of the cube axes.[2,3] When annealing the material at temperatures between $Tc - 17°$ and Tc, the width of the interface gradually increases;[4] this is obvious from Fig. 1b, which is made under similar imaging conditions as Fig. 1a. The antiphase boundary is 2 to 4 nm wide and at the interface the basic square *fcc* lattice of 0.19×0.19 nm can be recognized, indicating the presence of disordered material between the two ordered regions. This wetting behaviour of antiphase boundaries at temperatures close to the order-disorder transition was recorded on video during *in situ* heating inside the electron microscope[5] and was confirmed for furnace annealed bulk material, quenched prior to investigation. In the latter case a better temperature controle, down to 0.1°, is possible and allows a quantitative analysis of the wetting effect.

In the Cu-17% Pd, the low order-disorder transition temperature has allowed us to perform highly controlled heat treatments in the range between $Tc - 10°$ and Tc with temperature steps of 0.2°.[6] This experimental study has been undertaken in order to characterize the nature of the divergence of the width of the domain walls. Although the concentration of the alloy was chosen to avoid two-phase effects, the Cu-17% Pd system still presents a two phase field of about 1° wide; evidence is presented in Fig. 2. The presence of this two phase field hampers the boundary-width measurements as well as the analysis of interface instabilities in the immediate vicinity of Tc; on the other hand the measurements clearly allow to distinguish between the wetting regime and the two phase regime. In order to measure the exact width of the antiphase boundary a number of precautions have to be taken. However some problems persist: (a) the width of the interface is always measured at room temperature after quenching, (b) high resolution images invariably show a sharper interface than conventional dark field images. Moreover close to Tc the antiphase boundaries show a tendency to fluctuate dramatically

Figure 1. High resolution image along [001] of antiphase boundaries in the $Pt_{70}Cu_{30}$ alloy. a) when annealed well below Tc. b) When annealed just below Tc.

and they may indicate the initial stages of formation of a long period superstructure. This long period superstructure is only stable however at higher Pd content. The preliminary results of the change in boundary width as function of temperature are shown in Fig. 3. They show a good agreement with statistical mechanics calculations which predict that the width of the interface logaritmically diverges as $\log(Tc - T)$.[7]

When the palladium content is increased in the Cu-Pd system the $L1_2$ structure becomes unstable towards antiphase boundary formation and a one dimensional long period superstructure is formed. The average spacing between subsequent interfaces increases and is function of composition and temperature. This has recently been extensively studied in Refs. 3 and 8. Close to the disordering temperature the interfaces exhibit similar wetting phenomena as the individual boundaries, this is clear from the image of Fig. 4a which is from a Cu-19 at % Pd alloy annealed only 5° below the disordering temperature. The disordered region is 1–2 $L1_2$-unit cells wide and moreover the interfaces, which at lower temperature are strictly bound to (001) planes, now become more or less wavy. The inset, which is a processed image of the original HREM, accentuates the local changes at the interface. These effects are a general feature of the long period superstructures which are formed from a solid solution; a similar observation for

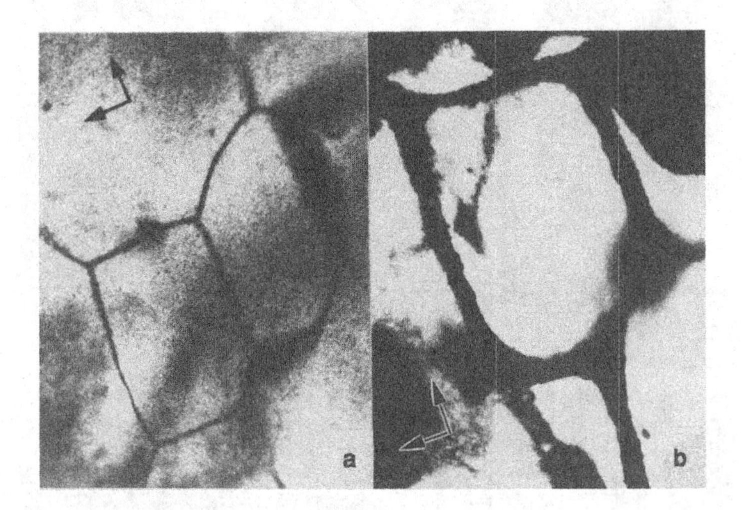

Figure 2. Dark field images of an alloy Cu-17%Pd annealed at two different temperatures. a) 4° below Tc. b) In the two phase field between the ordered and the disordered region.

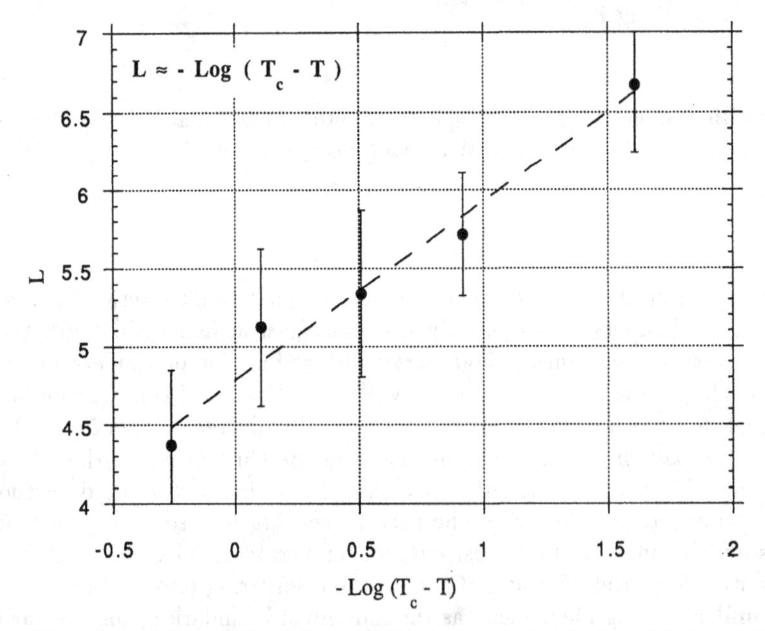

Figure 3. Variation of the width of the antiphase boundaries in the vicinity of the order-disorder temperature, as function of temperature.

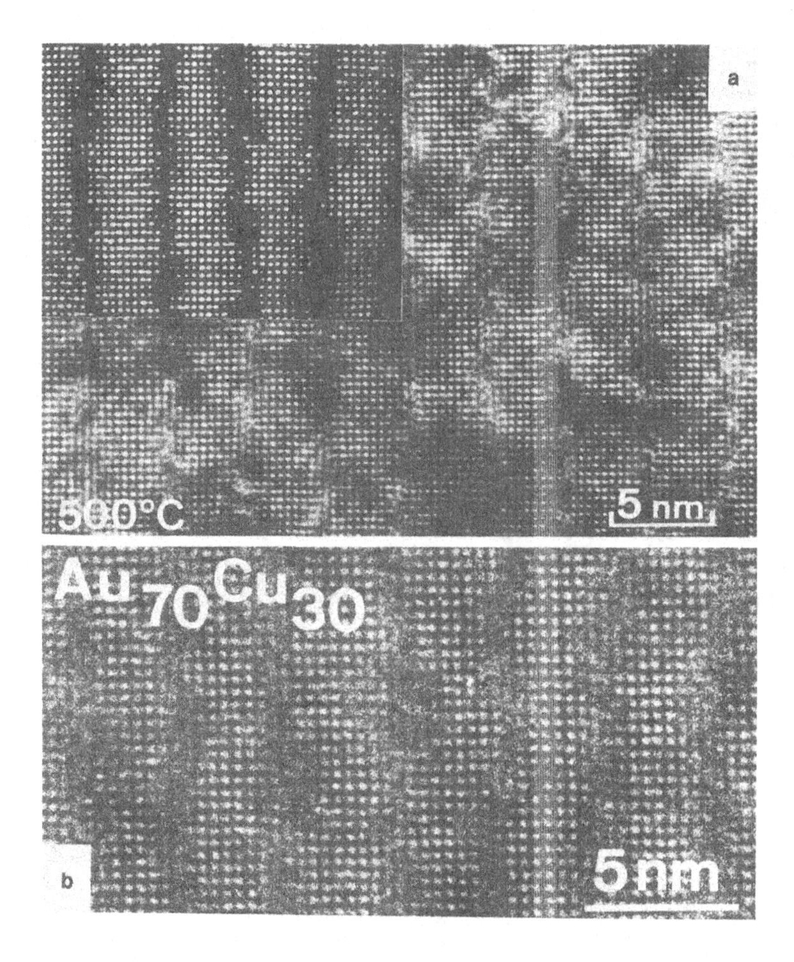

Figure 4. High resolution image along [010] of the incommensurate long period structures in $Cu_{81}Pd_{19}$ (a) and $Au_{70}Cu_{30}$ (b). Note the fluctuations in the APB positions.

$Au_{70}Cu_{30}$ is shown in Fig. 4b. In both images the bright dots are to be interpreted as the minority atom configuration; this has been proven by computer simulation of the experimental data.[9]

In conclusion we can say that electron microscopy techniques allow to study in detail, as well qualitatively as quantitatively, the complex behaviour of interfaces in the vicinity of the order-disorder transition. Wetting of antiphase boundaries has been observed in a number of alloys; the width of the interface has been shown to follow a logarithmic law as function of $(Tc - T)$.

III. Displacive Transformations : Austenite-Martensite Transition in Ni-Al

The austenite-martensite transition in Ni-rich Ni-Al has been reexamined using different electron microscopy techniques including HREM. Careful investigation of results

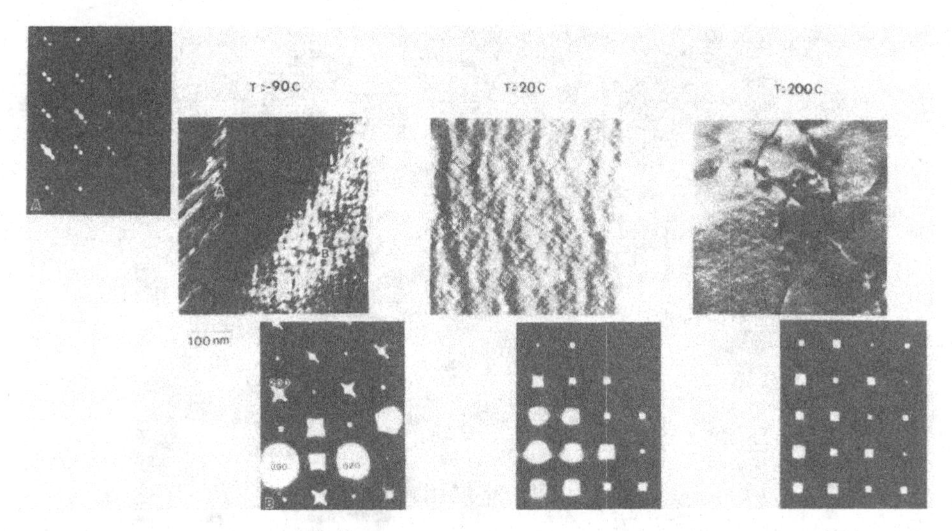

Figure 5. BF images and SAED zone patterns of the corresponding regions of a cooling sequence through Ms in a $Ni_{62.5}Al_{37.5}$ sample.

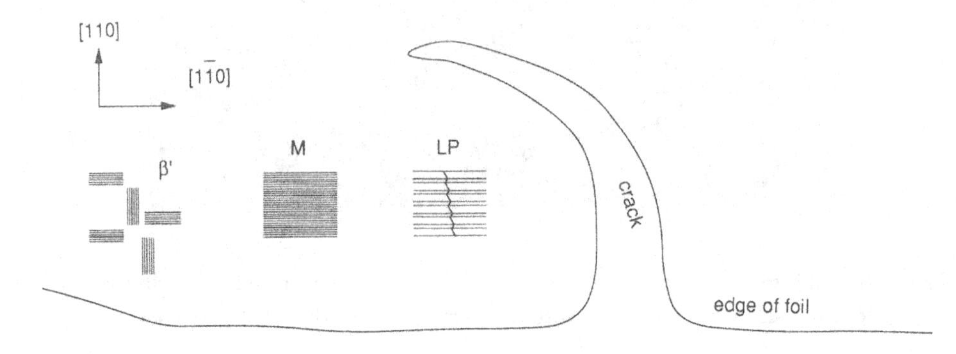

Figure 6. Schematical drawing of an asymmetric configuration showing the same structures as in Fig. 4.

obtained from selected samples enabled us to interpret images of precursor and transition stages, including the well known tweed structure, in terms of sequential steps in the transition sequence. One of the conclusions is that, as early as the precursor state, the material develops transverse displacement modulations that will finally lock in to yield the martensite structure.

The transition from the *bcc* based B2 austenite phase (CsCl structure) to the *fct* based $L1_0$ (possibly twinned) martensite phase can be followed in situ by cooling inside the electron microscope. In Fig. 5 direct images as well as diffraction patterns of a $Ni_{62.5}Al_{37.5}$ sample at different temperatures are presented. The high temperature (200° C) tweed pattern revealing 110 striations in the image and $\langle 110\rangle^*$ diffuse streaks in the diffraction pattern clearly exists far above M_s, although the contrast of the mod-

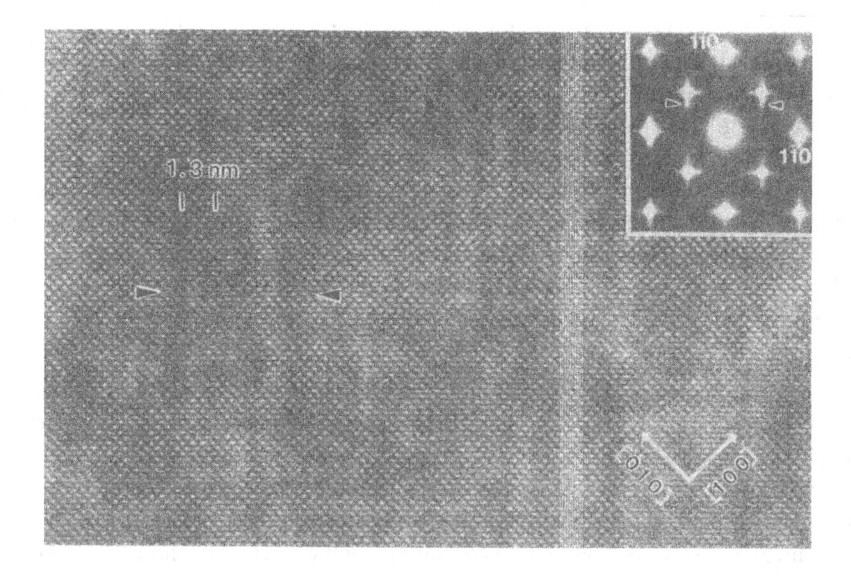

Figure 7. [001] HREM image of the precursor austenite phase.

ulation increases with decreasing temperature (20° C). When the intensity distribution in the diffuse streaks of the precursor tweed structure is measured faint satellites at $q^* = 0.1 \cdots 0.2\langle 110\rangle^*$ are found.[10] Elastic neutron scattering shows these peaks much more pronounced at $q^* = 0.15 \cdots 0.16\langle 110\rangle^*$ corresponding to a wavelength of 1.35 nm.[11] These satellites exist in all of the six $\langle 110\rangle^*$ variants of the diffuse streaking. Below M_s and inside the final martensite plate (A in Fig. 5 at −90° C) the diffraction pattern clearly shows the long period stacking of the 7R multiply twinned martensite.[12] At the transition region (B at −90° C) between the growing martensite plate and the austenite matrix a heavily-faulted quasi-long period structure is observed. In the corresponding diffraction pattern the intensity in the diffuse streaks is piling up at diffuse satellite positions $q^* \simeq 0.2\langle 110\rangle^*$, indicating a better correlated modulation with a shorter wavelength. From Fig. 5A it is seen that this sharpening occurs in only one of the six possible $\langle 110\rangle^*$ directions, *i.e.* locally one variant grows at the expense of the others. Finally, when the 7R product phase forms, rows of equally spaced reflections become visible. This was first observed by Reynaud in 1977.[13]

These results were obtained by cooling a Ni-Al sample with the appropriate composition through its thermoelastic M_s temperature. However a similar transition can be induced by applying a proper stress configuration.[12] In practice, when a Ni-Al sample with an M_s close but still below room temperature (higher Ni content) is thinned for observation in a HREM instrument, strong but local stress fields around micro-cracks in the thin foil can induce the transformation.

The different transition stages can be observed under HREM conditions; the overall field of view is schematically given in Fig. 6. On the left-hand-side the precursor austenite phase, labelled β', is observed. In Fig. 7a typical [001] HREM image of this region is presented:[14] a patchwork of contiguous domains of approximately 3–5 nm diameter can be observed, the contrast inside each domain being modulated parallel with (110) or $(1\underline{1}0)$ (the other four {110} possibilities are extinct) and with a wavelength slightly incommensurate with the *bcc* lattice and corresponding with the satellites observed in the diffuse streaks. It was concluded that the distortion of the perfect *bcc* lattice can best be described by superimposing a transverse $(110)[1\bar{1}0]$ sine modulation onto a uniform $[1\bar{1}0]$ shift of the (110) planes of the austenite *bcc* lattice.[14] Filling space with different variants of this construction yields the proper low magnification tweed contrast[15] while the average symmetry remains cubic. The present multiple-beam phase contrast modulation is called "micro-tweed"[14] and should not be confused with the original two-beam tweed contrast. The origin of this patchwork of small distorted domains can most probably be related to the existence of a large number of point-like defects inducing a local strain field with tetragonal symmetry to which the surrounding matrix responds following its soft modes.[11]

The central region in Fig. 6 exhibits a one-dimensionally long period modulated structure with a larger correlation length than the micro-modulations. It is clear that out of all variants of the micro-modulation existing in the precursor phase, one is chosen which gradually increases its correlation both along and perpendicular to the corresponding wave-vector when moving closer to the micro-crack. Selected area computer diffraction (SACD) shows that the uniform distortion observed in the micro-modulated domains still exists in the present long period modulated structure. As indicated in Fig. 8 the direction of distortion can be positive ($\theta > 90°$) or negative ($\theta < 90°$) in the precursor state (depending on the chosen domain), whereas one of both is singled out in the long period case. No difference could be measured between the amount of distortion in the micro-modulated domains and in the long period structure. On the other hand, the periodicity of the modulation shows a tendency of being larger in the separate micro-modulated domains, where it varies between 6 and 8 {110} planes, than in the long period modulated structure where it reaches an average of only 5.5 {110} planes. In the small domains one thus finds a periodicity close to that of the 7R martensite structure, whereas the wavelength of the long period modulated structure clearly deviates from this value.

From these experimental results it can be concluded that when moving from the premartensitic region into the long period transition structure the local atomic arrangements must undergo some well defined alterations. The stress configuration near the micro-crack apparently dedicts which, out of the six possible transverse $110\langle 1\bar{1}0\rangle$ distortions will be preferred for the formation of the long period modulated structure with the increased correlation lengths. In the thin foil seen in Fig. 8a; the distortions in the long period modulated structure belong to the $(110)[\bar{1}10]$ variant, with a displacement to the left implicite in this choice of indices. The external stress field overcomes the different orientations of the local and smaller stress fields inferred in the centers of the domains.

It can thus be expected that the domains belonging to the $(110)[\bar{1}10]$ variant will grow at the expense of domains distorted in one of the other variants. Since the uniform distortions inside the micro-modulated domains are small, those domains belonging to the $(110)[1\bar{1}0]$ variant will probably also grow, but after flipping their uniform distortion from negative to positive. Since there is no correlation between domains of the same variant in the precursor phase[14] some conflicting situations will arise when domains with out-of-phase micro-modulations connect during this growth: an example hereof is seen at the marker A in Fig. 8. Also, as found from SACD patterns, the periodicity of the modulation decreases when moving into the long period modulated structure. Two different mechanisms by which the structure achieves this decrease in wavelength are observed in Fig. 8. The first is a bifurcation at B while the second is a dislocation type arrangement at C, both enlarged in 8b and 8c, respectively. The continuous nature

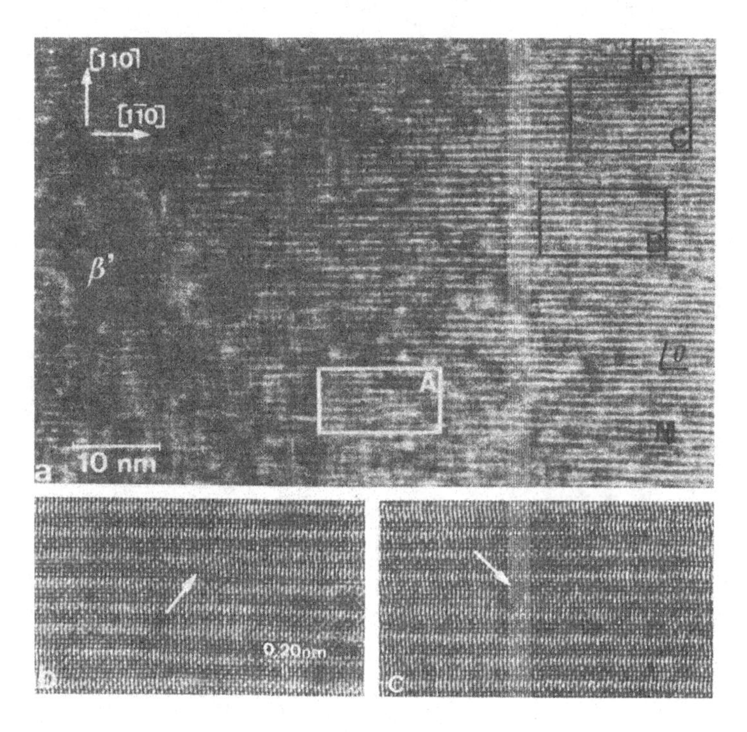

Figure 8. [001] HREM image of the long period transition structure with indications of some specific defects.

of the shifting observed in nearly all the crests indicates that the modulation is not strongly coupled with the basic lattice explaining the possibility for the material to acquire a range of incommensurate modulations as measured locally at the different sites in Fig. 8.

Only on the very edge of the foil the amount of stress is sufficient to fully transform a small portion of the foil into the 7R phase. The wavelength (or the number of close packed planes in the 7R monoclinic unit cell) must now increase from 5.5 to 7 planes. Similar but opposite behaviour as for the previous decrease in wavelength can be suggested: larger regions of micro-twinned martensite indeed show similar defects, such as steps in twin habit planes and small twin bands acting as dislocation twins (arrow in Fig. 8). Also, the modulation wave must now become locked in on close packed planes so as to form the stable $(5\underline{2})$ stacking. Fig. 9 shows another HREM image of this structure which is taken even closer to the crack, *i.e.*, in a region accommodating stronger stress. It is clear, however, that the perfect $(5\underline{2})$ stacking sequence, as proposed by Martynov *et al.*,[12] does not persist in large regions. In the present stress-induced situation this is probably due to a change in the amount or direction of the stress when looking at different locations around the crack. However, recent neutron scattering experiments on the thermoelastic transition in bulk material revealed superreflections that can not be attributed to the perfect $(5\underline{2})$ stacking,[16] thus a more general deviation behaviour might also exist.

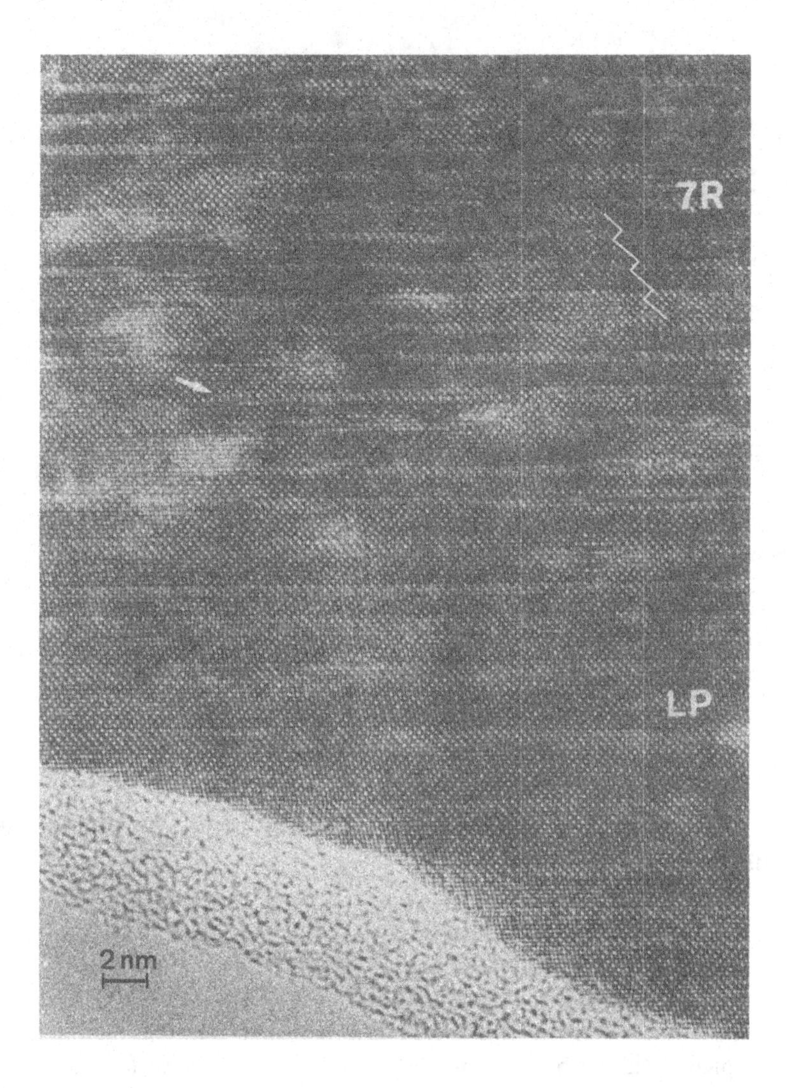

Figure 9. HREM image of long period (LP) multiply twinned martensite with a 7R region indicated.

In samples with a higher Ni content the distortion of the martensite relative to the austenite increases ($c/a \rightarrow 1$). Meanwhile the density of $(111)_{fct}$ twins decreases but no new long period stackings were observed.[17]

In conclusion we can say that from the present HREM investigation new detailed information on the atomic structures of precursor and transition states of martensitic transition in Ni-Al has been collected. It can be concluded that the precursor phase consists of contiguous domains (3–5 nm in diameter) with each domain being modulated by an incommensurate transverse {110}¡1$\underline{1}$0¿ displacement wave superimposed onto a uniform shift of the same type. As the transformation continues, "external" factors such as strain fields with higher amplitudes arising from, e.g., dislocation tangles, grain boundaries etc., determine which of the six variants will grow and develop the microtwinncd martensite structure. Before this phase is formed, an intermediate long period modulated phase with a shorter wavelength appears. Its modulation will then lock in on the *fcc*-like close packed planes to construct the 7R stacking or multiply twinned $L1_0$. Two mechanisms by which the wavelength of the modulation can de increased or decreased are described.

Acknowledgements

The authors (L.T, N.S and G.V.T.) like to thank the National Centre for Electron Microscopy at the Lawrence Berkeley Laboratory, Berkeley, (U. S. A.) for the use of the facilities. Part of this work was supported by the U. S. Department of Energy contract no. W-7405-ENG-48.

References

1. C. Leroux, M. C. Cadeville, and R. Kozubski, *J. Physics: Condensed Matter* **1**, 6403 (1989).
2. D. Broddin, C. Leroux, and Van Tendeloo G. *Proceedings of MRS Spring Meeting*, San Francisco 1990, in press.
3. D. Broddin, G. Van Tendeloo, J. Van Landuyt, S. Amelinckx, R. Portier, M. Guymont, and A. Loiseau, *Phil. Mag.* **A54**,395 (1988).
4. C. Leroux, A. Loiseau, M. C. Cadeville, D. Broddin, and G. Van Tendeloo, *J. Physics: Condensed Matter* **2**, 3479 (1990).
5. C. Leroux and A. Loiseau. ONERA film nr 1202 (1989).
6. C. Ricolleau, A. Loiseau, and F. Ducastelle, *Phase Transitions*, in press.
7. A. Finel, V. Mazauric, and F. Ducastelle, *Phys. Rev. Lett.* **65**, 1016 (1990).
8. D. Broddin, G. Van Tendeloo, J. Van Landuyt, S. Amelinckx, and A. Loiseau, *Phil. Mag.* **B57**, 31 (1988).
9. D. Van Dyck, G. Van Tendeloo, and S. Amelinckx, *Ultramicroscopy* **10**, 263 (1982).
10. S. M. Shapiro, J. Z. Larese, Y. Noda, S. C. Moss, and L. E. Tanner, *Phys. Rev. Lett.* **57**, 3199 ((1986).
11. L. E. Tanner, D. Schryvers, and S. M. Shapiro, *Mat. Sci. and Eng.* **A127**, 205 (1990).
12. V. V. Martynov, K. Enami, L. G. Khandros, S. Nenno, and A. V. Tkachenko, *Phys. Met. Metall.* **55**, 136 (1983).
13. F. Reynaud, *Scripta Metall.* **11**, 765 (1977).
14. D. Schryvers, L. E. Tanner, *Ultramicroscopy* **32**, 241 (1990).
15. W. Bell, private communication.
16. S. M. Shapiro, B. X. Yang, G. Shirane, Y. Noda, and L. E. Tanner, *Phys. Rev. Lett.* **62**, 1298 (1989).
17. D. Schryvers, B. De Saegher, and J. Van Landuyt, *Mat. Res. Bull.* **26**, 57 (1991).

Thermodynamics of Surfaces and Interfaces

François Ducastelle

Office National d'Etudes et de Recherches Aérospatiales
(ONERA)
BP72 92322 Châtillon Cedex
France

I. Introduction

Order-disorder transitions are frequent in alloys. The homogeneous ordered phase characterized by long-range order parameters which vanish in the disordered phase. In fact, local order parameters can also be defined on a scale large enough compared to the size of the relevant unit cells, and they vary in space in the presence of defects.

We shall consider here planar defects for which well defined thermodynamical quantities can be introduced. The simplest example is that of a surface, the problem being to understand how the bulk ordered phase is perturbed at the surface. Another example is that of interfaces between different ordered domains. It turns out that interesting phenomena occur when approaching the order-disorder transition from below. This is the main subject matter of this short and elementary review.

II. Order-disorder transitions

We first recall some elementary facts and definitions concerning order-disorder on fixed lattices.

II.1 Simple Examples

The B2 (CsCl) and $L1_2$ (Cu_3Au) ordered structures provide us with canonical examples which are very useful to illustrate many arguments.

In the case of B2-ordering, the *bcc* lattice is split into two interpenetrating sublattices (Fig. 1). The order parameter ϕ is then related to the concentration of these sublattices, $c_1 = (1+\phi)/2$, $c_2 = (1-\phi)/2$ for an alloy of mean concentration $c = 1/2$. It is clear that

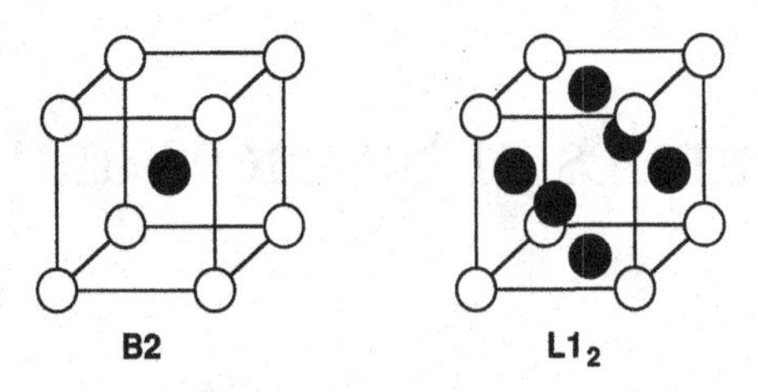

Figure 1. B2 (CsCl) and L1$_2$ (Cu_3Au) structures.

$-1 \leq \phi \leq +1$ and that changing ϕ into $-\phi$ is equivalent to exchanging the sublattices, or also to make a translation of vector [111]/2.

The L1$_2$ structure is also shown in Fig. 1. Here the *fcc* lattice should be split into four simple cubic sublattices with concentrations $c_0 = (1+3\phi)/4$, $c_1 = c_2 = c_3 = (1-\phi)/4$ but changing ϕ into $-\phi$ no longer preserves the structure. Actually ϕ should satisfy $-1/3 \leq \phi \leq 1$. This asymmetry will be explained later.

II.2 Concentration Waves and Order Parameters

Consider now the general situation where the concentration c_n at site n depends on n and perform a Fourier analysis [1–4]

$$c_n = c + \sum_{\mathbf{k}} \phi_{\mathbf{k}} e^{i\mathbf{k}\cdot\mathbf{n}}.$$

For simplicity we assume here a disordered phase with a single atom per unit cell, e. g. *bcc* or *fcc*. The vectors $\mathbf{k}$ in the sum are defined modulo a vector of the reciprocal lattice and can therefore be taken within the first Brillouin zone.

It is very easy to verify that the B2 structure is obtained using a single wave, for example $\mathbf{k} = \frac{2\pi}{a}$ [100]. Actually the other equivalent vectors such as $\mathbf{k} = \frac{2\pi}{a}$ [010] are equal to it modulo vectors belonging to the reciprocal lattice of the *bcc* structure such as $\mathbf{k} = \frac{2\pi}{a}$ [1$\bar{1}$0]. Comparing with the definition of ϕ given in II.1, it is clear that $\phi_{\mathbf{k}}$ is just equal to $\phi/2$.

In the case of the L1$_2$ structure, three concentration waves are necessary, and c_n is given by

$$c_n = c + \frac{1}{4}(\phi_1 e^{i\mathbf{k}_1\cdot\mathbf{n}} + \phi_2 e^{i\mathbf{k}_2\cdot\mathbf{n}} + \phi_3 e^{i\mathbf{k}_3\cdot\mathbf{n}}),$$

with

$$\mathbf{k}_1 = \frac{2\pi}{a}[100];\ \mathbf{k}_2 = \frac{2\pi}{a}[010];\ \mathbf{k}_3 = \frac{2\pi}{a}[001].$$

The perfect homogeneous L1$_2$ structure actually corresponds to $\phi_1 = \phi_2 = \phi_3 = \phi$. It is however useful to consider the set (ϕ_1, ϕ_2, ϕ_3) as a vector Φ. The four possible L1$_2$ variants are then found to be associated with vectors proportional to (111), (1$\bar{1}\bar{1}$), ($\bar{1}$1$\bar{1}$) and ($\bar{1}\bar{1}$1). From this point of view it is natural to consider Φ as a three-dimensional long-range order parameter, the previous vectors defining the vertices of a tetrahedron.

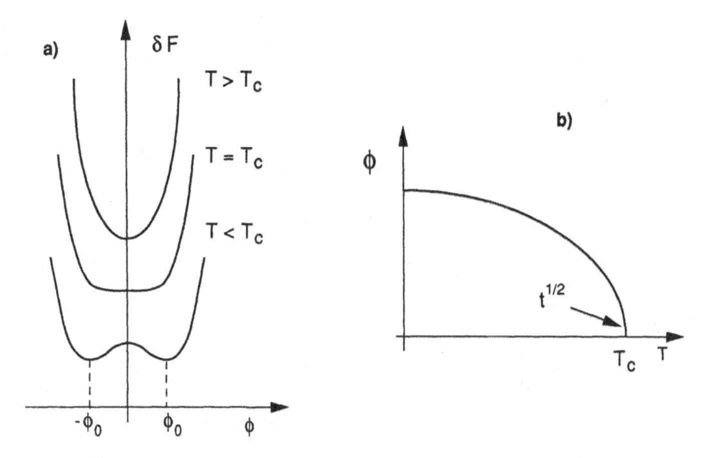

Figure 2. a) Free energy and b) order parameter for the ϕ^4 model, Eq. (1).

It is then clear that exchanging variants is equivalent to exchanging vertices, which cannot be obtained here through the symmetry $\Phi \rightarrow -\Phi$.

II.3 Landau Theory

Within Landau theory, we compare the free energy $F(\{\phi_{\mathbf{k}}\})$ of an ordered state with that of the disordered state and perform an expansion as a function of the $\phi_{\mathbf{k}}$

$$\begin{aligned}\delta F &= F(\{\phi_{\mathbf{k}}\}) - F(\{\phi_{\mathbf{k}} = 0\}) \\ &= \frac{1}{2}\sum_{\mathbf{k}} r_{\mathbf{k}}|\phi_{\mathbf{k}}|^2 + \sum_{\mathbf{k}_1+\mathbf{k}_2+\mathbf{k}_3=\mathbf{K}} w_{123}\phi_{\mathbf{k}_1}\phi_{\mathbf{k}_2}\phi_{\mathbf{k}_3} + \cdots .\end{aligned}$$

The sums over $\mathbf{k}$ are limited to the vectors necessary to describe the structure of interest. In the case of B2 ordering, δF cannot contain odd powers of ϕ because of the symmetry $\phi \rightarrow -\phi$ and δF is given by

$$\delta F = \frac{1}{2} r\phi^2 + u\phi^4 + \cdots . \tag{1}$$

The sign of r controls the stability of the disordered phase. In this phenomenological theory we shall therefore assume that r is proportional to $t = T_c^* - T$. The equilibrium order parameter is given by $\partial\delta F/\partial\phi = 0$ and if $u > 0$ a standard discussion shows that ϕ vanishes continuously at $T_c = T_c^*$ according to a $t^{1/2}$ law (Fig. 2). Modern theories of phase transitions have shown that such a mean field law should be modified, but we shall not enter into these complications here.

Consider now an homogeneous $L1_2$ structure $\Phi = \phi(111)$. The free energy difference now contains a cubic term

$$\delta F = \frac{1}{2} r\phi^2 + w\phi^3 + u\phi^4 + \cdots , \tag{2}$$

and this necessarily implies a first-order transition above T_c^*, *i. e.* ϕ is discontinuous at the transition (Fig. 3).

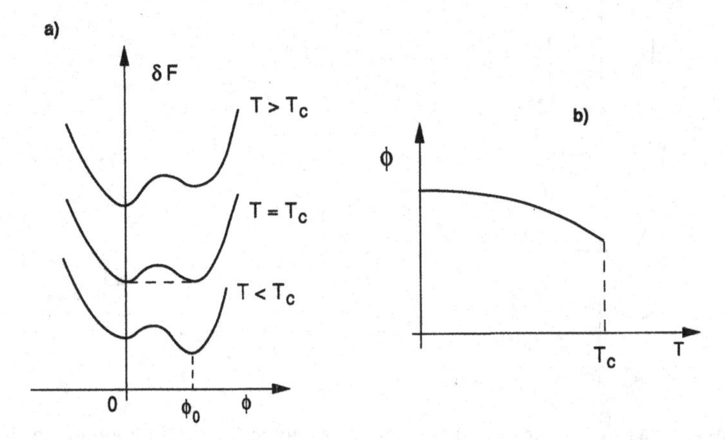

Figure 3. a) Free energy and b) order parameter in the case of a first order transition.

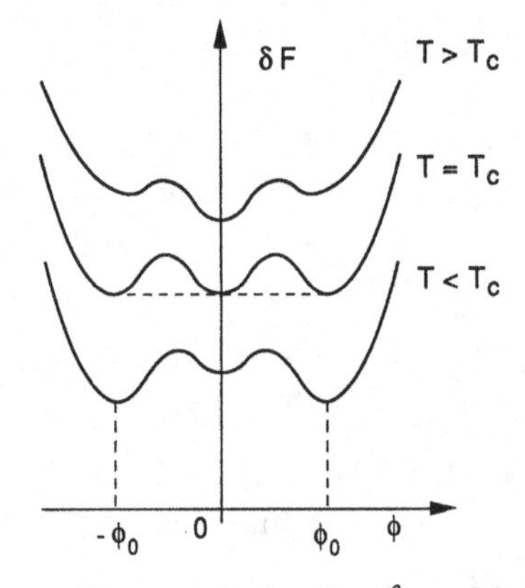

Figure 4. Free energy for the ϕ^6 model, Eq. (3).

For our purposes the main difference between first order (discontinuous) and second order (continuous) transitions is that the ordered and the disordered phase can coexist in the first case, which is clearly impossible in the second one.

It is also useful to consider a model displaying a first order transition with a scalar (one-dimensional) order parameter ϕ. Assume that the coefficient in front of ϕ^4 is negative. One should then introduce a positive term of order six

$$\begin{aligned} \delta F =& \frac{1}{2} r\phi^2 + u\phi^4 + v\phi^6 + \cdots , \\ & u < 0; \;\; v > 0, \end{aligned} \tag{3}$$

and this yields the desired symmetric first order transition (Fig. 4).

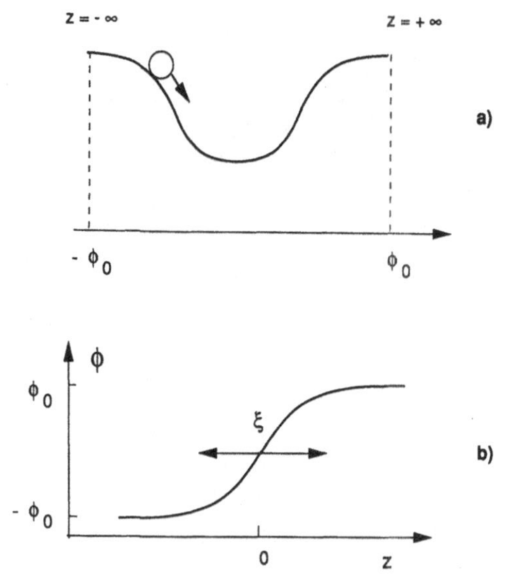

Figure 5. a) "Potential energy" and b) profile $\phi(z)$ for a bulk continuous transition.

III. Surfaces and interfaces

We shall principally discuss interfaces. Assume then a symmetric scalar model with just two variants characterized by two opposite values $\pm\phi_\circ$ of the order parameter. This should necessarily occur in practice if an alloy is quenched from the disordered phase since ordering breaks the symmetry of the disordered phase. For instance in the case of B2 ordering, the $\frac{1}{2}$ [111] translation is lost and both types of variants should nucleate at random. Such interfaces associated with a "translational" defect are generally called antiphase boundaries.

From a thermodynamical point of view, a planar domain wall forced by appropriate boundary conditions (here $\phi \to \pm\phi_\circ$ when $z \to \pm\infty$, where z is the coordinate normal to the interface) is stable. Then, by comparing with an homogeneous system, excess interface free energies, entropies, etc, can be defined properly.

III.1 Landau Ginzburg Theory

The problem is to determine the profile $\phi(z)$. As pointed out in the introduction, the definition of $\phi(z)$ requires some coarse grained average. It is therefore assumed that $\phi(z)$ does not vary too rapidly. As a consequence a first local contribution to the free energy is simply the energy calculated for the local value of the order parameter $\phi(z)$. On the other hand, the variation of $\phi(z)$ should cost something, and the simplest approximation is to include first order derivatives (see e. g. Refs. 2 and 5, and references therein)

$$\delta F \approx \int dz \left[\delta F_L(\phi(z)) + \frac{1}{2} m \left(\frac{d\phi}{dz} \right)^2 \right],$$

where δF_L is the local free energy difference and m is a positive stiffness constant.

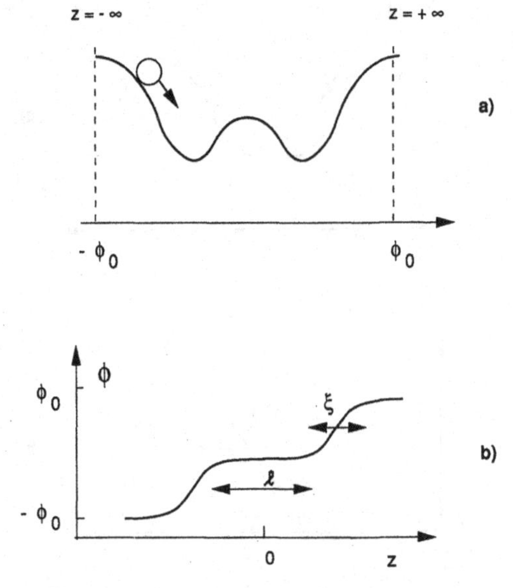

Figure 6. a) "Potential energy" and b) profile $\phi(z)$ for a bulk first order transition.

We have to minimize δF with respect to the profile $\phi(z)$. The corresponding Euler-Lagrange equations are in fact completely similar to those used in classical mechanics if we identify $\frac{1}{2}m(d\phi/dz)^2$ with a kinetic energy, $-\delta F_L(\phi)$ with a potential energy and the quantity to be integrated with a Lagrangian. Within this analogy, z plays the part of time, ϕ of the position, and m of the mass. However since the "potential" energy is the opposite of $-\delta F_L$, the equilibrium values of ϕ correspond here to maxima and not to minima of this potential energy.

III.2 Second Order Transitions

We assume that we are in the ordered phase $T < T_c$, the part of the curve $-\delta F_L(\phi)$ explored along the profile being that between $-\phi_o$ and $+\phi_o$. This yields the potential energy shown in Fig. 5a and our problem is to describe the motion of a particle starting at $-\phi_o$ at "time" $z = -\infty$ and arriving at $+\phi_o$ at time $z = +\infty$. This obviouly yields a profile of the type characterized by a length ξ equal to the correlation length of the bulk. In fact, for the ϕ^4 model defined in Eq. (1) an exact solution is available[6]

$$\phi = \phi_o \tanh(z/\xi),$$

with $\phi_o \approx t^\beta$, $\xi \approx t^{-\nu}$ when $t \to 0$. Within the present mean field theory, $\beta = \nu = 1/2$. As a result the amplitude of the profile vanishes when $T \to T_c$ (the interface disappears) while its width diverges.

III.3 First Order Transitions

Consider the potential energy corresponding to the ϕ^6 model. There is now an intermediate hill at $\phi = 0$ corresponding to the metastable disordered phase (Fig. 6a). It is then clear that the particle will slow down there so that the profile should look like

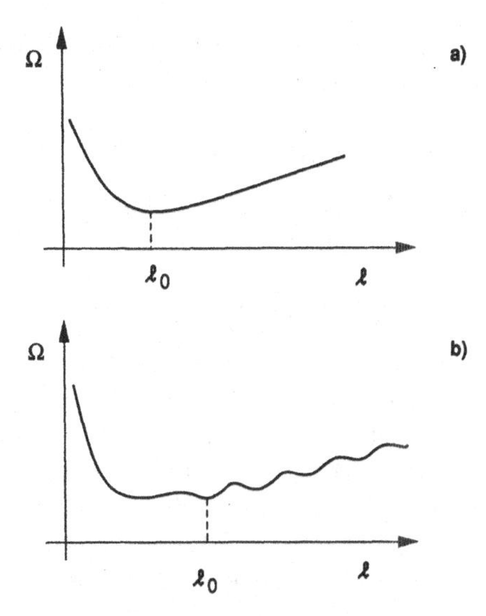

Figure 7. Schematic form of the functional $\Omega(\ell)$. a) Continuous model; b) in the presence of a lattice. The mean slope for large values of ℓ is proportional to t.

the one shown in Fig. 6b. A straightforward calculation[5,7,8] shows that the width ℓ of the region where $\phi \approx 0$ diverges when $t \to 0$, $\ell \approx \log(1/t)$. Actually when $t = 0$ the hill at $\phi = 0$ is exactly at the same height than the hills at $\pm\phi_o$ so that the mechanical particle starting from $-\phi_o$ will spend an infinite time at $\phi = 0$.

This is what is called (complete) wetting in the present context: the interface variants splits into two order-disorder interfaces. When $t \to 0$, the disordered phase wets the antiphase boundary (Fig. 6b). The logarithmic law is not modified when using more sophisticated renormalization group theories.[9]

It is clear that wetting is unavoidable for any first order transition with a scalar order parameter, but this is not necessarily so for a three-dimensional order parameter as in the case of $L1_2$: one can then imagine a path around the intermediate hill.

III.4 Surfaces

In the case of a surface, we have different boundary conditions. The profiles are the same as before but the position of the surface on the z-axis has to be determined from these conditions. Many different behaviors can then be imagined, the equivalent of the wetting behavior mentioned previously being still called wetting or surface-induced disorder, or surface-induced melting.[10–13]

III.5 Phenomenological Theory and Layering Effects

If wetting is assumed to occur, it is fairly simple to guess the general form of the free energy $\Omega(\ell)$ describing the interaction between the two order-disorder interfaces separated by a distance ℓ. There is an attractive part $A \cdot \ell$ where A is the free energy difference between the metastable disordered phase and the stable ordered one, but there is also a repulsive part related to the overlap of the two partial order-disorder interfaces which is of the form $\exp(-\ell/\xi)$ where ξ is the bulk correlation length which remains finite at the first order transition. Thus

$$\Omega(\ell) \approx A\ell + B\exp(-\ell/\xi),$$

and since A is proportional to t when $t \to 0$ we obtain for the equilibrium position ℓ_o (see Fig. 7a)

$$\ell_o/\xi \approx \log(1/A) \approx \log(1/t).$$

The previous Landau-Ginzburg theory assumes a continuous medium. In the presence of a lattice, friction effects are expected, the energy of the order-disorder interface depending on its position with respect to the lattice. This provides an additional periodic term $C\cos(\ell/a)$ in $\Omega(\ell)$, where a is the periodicity along the z-axis, and from the graph of $\Omega(\ell)$ one easily deduces that it necessarily induces first order layering transitions close enough to T_c. However the range in temperature where this effect can be observed depends crucially on the amplitude C which itself varies exponentially with ξ/a.

IV. Experiments

We mention very briefly a few recent experiments showing wetting effects as well as the expected logarithmic law.

(i) Wetting of antiphase boundaries has been observed in several $L1_2$ alloys,[14] $CoPt_3$ (Refs. 15 and 16) and Cu_3Pd (Refs. 17 and 18), using electron microscopy. Careful investigations for Cu_3Pd are consistent with a logarithmic law. This is in agreement with the microscopic calculations discussed in Sec. V.2.

(ii) In the case of Cu_3Au, X-ray experiments at grazing incidence reveal a surface induced disorder with again a logarithmic law.[19,20]

(iii) Finally, although the physical system is different, the theory for surface melting is completely identical, and experiments on Pb are described in detail in Refs. 21 and 13.

It is worth saying that logarithmic laws are very difficult to detect; the divergence is very weak so that the results would certainly not be incompatible with power laws $1/t^n$ with small values for n.

V. Microscopic Theories

V.1 Ising Model for Alloys

Ordering of a binary alloy on a fixed lattice can be described by an Ising model hamiltonian

$$H = -\sum_{\langle nm \rangle} J_{nm}\sigma_n\sigma_m - \sum_n h_n\sigma_n,$$

where the spin-like variable σ_n is equal to $+1$ or -1 depending on site n being occupied by an A or B atom. The field h_n is related to chemical potential differences which can depende on n in inhomogeneous systems.

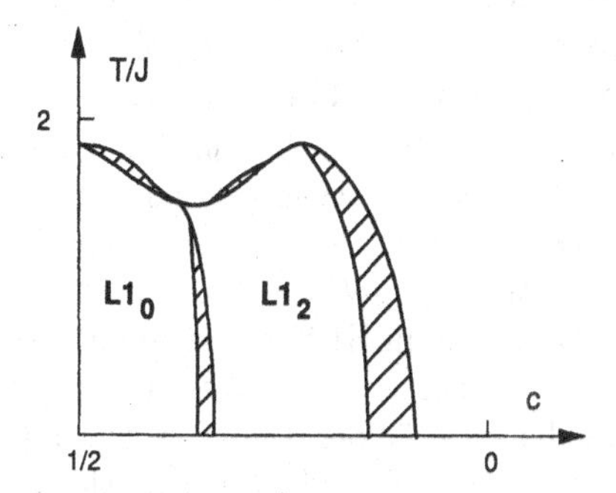

Figure 8. Schematic phase diagram of the *fcc* lattice with first neighbor interactions J calculated within the tetrahedron-octahedron CVM approximation; more accurate Monte Carlo and CVM calculations indicate that in fact the triple point is about $T/J \approx 1$; see Refs. 22 and 23.

The above hamiltonian can be justified from the electronic structure of the alloy using the so-called generalized perturbation method (see e.g. Ref. 24). For an antiphase boundary, J_{nm} and h_n can be taken to be identical to their bulk values and therefore J_{nm} only depends on $m-n$ and h_n is equal to its bulk value h_b.

In the case of a surface, the local electronic structure is generally strongly perturbed, so that the surface field h_s differs from the bulk one. In most cases it is a reasonable approximation to assume that $h_s - h_b$ is related to the difference in surface tensions of pure A and pure B. As for the pair interactions, they are generally found larger than in the bulk as soon as a site of the pair lies at the surface.[25,26]

V.2 FCC Lattice with First Neighbor Interactions

The Ising model cannot be solved in three dimensions, but accurate mean field methods such as the cluster variation method (CVM) or Monte Carlo simulations are now available. In the case of antiferromagnetic first neighbor interactions J (*i. e.* negative J favouring heteroatomic pairs) the phase diagram obtained from these methods is schematically shown in Fig. 8. Some difficulties and controversies have remained for a long time but most of them seem now to be solved (for recent discussions, see Refs. 22 and 23, and references therein).

These methods can be extended to inhomogeneous systems. The numerical calculations are obviously much heavier than for homogeneous systems, but several results are now available. We just describe here those concerning the antiphase boundaries in the $L1_2$ structure (for studies on surfaces, see the references 27-33).

The first detailed calculation is due to Kikuchi and Cahn.[34] They have calculated the profiles and free energies of the order-disorder interfaces as well as those of non-conservative [100] antiphase boundaries. Sanchez *et al.*[35] on the other hand have studied [111] antiphase boundaries. The accuracy of their calculations however was not sufficient to show wetting effects without ambiguity, but layering effects were already perceptible in the work by Kikuchi and Cahn.[34] In both cases the tetrahedron CVM was used. Recent improved calculations by Finel *et al.*[36] within the same approximation have shown indeed an infinite sequence of layering transitions in a temperature range ΔT below T_c of the order of $\Delta T/T_c \approx 10^{-1}$ with an average logarithmic behavior. These microscopic calculations do show that wetting occurs in $L1_2$ alloys with short-range order interactions. Let us recall that this was not obvious *a priori* in the case of a three-dimensional order parameter (see III.2).

More recent tetrahedron-octahedron calculations and Monte Carlo simulations by Finel[37] and by Finel and Mazauric[38] respectively have shown that in fact the order-disorder interface (or in other words that the correlation length at coexistence) was much larger than that deduced from the Bragg-Williams or tetrahedron CVM approximations. As a consequence (see the discussion in III.5) the layering effects practically disappear, *i. e.* they are invisible with an accuracy in the free energy calculations about $10^{-10}J$. Thus, contrary to what was believed before, the tetrahedron CVM approximation may not always be sufficiently accurate to describe ordering effects on the *fcc* lattice.

V.3 Antiphase Boundaries and Dislocation Dissociation

In the presence of order, the shortest lattice translations are generally lost and many perfect dislocations of the disordered lattice introduce order faults in the ordered phase. As a result there is a tendency to have pairs of so-called superpartial dislocations separated by an antiphase boundary.[39,40] The lower the (free) energy of the antiphase boundary, the larger the separation. This provides a method to determine antiphase boundary energies. On the other hand, provided the atoms remain located at the sites of the averaged fixed lattice, these energies can be determined from CVM or Monte Carlo calculations using the Ising model introduced previously if the pair interactions J_{nm} are known.

An efficient method for determining these interactions from experiment is to measure the local order (more precisely the short-range order parameters) in the disordered phase, using diffuse X-ray or neutron scattering and then to perform "inverse" CVM or Monte calculations (see Ref. 41 and the references therein).

Thus, it is in principle possible to relate quantitatively the observed dislocation dissociations to the diffuse scattering experiments.

Such a program has been recently achieved on the DO_{22} compound Ni_3V . From neutron diffuse scattering and CVM calculations, the interactions up to fourth neighbors determined.[42] The calculated free energies as a function of temperature are then found to be in excellent agreement with those determined from weak beam electron microscopy observations at several temperatures.[43]

Acnowledgements

Many results mentioned in this paper are due to several colleagues: R. Caudron, A. Finel, A. François, A. Loiseau, V. Mazauric, C. Ricolleau, F. Solal, and R. Tétot.

References

1. A. G. Khachaturyan, *Progr. Mater. Sci.* **22**, 1 (1978).
2. A. G. Khachaturyan, *The Theory of Structural Transformations in Solids*, Wiley, New York, 1983.
3. D. de Fontaine, *Solid State Physics* **34**, 73 (1979).

4. F. Ducastelle, *Order and Phase Stability in Alloys*, North-Holland, Amsterdam, to be published.
5. K. Binder in *Phase Transitions and Critical Phenomena*, Vol. 8, edited by C. Domb and J. L. Lebowitz, Academic Press, New York, 1983. p. 2.
6. J. W. Cahn and J. C. Hilliard, *J. Chem. Phys.* **28**, 258 (1958).
7. B. Widom, *J. Chem. Phys.* **68**, 3878 (1978).
8. J. Lajzerowicz, *Ferroelectrics* **35**, 219 (1981).
9. S. Dietrich in *Phase Transitions and Critical Phenomena*, Vol. 12, edited by C. Domb and J. L. Lebowitz, Academic Press, New York, 1988. p. 2.
10. R. Lipowsky, *Phys. Rev. Lett.* **49**, 1575 (1982).
11. R. Lipowsky and W. Speth, *Phys. Rev.* **B28**, 3983 (1983).
12. R. Lipowsky, *Ferroelectrics* **73**, 69 (1987).
13. B. Pluis, D. Frenkel, and J. F. Van der Veen, *Surface Sci.* **239**, 282 (1990).
14. D. G. Morris, *Phys. Stat. Sol. (a)* **32**, 145 (1975).
15. C. Leroux, A. Loiseau, M. C. Cadeville, D. Broddin, and G. Van Tendeloo, *J. Phys. Condens. Matter* **2**, 3479 (1990).
16. C. Leroux, A. Loiseau, M. C. Cadeville, and F. Ducastelle, *Europhys. Lett.* **12**, 155 (1990).
17. C. Ricolleau, A. Loisseau, and F. Ducastelle, *Phase Trans.* **30**, 243 (1991).
18. G. Van Tendeloo, this volume Chapter 14.
19. H. Dosch, L. Mailänder, A. Lied, J. Peisl, F. Grey, R. L. Johnson, and S. Krummacher, *Phys. Rev. Lett.* **60**, 2382 (1988).
20. H. Dosch, L. Mailänder, H. Reichert, A. Lied, J. Peisl, and R. L. Johnson, *Phys. Rev.* **B43**, 13172 (1991).
21. B. Pluis, A. W. Denier van der Gon, J. F. Van der Veen, and A. J. Riemersma, *Surface Sci.* **239**, 265 (1990).
22. R. Kikuchi, *Progr. Theor. Phys. Supp.* **87**, 69 (1986).
23. R. Tétot, A. Finel, and F. Ducastelle, *J. Stat. Phys.* **61**, 121 (1990).
24. F. Ducastelle in *Alloy Phase Stability*, edited by G M Stocks and A Gonis, Nato-Asi Series E, Vol. 163, 1989. p. 293.
25. G. Tréglia, B. Legrand, and F. Ducastelle, *Europhys. Lett.* **7**, 575 (1988).
26. F. Ducastelle, B. Legrand, and G. Tréglia, *Progr. Theor. Phys. Supp.* **101**, 159 (1990).
27. J. M. Sanchez and J. L. Morán-López, *Phys. Rev.* **B32**, 3534 (1985).
28. J. M. Sanchez and J. L. Morán-López, *Phys. Rev. Lett.* **58**, 1120 (1987).
29. D. M. Kroll and G. Gompper, *Phys. Rev.* **B36**, 7078 (1987).
30. G. Gompper and D. M. Kroll, *Phys.Rev.* **B38**, 459 (1988).
31. X. M. Zhu and H. Zabel, *Acta Cryst.* **A46**, 86 (1990).
32. Y. Teraoka, *Surface Science* **232**, 193 (1990).
33. W. Schweika, K. Binder, and D. P. Landau, *Phys. Rev. Lett.* **65**, 3321 (1990).
34. R. Kikuchi and J. W. Cahn, *Acta Metall.* **27**, 1337 (1979).
35. J. M. Sanchez, S. Eng, Y. P. Wu, and J. K. Tien, MRS Symp. Proc. Vol. 81, 57 (1987).
36. A. Finel, V. Mazauric, and F. Ducastelle, *Phys. Rev. Lett.* **65**, 1016 (1990).
37. A. Finel, *Proceedings of the European Workshop on Ordering and Disordering in Alloys*, Grenoble, July 10–12, 1991, Elsevier.
38. A. Finel and V. Mazauric, *Proceedings of the European Workshop on Ordering and Disordering in Alloys*, Grenoble, July 10–12, 1991, Elsevier.
39. M. J. Marcinkowski, N. Brown, and R. M. Fisher, *Acta Metall.* **9**, 129 (1961).

40. G. Vandershaeve in *L'Ordre et le Désordre dans les Matériaux*, Les Editions de Physique, Les Ulis, 1984. p. 135.
41. T. Priem, B. Beneu, C. H. De Novion, A. Finel, and F. Livet, *J. Phys. France* **50**, 2217 (1989).
42. R. Caudron R, M. Sarfati, A. Finel, and F. Solal, in MRS Symposium Proceedings, Vol. 166, (1990) p. 243; actually improved analyses (to be published) have been developed since then.
43. A. François and A. Finel, private communication; to be published.

Spatial Ordering in Bimetallic Nanostructures

J. L. Morán-López and J. M. Montejano-Carrizales

Instituto de Física
Manuel Sandoval Vallarta
Universidad Autónoma de San Luis Potosí
78000 San Luis Potosí, SLP
Mexico

Abstract

A systematic study of ordering and segregation in bimetallic nanoclusters is presented. Icosahedral and cubo-octahedral clusters are considered. The spatial distribution of the two components in the cluster is obtained by calculating the free energy within the regular solution model. Depending on the heat of mixing and on the difference in the cohesive energies of the components, and as a consequence of the large surface to volume ratio, one obtains a variety of phases characteristic only of the nanosystems. By comparing the total energies of the icosahedron and the cubo-octahedron the equilibrium shape is obtained as a function of the cluster size and of the interaction energies. The theory is applied to CuNi nanoclusters. The effect of chemisorption of O and H on the spatial atomic distribution is also discussed.

I. Introduction

The study of the properties of clusters in the nanometer scale is an interdisciplinary enterprise that involves physicists and chemists. The understanding of cluster properties, ranging sample preparation to specific applications in catalysis, optics, and electronics, has become a very active field of research in the last decade. In general, it is observed that many of the properties of small clusters are far from linear interpolations between the atomic or molecular world and the solid state. This fact had led

Structural and Phase Stability of Alloys
Edited by J.L. Morán-López *et al.*, Plenum Press, New York, 1992

to recognize that nanoclusters belong to a new phase of matter. In addition it is expected that these systems posses special properties with a wide range of applications in materials science.

Metal clusters with a small number of atoms usually adopt compact shapes. The icosahedron (I) and the cubo-octahedron (C) are the most common examples.[1,2] Clusters with less than 150~200 atoms crystallize in the form of icosahedra. Larger clusters become unstable and transform to the cubo-octahedron structure. That change in structure is expected since, in contrast to the icosahedron, the cubo-octahedron is formed by a subset of points that belongs to a crystalline lattice (the face-centered-cubic lattice).

A characteristic unique of small clusters is the very high ratio (called dispersion D) of the number of the surface atoms to the total number of atoms in the cluster. This fact makes these systems ideal as catalysts since a large number of atoms at the surface can be exposed to chemical reactions. That property was recognized early and metallic clusters are now widely used in the petroleum industry.[3,4] In particular, nanoclusters made of catalytically active and inactive components (like those of group VIII and of group IB elements), have been used to elucidate the role of geometric and electronic effects in catalysis.[5] We feel, however, that a necessary step towards the understanding of their catalytic behavior is the knowledge of the spatial distribution of the two chemical species in the cluster.

A phenomenon that may be the most important in determining the spatial atomic distribution in the bimetallic clusters is surface segregation. It is well established[6–8] that near the surface of macroscopic binary alloys the chemical composition and the spatial ordering may differ considerably from the bulk values. Thus, since the number of surface atoms in nanoclusters is very high compared to the total number of atoms in the system, the spatial arrangement is ruled, to a large extent, by the surface. The problem that we address here is how the two components arrange themselves spacially. We study in a detailed how the spatial atomic distribution depends on the size and the geometry of the cluster, on the interaction energies,[9] on the temperature,[10] and what is the effect of chemisorbed species on the surface of the cluster.[11]

We calculate, within the regular solution model,[12] the equilibrium spatial distribution of atoms in bimetallic icosahedral and cubo-octahedral clusters with 55 and 147 atoms. The general geometrical aspects of the I-and C-clusters are given in Sec. II. The ground state as a function of the interaction energies and of the number of A (N_A) and B ($N_B = N - N_A$) atoms is discussed in Sec. III. The equilibrium shape adopted by the clusters, icosahedral or cubo-octahedral, is addressed in Sec. IV. Finite temperature results for CuNi are discussed in Sec. V and the effect of H and O chemisorption on this system is discussed in Sec. VI. Finally the results are summarized in Sec. VII.

II. Geometrical Aspects

A cubo-octahedron is a geometrical figure with 6 square- and 8 triangular-faces, 24 edges and 12 vertices. On the other hand, the icosahedron possesses 20 triangular-faces, 30 edges and 12 vertices. The icosahedra are obtained by distorting the cubo-octahedra.[13] In contrast to the cubo-octahedron, where all the nearest neighbor sites are equidistant, in the icosahedron, the distance between the nearest neighbor atoms at the surface of the cluster is 1.05 larger than the distance between the atoms at the surface and their nearest neighbors in the core.[13] The smallest cluster (order one)

Table I

ν P	N_i	i\j	0	1	2	3	4	5	6	7	8
	1	0	0	12							
1 V	12	1	1	4	2	4	1				
2 S	6	2		4	0	4	0	4			
E	24	3		2	1	2	2	2	1	2	
V	12	4		1	0	4	0	2	0	4	1
3 S	24	5			1	2	1	2	0	2	0
T	8	6				3	0	0	0	6	0
E	48	7				1	1	1	1	2	1
V	12	8					1	0	0	4	0

is formed by 13 atoms; one in the center (called zeroth shell) and 12 located at the vertices (first shell). We consider all the atoms equidistant from the center of the cluster to form a shell. The next cluster size (order 2) is formed by covering the surface of the previous cluster. This can be accomplished by adding 42 atoms. In the cubo-octahedron they are distributed in three shells and in the icosahedron they occupy only two shells.

If we call ν the order of the cluster (cubo-octahedral or icosahedral), the total number of atoms is $N = 10\nu^3/3 + 5\nu^2 + 11\nu/3 + 1$, and the number of atoms at the surface is $N_S = 10\nu^2 + 2$. The dispersion D is defined as: $D = N_S/N$. It is worth noticing that for clusters containing approximately 10^3 and 10^4 atoms (ν= 6 and 14), 39% and 20% of the atoms are located at the surface, respectively. Obviously, even for these relatively large clusters, surface effects are very important in determining their properties.

The geometrical characteristics of the cubo-octahedron and icosahedron for $\nu = 3$ are summarized in Tables I and II, respectively. The meaning of the various columns is the following: ν denotes the order of the cluster, P is the position that the atoms occupy in the cluster (V = vertex, E = edge, S = square face, T = triangular face), and N_i is the number of sites forming the i-th shell. The numbers at the right hand side of the Tables are the number of nearest neighbors of an atom in the shell i located in shell j, and are denoted by Z_{ij}. In the Table for the icosahedron cluster, the nearest neighbors are located at the surface, which as mentioned above, are at a larger distance, are printed in bold.

The icosahedron and the cubo-octahedron with 147 atoms ($\nu = 3$) are illustrated in Fig. 1. The various shells that compose the surface are indicated by numbers and different shadowing.

Table II

ν P	N_i	i\j	0	1	2	3	4	5	6
	1	0	0	12					
1 V	12	1	1	**5**	5	1			
2 E	30	2		2	**4**	**2**	2	2	
V	12	3		1	**5**	**0**	0	5	1
3 T	20	4			**3**	0	**0**	**6**	**0**
E	60	5			1	1	**2**	**3**	**1**
V	12	6				1	**0**	**5**	**0**

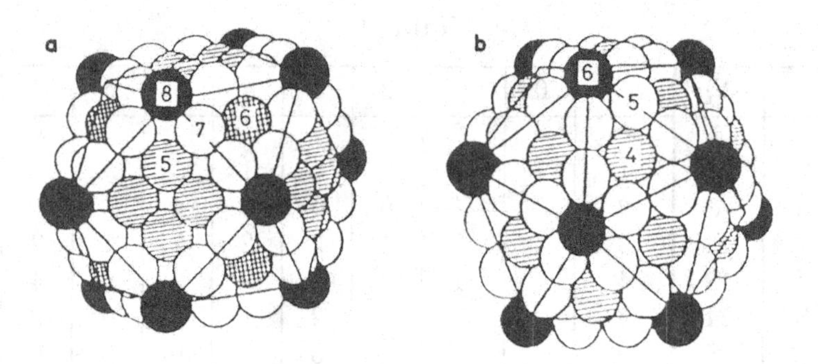

Figure 1. Cubo-octahedron (a) and icosahedron (b) with 147 atoms. In each cluster, equivalent sites on the surface have the same shading. The numbers represent the shells on which they lie.

III. The Ground State

Once the geometry of the cluster s has been defined, we proceed to calculate the equilibrium configuration in a bimetallic cluster with N_A atoms of type A and $N_B = N - N_A$ atoms of type B (note that vacancies are not allowed). The free energy of the system is calculated within the regular solution model.[12] We take into account only interactions between nearest neighbor atoms and they are denoted by $U_{ij}, i, j = A, B$. For the icosahedra, we assume that the ratio between the bond energies of atoms located at long and short distances is α. Due to the symmetry of the cluster, all the sites in a given shell are equivalent and the concentration of the component A in the i-th shell is defined by:

$$x_i = N_{A,i}/N_i, \tag{1}$$

where $N_{A,i}$ and N_i denote the number of A-atoms and the total number of sites in the i-th shell, respectively.

The internal energy of the system can be written in terms of the various concentrations as follows:

$$U = \frac{W}{2}\sum_{ij} N_i Z_{ij} x_i x_j + \frac{1}{2}(\Delta - W)\sum_{ij} N_{A,i} Z_{ij} + \frac{U_{BB}}{2}\sum_{ij} N_i Z_{ij}, \tag{2}$$

where the values of Z_{ij} are given in Tables I and II, and where $W = U_{AA} + U_{BB} - 2U_{AB}$ is the heat of mixing and $\Delta = U_{AA} - U_{BB}$ is identified as the difference in the cohesive energies of the pure components. To take into account the difference in the bond energies of the atoms located at larger distances in the icosahedra, we multiply the Z_{ij} printed in bold in Table II by the parameter α.

In macroscopic binary alloys, where boundary effects do not play any important role, the energy of the system is minimized at low temperatures, either by developing long-range order structures or by forming separate phases. Within the regular solution model that behavior is ruled by the sign of W: positive (negative) values correspond to ordering (segregating) alloys. Furthermore, at the surface of macroscopic systems, and due to the fact that the bonds of the surface atoms are broken, the concentration and the spacial order in that region differ from the bulk values. The parameter that

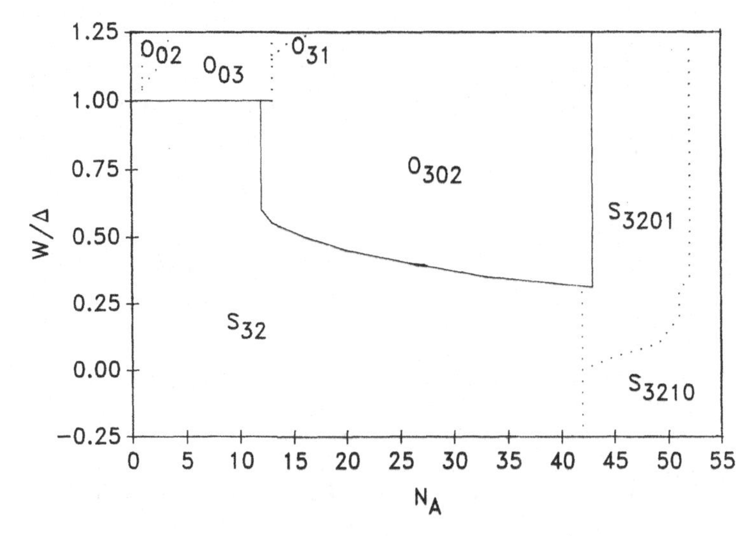

Figure 2. Phase diagram in the N_A *vs.* W/Δ parameter space of a icosahedral 55-atom bimetallic cluster. The phases are ordered (O) or segregated (S) and the subindices denote the order in which the B-atoms in those shells get substituted by A-atoms.

regulates the surface properties, within the regular solution model, is Δ. For example, in random alloys with a nominal concentration x, the concentration of the component A at the surface exceeds the bulk concentration for values of $\Delta > \Delta_c = W(1-2x)$.

In small clusters, due to the large dispersion, the atomic spatial distribution in the whole system is determined by both parameters, W and Δ, and one obtains phases that are characteristic of small clusters uniquelly. We show in Fig. 2 the phase diagram in the N_A *vs.* W/Δ parameter space of 55-atom clusters with icosahedral shape. These results correspond to $\alpha = 0.9$. Let us notice that the surface of the icosahedron is formed by the 2nd and 3rd shells with 30 and 12 sites, respectively.

In nanosystems, the phases are not sharply defined. However, one can still recognize general features. In the icosahedron with 55-atoms one finds only two phases, ordering (O) and segregating (S). For values of W/Δ near to zero, all the A-atoms segregate to the surface. Starting with a cluster made up of only B-atoms ($N_A = 0$), and by substituting B- by A-atoms, the lowest-energy configuration is obtained when the A-atoms occupy the outer (3rd) shell. Once that shell is fully occupied by A-atoms ($N_A = 12$), and by further substituting B- by A-atoms, the state of lowest energy is obtained by locating the A-atoms in the 2nd shell. Just after the surface is completely covered by A-atoms, the core B-atoms begin to get substituted by A-atoms. The opposite behavior (ordering) is observed for particles with $W/\Delta > 1$. When an atom of type B is substituted by another of type A, the minimum energy is obtained when the impurity goes to the central site (zeroth shell). With increasing N_A, the ground state is obtained by locating the A-atoms first in the 3rd shell, then in the second and finally in the first. An alternative behavior is obtained for $0.5 < W/\Delta < 1.0$, where the first substitutional A-atoms occupy the 3rd shell and then the central site. The nomenclature used in the phase diagram, P_{ijkl} means: P = O (ordering), S (segregating) and $ijkl$ is the order in which the various shells, originally occupied with B-atoms, get substituted by A-atoms. The heavy lines separate the two phases

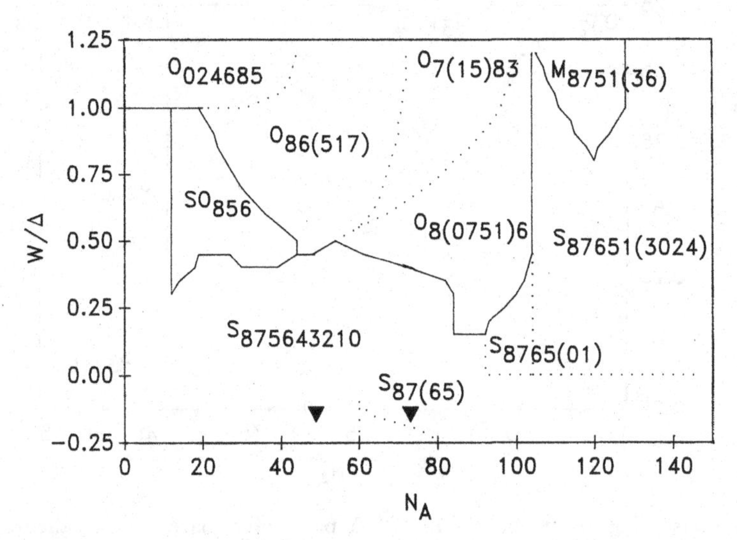

Figure 3. Phase diagram in the N_A *vs.* W/Δ parameter space of a cubo-octahedral 147 atom bimetallic cluster. The phases are ordered (O), segregated ordered (SO), mixed (M) or segregated (S). The subindices denote the order in which the B-atoms in those shells get substituted by A-atoms. The numbers in brackets mean that in those shells the B-atoms get substituted by A-atoms simultaneously. The symbols ∇, correspond to the parameters of CuNi system.

while the dotted lines indicate regions in a given phase where the substitutions follow a different sequence.

The phase diagram for the 147-atom cubo-octahedron is displayed in Fig. 3. The surface is formed by the 5, 6, 7 and 8-shells. Due to the larger number of shells, the number of ways of arranging the atoms increase sharply, generating thereby a larger variety of phases. The notation used in these figures is the same as the one discussed above and the numbers in brackets (mn) mean that in those shells the B-atoms get substituted by A-atoms simultaneously. The symbols ∇ correspond to the parameters of CuNi.

To visualize how the A-atoms occupy progresively the surface and core sites as the B-atoms get substituted by A-atoms, we plot in Fig. 4 the average concentration at the surface and in the core as a function of N_A, for W/Δ= 0.35, 0.85 and 1.2, for the cubo-octahedral particle. In the case of W/Δ= 0.35, the A-atoms segregate to the surface for almost any value of N_A. For W/Δ= 0.85 the surface also gets enriched with A-atoms but some of the core B-atoms are also sustituted by A atoms. In addition, one observes that drastic changes in the concentration occur when various of the A-atoms in the core are moved to the surface sites. Finally, in the case of W/Δ= 1.2, and for small values of N_A, the A-atoms occupy the core sites. However, the overall behavior is that the surface, in most of the N_A range, is enriched with the component A. For comparison, the average concentration in the whole cluster is also plotted (dotted line).

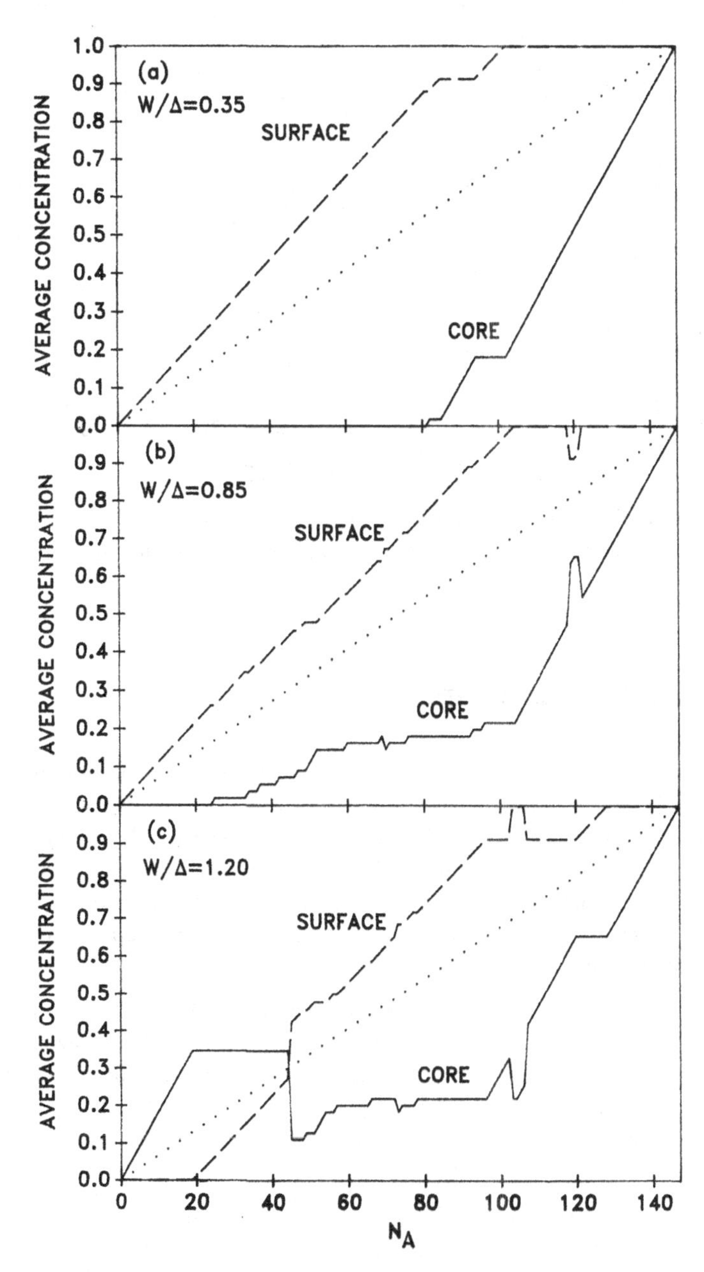

Figure 4. The average concentration at the surface and in the core as a function of N_A in a cubo-octahedral 147-atom cluster for $W/\Delta= 0.35$ (a), 0.85 (b) and 1.2 (c). The dotted line is the average concentration in the whole cluster.

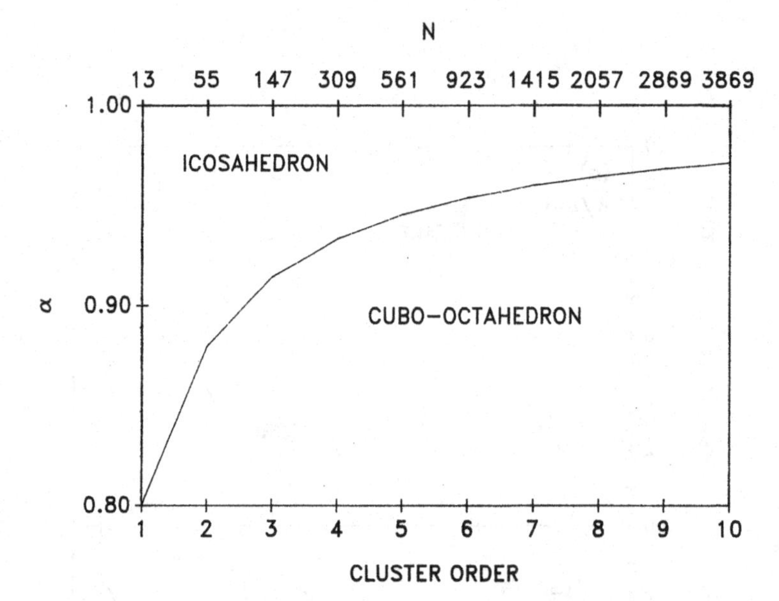

Figure 5. Phase diagram in the α *vs* cluster order parameter space of the stable cluster structures. The upper scale is the total number of atoms in the cluster.

IV. The Equilibrium Shape

It has been observed experimentally[1,2] that the equilibrium shape of small clusters depends on their size. Small icosahedral clusters transform to cubo-octahedral as the number of atoms increases. Based on electronic structure calculations, carried out within the tight-binding Hamiltonian,[14] it has been found that the most stable shape in transition-metal particles with a number of atoms larger than 150 is the cubo-octahedral. The shape stability in single component clusters depends on the value of the ratio, α, of the long- to short-bonds in the icosahedra.

We present in Fig. 5 the phase diagram for clusters made up of a single component. The figure shows how the critical value of α depends on the cluster size. We used two scales in the x-axis: the cluster order, ν, (lower scale) and the total number of atoms in the cluster N (upper scale). One clearly sees from this figure that the icosahedron phase field gets strongly reduced as the cluster size increases. For α=0.9 the instability between the two structures is $\sim$130 atoms.

In bimetallic clusters, the equilibrium shape also depends on the number of A-atoms, N_A, and on the cohesive energy of the B-component. We present in Fig. 6 the phase diagram for a cluster with 147 atoms in the α *vs.* W/Δ parameter space. The parameters used are: N_A= 73 and U_{BB}/Δ= 3.93. The equilibrium phases are the cubo-octahedron (to the left-hand side of the curve), and the icosahedron (to the right-hand side). There are various phases within the icosahedron and cubo-octahedron regions. The notation used here is similar to the one used above, althought we have added the symbol C (I) for the cubo-octahedral (icosahedral) structures.

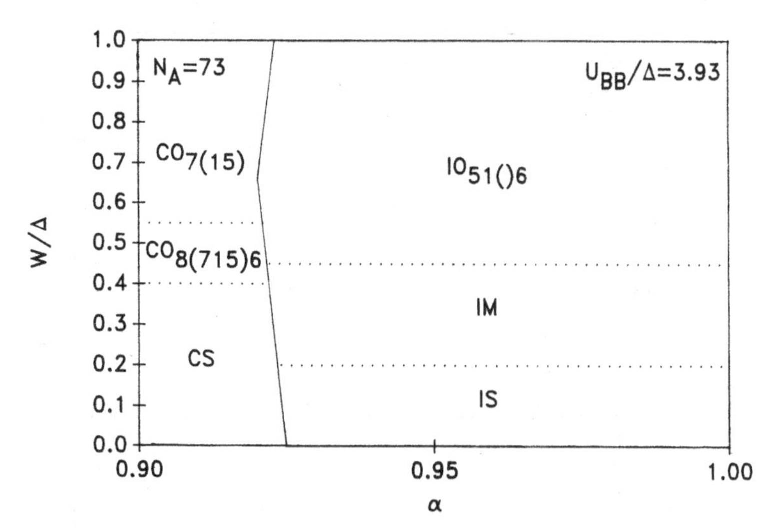

Figure 6. Phase diagram in the α *vs.* W/Δ parameter space of 147-atom clusters with $N_A = 73$, and $U_{BB}/\Delta = 3.93$. The equilibrium phases are the icosahedral ordered(IO), mixed (IM) and segregated (IS), and cubo-octahedral ordered (CO) and segregated (CS).

V. Finite Temperature Effects

The equilibrium configuration at finite temperatures is obtained by minimizing the free energy $F = U - TS$ with the constraint that the total number of A- and B-atoms remain constant. The configurational entropy is given by the expression:

$$S = \sum_i ln \frac{N_i!}{(N_i - N_{A,i})! N_{A,i}!}, \tag{3}$$

where the sum is carried out over all the shells around the central atom.

An extensive analysis of icosahedral and cubo-octahedral clusters of various sizes can be found in Ref. 10. Here we present only some illustrative cases. In Fig. 7 we present how the A-population $N_{A,i}$ depends on the total number of A-atoms in the cluster, N_A, for three different temperatures. The parameters used in the calculation, $W = 4$, $\Delta = 2$, and $\alpha = 0.9$, correspond to a system with a large tendency to form ordered patterns. The low temperature behaviour is characterized by arrangements in which the bonds of AB-type are maximized. For small values of N_A this is accomplished by locating the A-atoms in the 0 and 2 or in the 1 and 3 shells. One can notice that at T = 0 (Fig. 7a), drastic changes in the shell population occurs when the minimum of energy is obtained by switching between those arrangements. At high temperatures and for small N_A, the A-atoms are segregated to the surface (shells 2 and 3).

We applied the theory to the CuNi system. Based on the data of the cohesive energy of the pure elements,[15] and on the alloy phase diagram,[16] we estimated the value of Δ and W. Obviously this is a rough estimate since the electronic structure of small cluster differs from the one of macroscopic systems.[14,17–19] We expect, however, that this first order approximation contains the essential features.

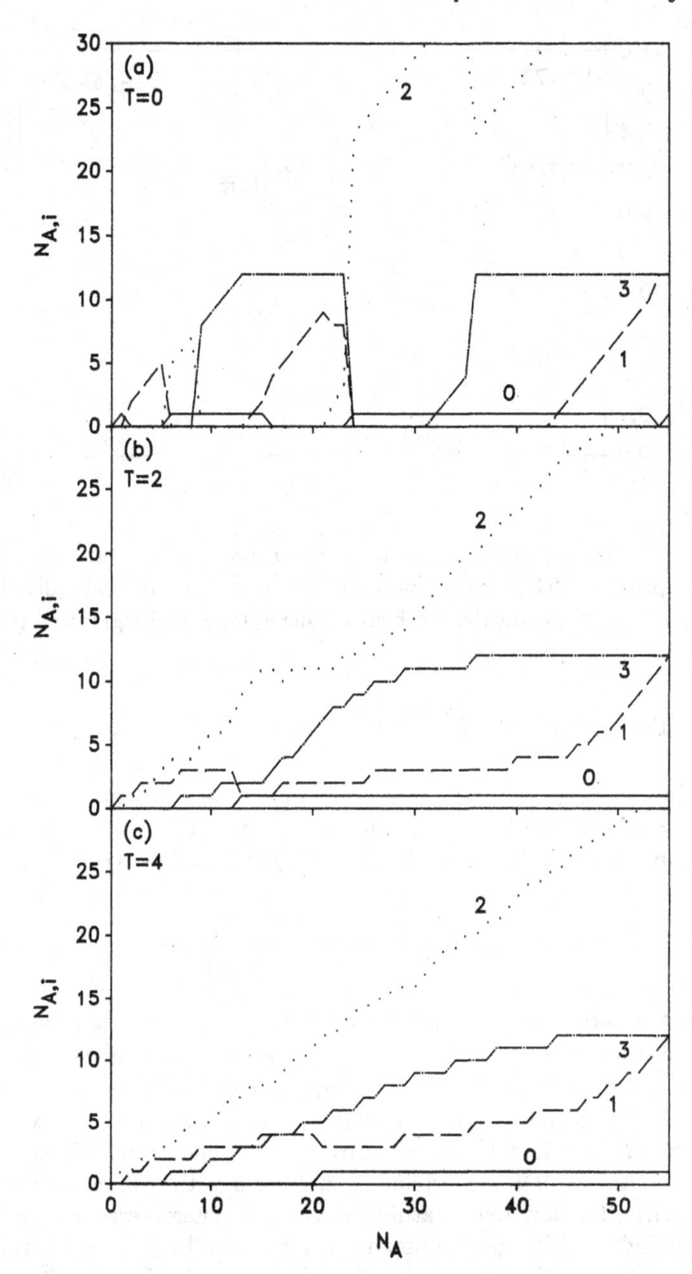

Figure 7. The number of A-atoms at the various shells $N_{A,i}$, $i = 0 - 3$, as a function of the total number of atoms A, N_A, for icosahedral particles made of 55-atoms. The various temperatures are given in units of k. The parameters used are $W = 4$, $\Delta = 2$, and $\alpha = 0.9$.

The CuNi alloys are some of the most studied systems with respect to the bulk phase diagram,[16] surface segregation,[20–24] catalytic properties,[25,26] etc. The CuNi alloys do not mix at low temperatures. This system crystalizes with a *fcc* structure and presents a miscibility gap that goes up to 400 °C. From the viewpoint of surface segregation, CuNi alloys represent a prototype system for both experimentalists and theorists. Equilibrium segregation experiments,[20–23] show that copper segregates to

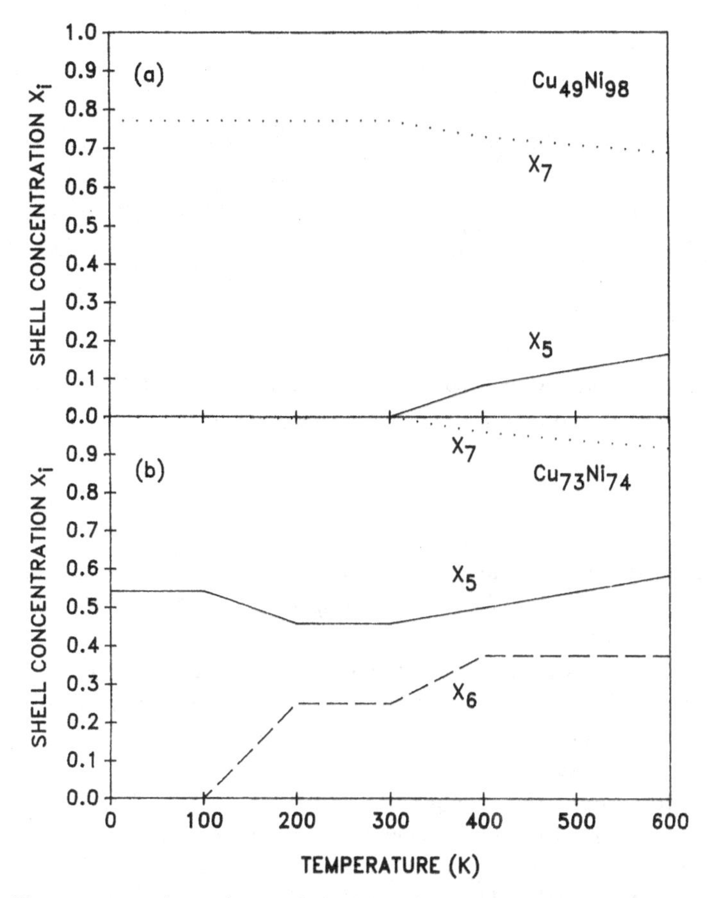

Figure 8. Temperature dependence of the concentration at the surface shells of the 147 cubo-octahedral particles $Cu_{49}Ni_{98}$ and $Cu_{73}Ni_{74}$.

the surface. On the other hand, it has been observed that chemisorbed atoms like S and O enhance the Ni concentration at the surface,[23] and that Ni enrichment resulted from radiation experiments.[24] However, all theoretical models predict the segregation of copper to the surface.

In our model, we estimate the parameter $W/\Delta = -0.146$. The 147-cubo-octahedra, $Cu_{49}Ni_{98}$ and $Cu_{73}Ni_{74}$, are marked in the ground state phase diagram (Fig. 3). The temperature dependence of the surface shells, $x_i, i = 5{,}6$ and 7 is shown in Fig. 8. The core is formed by Ni atoms and do not mix with the Cu atoms at any of the calculated temperatures. Furthermore, the outermost layer (8) remains occupied by Cu atoms in all the temperature range. The only effect of temperature is to mix the atoms at the 5, 6 and 7 shells.

VI. Chemisorption Effects

Catalytic processes are complex phenomena that, of necessity, involve the adsorption of atoms at the surface of the cluster. Due to the fact that the interactions between

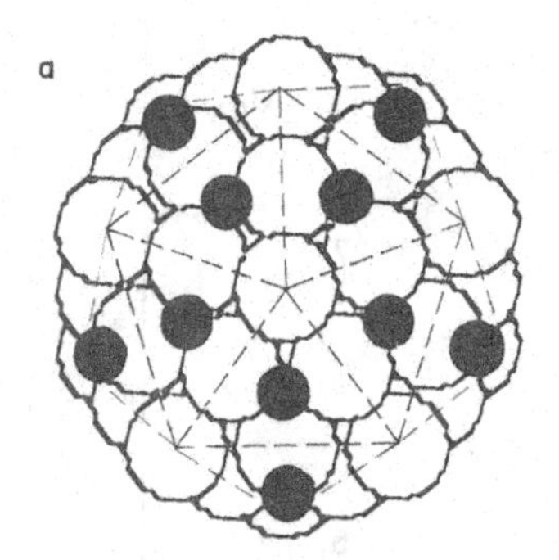

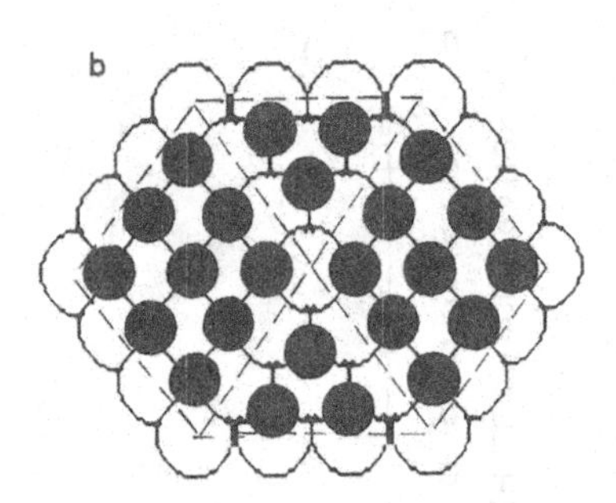

Figure 9. Chemisorbed oxygen (black circles) on triangular and four-coordinated sites on a) 55-atom icosahedral and b) 147-atom cubo-octahedral clusters.

the chemisorbed species and the cluster components may be very different, the atomic distribution in the presence of chemisorbed species will change as compared to the case of a clean cluster surface.

For example, it has been observed that chemisorption on macroscopic surfaces changes the surface concentration measured in clean surfaces.[27,28] In some cases the segregation is even inverted. Due to the high dispersion, we expect more drastic effects in bimetallic nanostructures.

To take into account the chemisorbed atoms we assume that they cover all the surface of the cluster. However, the chemisorption position depends on the atom under consideration and on the geometry of the surface. For example in the case of O on Ni, it is found that in the (111) surface O is chemisorbed on bridge and triangular sites.[29,30] On the other hand, in the (100) plane O is chemisorbed in four-coordinated central sites.[29] In the case of H it is observed that it is chemisorbed mainly on top of single-coordinated sites.[31] We assume that H is chemisorbed on top positions and that in the case of O it is chemisorbed on triangular and four-coordinated sites. In Fig. 9a we illustrate the O (black circles) chemisorption on a 55-atom icosahedral cluster, and in Fig 9b on a 147-atom cubo-octahedral cluster.

The equilibrium configuration now is obtained by minimizing the free energy of the system, which in the presence of chemisorbed species contains the additional term

$$U_{chem} = \sum_{ss} N_i Y_{ai} x_i (\xi_{aA} - \xi_{aB}) + \sum_{ss} N_i Y_{ai} \xi_{aB}, \tag{4}$$

where Y_{ai} is the coordination of the chemisorbed atom to the i-th shell located at the surface of the cluster. The interactions between the atoms located at the surface and the chemisorbed atoms a are denoted by ξ_{ak}, $k = A, B$. The sum in Eq. (4) is carried only over the shells that make the cluster surface. The number of surface shells of an icosahedral nanocluster containing a total number of atoms $N = 13$, 55, and 147 are 1, 2, and 3 respectively. On the other hand, for the same number of atoms, a cubo-octahedral nanocluster has 1, 3, and 4 respectively. The minimization of the free energy is performed with the additional constraint that the chemisorbed species do not diffuse into the cluster.

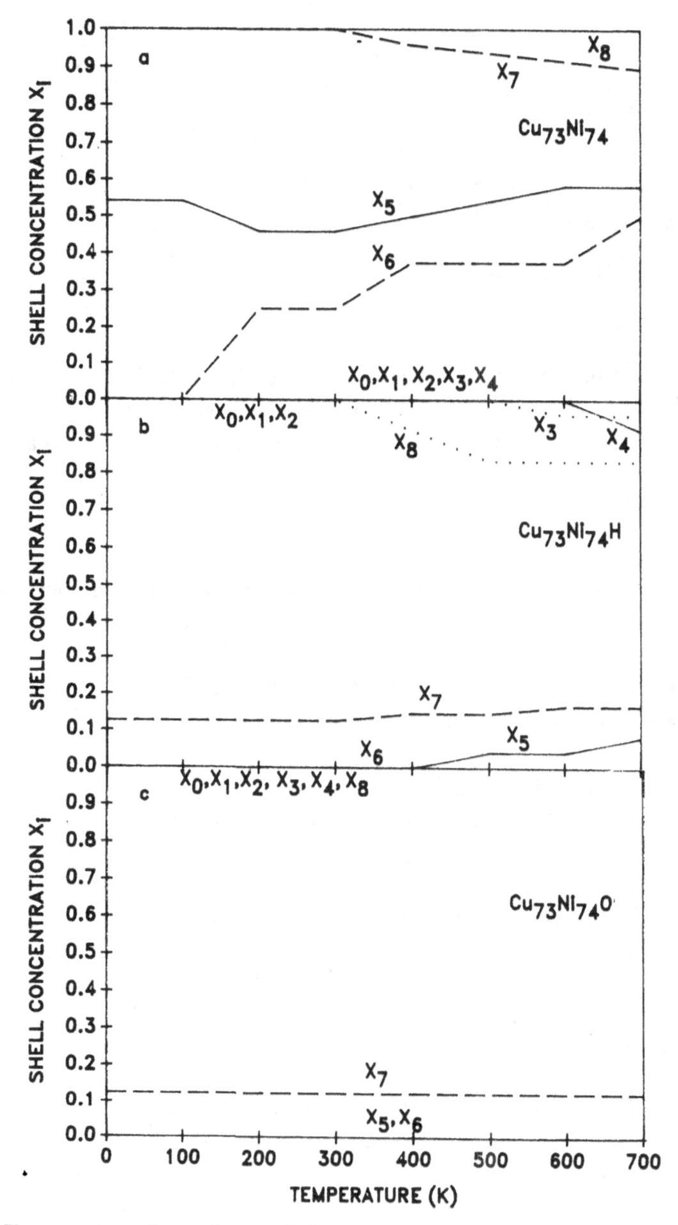

Figure 10. Temperature dependence of the concentration at the various shells, of the 147 atom cubo-octahedral particle $Cu_{73}Ni_{74}$, in the cases of: a) clean surface, b) H-chemisorption, and c) O-chemisorption.

We applied the theory to the O- and H-chemisorption effects on the atomic distribution of CuNi cubo-octahedral clusters. In this system the difference in adsorption energies for H-chemisorption on Cu and Ni is 0.584 eV, and it is 2.822 eV for O-chemisorption. The temperature dependence of the Cu concentrations x_i in the various shells for the 147-atom cluster $Cu_{73}Ni_{74}$ are presented in Fig. 10. The clean surface case is presented in Fig. 10a and the H- and O-chemisorption results are shown

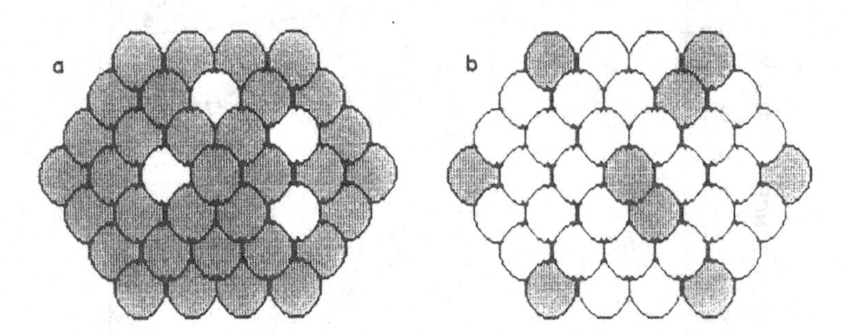

Figure 11. Equilibrium spatial distribution at T=300 K, of the $Cu_{73}Ni_{74}$ cubo-octahedral cluster a) without chemisorbed species, b) with H- and O-chemisorption. The chemisorbed atoms are not shown for clearness. Filled (empty) circles represent Cu (Ni) atoms.

in Figs. 10b and 10c, respectively. In the absence of chemisorbed species one finds a strong surface segregation of Cu atoms. On the other hand, we see that for H- and O-chemisorption, the core consists of Cu atoms, and the rest of them occupy partially shell 7 and shell 8. As the temperature rises, a small decrease in shells 3, 4, and 8, is accompained by a small increase in shells 5 and 7, in the case of H-chemisorption. No changes are observed in the case of O-chemisorption.

By comparing Figs. 10b and 10c, we can see that at $T = 300$ K the equilibrium spatial distribution is the same for H- and O-chemisorption. In Fig. 11 we show the equilibrium spatial distribution at $T = 300$ K for the 147-atoms cubo-octahedral cluster without (Fig. 11a) and with H- and O-chemisorbed at the surface (Fig. 11b). For clarity, the chemisorbed atoms are not shown. Filled (empty) circles represent Cu (Ni) atoms.

VII. Summary

The knowledge of the spatial atomic distribution in bimetallic nanoclusters is a fundamental step towards the understanding of their properties. Within a simple model, we studied in detail the ground state arrangements as a function of various parameters.

Here, we analysed first the geometrical aspects of two compact structures, the icosahedron and the cubo-octahedron. Then, we calculated the ground state as a function of the cluster size, the geometrical shape, and the interaction energies. In the regular solution model the parameters are the heat of mixing and the difference in the cohesive energies of the two components. We have found that, due to the large dispersion, surface segregation plays a decisive role in determining the equilibrium arrangement.

As a consequence of the two distances between nearest neighbors in the icosahedron, that structure becomes unstable in large clusters. This effect was modeled by a parameter α, defined as the ratio between the weak and strong interactions. The equilibrium shape was calculated under various circumstances and the phase diagrams were obtained.

We studied the temperature dependence of the concentration at the various shells. We applied the theory to the particular case of CuNi nanostructures, where it is seen that there is a strong surface segregation of Cu atoms that prevails to temperatures much higher that in macroscopic systems.

We studied the spatial atomic ordering in the presence of chemisorbed species. The contribution to the internal energy comming from chemisorbed species was calculated by assuming that the chemisorbed species cover all sites on the surface. Hydrogen was assumed to be chemisorbed in the top position and oxygen was assumed to be chemisorbed on triangular or four-coordinated sites depending on the geometry of the faces of the cluster.

We found that chemisorption may invert the surface segregation observed in clean clusters. When the theory is applied to the chemisorption of H and O on CuNi nanoclusters, we obtained that chemisorption inverts the segregation of Cu characteristic of clean surfaces.

Acknowledgements

This work was partially supported by the Dirección General de Investigación y Superación Académica, Secretaría de Educación Pública under Grant No. C90–07–0383.

References

1. M. Yacamán, in *Catalytic Materials, Relationship Between Structure and Reactivity*, edited by T.E. Whyte, R.A. Betta, E.G. Deroduane, and R.T.K. Baker (American Chemical Society, New York, 1983), p. 341.
2. A. Renou and M. Gillet, *Surface Sci.* **106**, 27 (1981).
3. J.H. Sinfelt, *Rev. Mod. Phys.* **51**, 569 (1979).
4. V. Poneç, *Adv. Catal.* **32**, 149 (1983).
5. J.H. Sinfelt *Acc. Chem. Res.* **10**, 15 (1977).
6. F.L. Williams and D. Nason, *Surface Sci.,* **45**, 377 (1974).
7. J.L. Morán-López and L.M. Falicov, *Phys. Rev. B***18**, 2542 (1978).
8. A.R. Miedema, *Z. Metallk.* **69**, 455 (1978).
9. J.M. Montejano-Carrizales and J.L. Morán-López, *Surf. Sci.* **239**, 169 (1990).
10. J.M. Montejano-Carrizales and J.L. Morán-López, *Surf. Sci.* **239**, 178 (1990).
11. J.M. Montejano-Carrizales and J.L. Morán-López, *Surf. Sci.* in press.
12. R.A. Swalin, *Thermodynamics of Solids* (Wiley, New York, 1972).
13. A.L. Mackay, *Acta Cryst.* **15**, 916 (1962).
14. M.B. Gordon, F. Cyrot-Lackmann, and M.C. Desjonqueres, *Surf. Sci.* **80**, 159 (1979).
15. C. Kittel, *Introduction to Solid State Physics* (Wiley, New York, 1971).
16. J. Vrijen and S. Radelaar, *Phys. Rev. B***17**, 409 (1978).
17. D. Tománek, S. Mukherjee, and K.H. Bennemann, *Phys. Rev. B***28**, 665 (1983).
18. S.N. Khanna, J.P. Butcher, and J. Buttet *Surf. Sci.* **127**, 165 (1983).
19. S. Mukherjee, J.G. Pérez-Ramírez, and J.L. Morán-López, in *Physics and Chemistry of Small Clusters*, eds. P. Jena, B.K. Rao, and S.N. Khanna (Plenum, New York, 1987) p. 451.
20. T. Sakurai, T. Hashizume, A. Jimbo, A. Sakai, and S. Hyodo, *Phys. Rev. Lett.,* **55**, 514 (1985).
21. N.Q. Lam, H.A. Hoff, H. Wiedersich, and L.E. Rehn, *Surf. Sci.,* **149**, 517 (1985).
22. H. Shimizu, M. Ono, and K. Nakayama, *Surf. Sci.,* **36**, 817 (1973).

23. L.E. Rehn, H.A. Hoff, and N.Q. Lam, *Phys. Rev. Lett.*, **57**, 780 (1986).
24. L.E. Rehn, in: *Metastable Materials Formation by Ion Implantation*, Eds. S.T. Picraux and W.J. Choyke (Elsevier, Amsterdam, 1982), p. 17
25. W.M.H. Sachtler and P. van der Plank, *Surf. Sci.* **18**, 62 (1969)
26. J.H. Sinfelt, J.L. Carter, and D.J.C. Yates, *J. Catal.*, **24**, 283 (1972).
27. S. Modak and B.C. Khanra, *Chem. Phys. Lett.* **134**, 39 (1987).
28. S. Modak and B.C. Khanra, *J. Phys. C: Solid State Phys.* **18**, L897 (1985).
29. J.E. Demuth and T.N. Rhodin, *Surf. Sci.* **45**, 249 (1975)
30. H. Conrad, G. Ertl, J. Kuppers, and E.E. Latta, *Surf. Sci.* **57**, 475 (1976)
31. J.A. Dalmon, G.A. Martin and B. Imelik, *Japan J. Appl. Phys. Suppl.* **2**, 261 (1974).

Summary Thoughts

D. G. Pettifor

Department of Mathematics
Imperial College of Science, Technology and Medicine
London SW7 2BZ
England

I. The Challenge

The first speaker[1] at this meeting threw down the challenge to the theorists in the audience that they should "calculate properties to 2% accuracy or else...".

II. Very Simple Situations

We have seen, of course, that accuracies approaching this can only be achieved for very simple situations.

For non magnetic *elements* structural energy differences between *fcc, bcc, hcp*, and diamond lattices can be calculated within Local Density Functional (LDF) theory as a matter of routine at absolute zero providing values that agree very well with experiment where known, such as for example *hcp-bcc* titanium or diamond-β-Sn silicon. However, where the metastable structures are not accessible directly to experiment, we saw that large discrepancies approaching nearly a factor of four can be found between the computed LDF values and the Calphad fitted values used for predicting binary and ternary phase diagrams. This reflected[2] the large uncertainty in the melting temperature and heats of fusion of the metastable phases, quantities which theorists are unable to predict except for the simplest of elements. Hopefully this is an area which will receive attention during the next few years. In the meantime LDF aficionados should predict the shear contents of metastable phases in addition to the usual binding energy curves as many are probably mechanically unstable. As a consequence the assignment of positive melting temperatures by the Calphad fitting procedure to such metastable lattices would be seen to describe effective melting temperatures which are useful only in the immediate concentration vicinity where the metastable structure is competing for dominance.

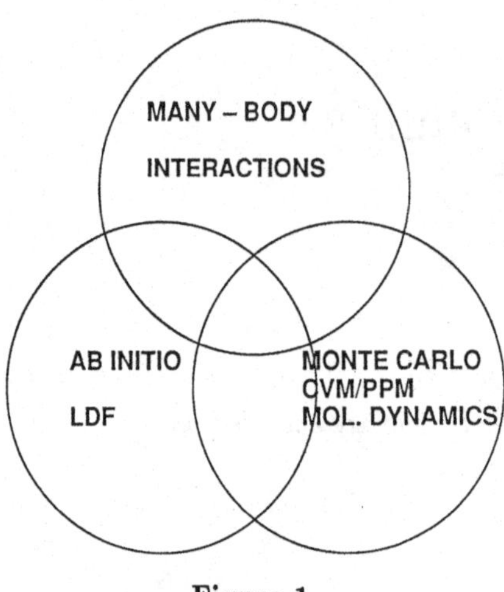

Figure 1

For *binaries* numerous intermetallic phase diagrams with respect to the *fcc* and *bcc* lattices were shown during the meeting. It appears that in all likelihood the result of first principles concentration dependent pairwise potentials or the Connolly-Williams cluster interactions will give similar results in most situations provided size effects and charge self-consistency are not important in the former and convergence has been obtained in the latter. However, we heard little about line compounds. All the transition metal tri-aluminides TAl_3 are line compounds which seems to indicate size rather than electronic factors as the origin. The present approximate inclusion of size effects would not, for example, predict the known line compound behaviour of $ScAl_3$ with the $L1_2$ crystal structure. A scandium (or titanium) atom in an all aluminium environment would have a much larger effective atomic volume due to the loss of the strong attractive d bond contribution which is present in the pure transition-metal. We saw that relaxation effects could be successively included within isovalent semi-conductors using perturbation theory with respect to the virtual crystal.[3] But in this case the hoped for ordering of the Si-Ge system was not obtained, indicating the importance of kinetic effects in the observed ordering within epitaxially grown silicon-germanium alloys. In the area of kinetics we heard that the Path Probability Method (PPM) had been successfully applied to spin systems and was being developed to treat alloys.[4]

III. Role of Theory

But, of course, the role of theory is not just to make quantitative predictions but also to provide concepts and models for understanding and guiding experiment. This was illustrated many times at this meeting where simple models explored, for example, ordering processes within microclusters, surfaces, interfaces, multilayers, and the bulk. However, I found the most impressive "confrontation between theory and experiment" (as François Ducastelle put it) concerned the wetting of Anti-Phase Boundaries.[5,6] In particular, the second, third, and fourth nearest neighbour pair interactions, which were obtained by fitting the short range order in Ni_3V obtained from neutron scattering, predicted the temperature dependence of the APB free energy that was in excellent agreement with that deduced from the APB domain widths between dissociated dislocations observed in the *electron* microscope.[5] To date first principles theory is unable to predict sufficiently reliable pair interactions in Ni_3V, highlighting yet again the need for improved treatments of size and charge-self-consistency within current theories.

IV. Complementarity Principle

The prediction of structural and phase stabilities is governed by a complementarity principle which may be represented diagrammatically as shown in Fig. 1. All three theoretical areas overlap and complement each other. It is possible, especially in Trieste using the Car-Parrinello technique, to proceed directly from first principles LDF calculations to Molecular Dynamics simulations. However, it is more usual to proceed from LDF via some effective many-body interactions to the prediction of structural and phase stability using CVM, PPM, Monte Carlo, or Molecular Dynamics. The organizers of the meeting are to be congratulated for bringing together experts in all three fields whose overlap and interactions have led to such a stimulating and successful week.

References

1. J. K. Tien (see Chapter 1).
2. A. P. Miodownik (see Chapter 5).
3. S. Baroni, S. de Gironcoli, and P. Giannozzi (see Chapter 9).
4. T. Mohri (see Chapter 7).
5. F. Ducastelle (see Chapter 15).
6. G. van Tenderloo and D. Schryvers (see Chapter 14).

Index

Printed in the United States
by Baker & Taylor Publisher Services